Praise for *On the Origin of the Deadliest Pandemic in 100 Years: An Investigation*

"Before I read this book, I thought the Wuhan lab leak hypothesis was just a wild conspiracy. Elaine Dewar makes a very strong case that it may well be true. She even goes farther, showing scientists are often sloppy with hazardous biological materials, that they engage in naive research with some dubious government actors, and, when caught, dodge responsibility by smearing their critics as enemies of science."

—Mark Bourrie, RBC Taylor Prize-winning author of *Bush Runner: The Adventures of Pierre-Esprit Radisson*

"In this age when reliable news is hidden in a cacophony of mis-information, dis-information, social media idiocy and the caterwauling of celebrity influencers, long-form journalism is coming back into its own. Elaine Dewar uses her formidable reporting skills to produce the story of the moment. The pandemic forced her to follow I.F. Stone's dictum to work exclusively from published, public sources, and her story is the stronger for it. This is, of course, a detective story, a whodunnit, and Dewar is unstinting in exposing the cover-ups, the political expediency, the deceits, and the sloppy work and judgements of usually diligent scientists along the route to her conclusions."

—Jonathan Manthorpe, author of *Restoring Democracy in an Age of*

"[Dewar] has spent the pandemic following the politics, the scientific research, the news coverage—and the money.... Her book casts a shadow over the wet market theory and points a finger at the Chinese government—and at some scientists and leading science journals for their single-minded support and promotion of this theory.... [It] reads almost like a detective novel."

—*Globe and Mail*

"Dewar gets us to an acceptable truth about the origin and the travels of this coronavirus, and the significant tardiness and incompetence of governments and science to protect the public.

—*Winnipeg Free Press*

"I devoured this book."

—Marsha Lederman, on Twitter

ELAINE DEWAR

On the Origin of the Deadliest Pandemic in 100 Years

An Investigation

BIBLIOASIS
Windsor, Ontario

FIRST EDITION
Second Printing, September 2021.

Library and Archives Canada Cataloguing in Publication

Title: On the origin of the deadliest pandemic in 100 years : an investigation / Elaine Dewar.
Names: Dewar, Elaine, author.
Description: Series statement: Field notes ; 4
Identifiers: Canadiana (print) 20210244755 | Canadiana (ebook) 20210244852 | ISBN 9781771964258 (softcover) | ISBN 9781771964265 (ebook)
Subjects: LCSH: COVID-19 (Disease)—China. | LCSH: COVID-19 Pandemic, 2020-—China.
Classification: LCC RA644.C67 D49 2021 | DDC 362.1962/414—dc23

Edited by Daniel Wells
Copyedited by Meghan Desjardins, Emily Donaldson, Theo Hummer, and John Sweet
Typeset by Vanessa Stauffer
Series designed by Ingrid Paulson

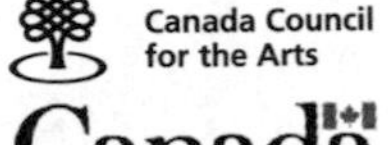

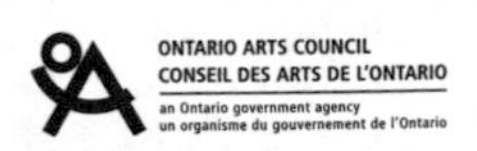

Published with the generous assistance of the Canada Council for the Arts, which last year invested $153 million to bring the arts to Canadians throughout the country, and the financial support of the Government of Canada. Biblioasis also acknowledges the support of the Ontario Arts Council (OAC), an agency of the Government of Ontario, which last year funded 1,709 individual artists and 1,078 organizations in 204 communities across Ontario, for a total of $52.1 million, and the contribution of the Government of Ontario through the Ontario Book Publishing Tax Credit and Ontario Creates.

PRINTED AND BOUND IN CANADA

For my beloved
Stephen Winston Dewar
1943–2019

Life is short
and Art long
opportunity fleeting
experimentations
perilous
and judgement
difficult.

–attributed to Hippocrates

Contents

Glossary of Terms

ACE2: Short for *angiotension-converting enzyme 2*, a particular type of receptor found on the surface of many types of cells in humans and in other mammals such as bats or mink or cats.

BSL-2, 3, 4 *laboratories:* Pathogens are studied in laboratories with different levels of containment related to the differing degrees of danger they represent to humans and animals.

BSL-2 *laboratory*: Work is done in a contained cabinet that sucks air away from the researcher to permit the safe study of Risk Group 2 pathogens. These cause illness but don't present a serious threat to humans or animals because vaccines and treatment methods are available.

BSL-3 *laboratory*: Work is done in a clean room separated from the environment by airlock. Researchers may shower in and out and wear protective suits that keep them safe from Risk Group 3 pathogens. These cause serious illness, but treatments are available so they don't pose a major public health threat.

BSL-4 laboratory: Work is done in a clean room behind an airlock. The whole system is separated entirely from the environment. Air is filtered, and all waste, including water, is sterilized before release. Researchers must wear specially designed suits with their own air supply while studying Risk Group 4 pathogens. These cause a high rate of mortality and no effective vaccines or treatments are available, which presents a pandemic risk if the pathogens are released. There are at least 42 known BSL-4s in the world, most in the United States.

chimera: A new virus made by recombining sequence information from two different viral strains.

COVID-19: The disease caused by the virus SARS-CoV-2.

dual use: A technology, pathogen, or toxin that may have a peaceful function, such as treating cancer, but which can also be repurposed as a weapon of terror or war.

first author: On scientific papers, the first author is the person who has done most of the work that results in the publication, often done under the supervision of a more senior scientist.

furin: An enzyme on the outside of certain cells that interacts with the receptor binding domain of a virus by cutting between specific amino acids, helping to activate it and aid the virus's entry into that host cell.

gain-of-function: An experiment that makes a virus more infectious, or more lethal, and/or gives it the capacity to infect new hosts.

genome: A sequence of genetic instructions written in ribonucleic acid or deoxyribonucleic acid, which guides the creation, behaviour, and replication of a virus or a living organism.

last author: Usually the most senior person in a laboratory or among a group of scientists working together who has supervised the research published in a scientific paper.

passaging: A method of speeding up the adaptation of a virus to a new host. This does not involve direct manipulation of the genome but takes advantage of the virus's natural tendency to mutate as it replicates.

receptor binding domain: The protein of a virus that binds to a receptor on a host cell.

receptor binding motif: The portion of the receptor binding domain that merges with the receptor of the host cell, permitting the virus's genetic information to enter the cell and the virus to be replicated.

rescue: The term of art for synthesizing a virus from its genome sequence.

SARS-*CoV-2*: The name given to the virus that causes COVID-19 by an international consortium of taxonomists.

synthetic biology: The creation of a virus or bacterium from its genome sequence. Small segments of the genome are manufactured then inserted into a bacteria and/or yeast, whose cellular organelles put the sequences together in the right order, eventually resulting in the reproduction of a functional virus or organism.

virion: A new virus made by an infected host cell.

virology: The study of viruses.

List of Characters

THIS BOOK DESCRIBES the work of scientists who became central players in the drama surrounding the origin of SARS-CoV-2. Several were born and raised in China, where traditional naming order puts the family name first, followed by the given name. In the last twenty years, Chinese scientists have forged partnerships with Western researchers and their work has frequently appeared in Western journals, where the traditional Chinese name order is reversed—the family name coming last instead of first. Both are shown below.

Basil Arif
Scientist emeritus of the Laboratory for Molecular Virology at the Great Lakes Forestry Centre in Sault Ste. Marie, Ontario. He is a long-time colleague of Zhihong Hu and Shi Zhengli of the Wuhan Institute of Virology and is an associate editor of *Virologica Sinica*, the Wuhan Institute of Virology's peer-reviewed journal.

Ralph S. Baric
A professor in the Departments of Epidemiology and Microbiology and Immunology at the University of North

Carolina, Chapel Hill. He is the leading expert on coronaviruses in the US and has worked with Shi Zhengli to produce chimerical coronaviruses that have pandemic potential.

Chen Wei (Wei Chen)
Born and educated in China. She is a major general in the People's Liberation Army, its leading bioweapons and Ebola expert. She tested her vaccine for Ebola at the National Microbiology Laboratory in Winnipeg thanks to Xiangguo Qiu (see below). She was decorated by Xi Jinping for her efforts in finding treatments for SARS-CoV-2. She created China's first SARS-CoV-2 vaccine.

Cheng K.D. (Keding Cheng)
Chinese born and a Chinese-, American-, and Canadian-educated expert in proteomics. He is the partner of Xiangguo Qiu. Both worked at the National Microbiology Laboratory in Winnipeg until they were marched out by the RCMP in July 2019.

Peter Daszak
UK-born and -educated parasitologist. He is president of EcoHealth Alliance, a US-based charity that funded some of Shi Zhengli's work on coronaviruses via grants won from US government agencies—the United States Agency for International Development (USAID) and the National Institutes of Health's National Institute of Allergy and Infectious Diseases (respectively NIH and NIAID).

DRASTIC
A loosely connected group with expertise in banking, data management, journalism, anarchism, virology, and genom-

ics. They found each other on Twitter as they each began to informally investigate the origin of SARS-CoV-2.

Richard H. Ebright
Molecular biologist and professor of chemistry and chemical biology at Rutgers in the US. He is a vocal critic of gain-of-function experiments and USAID's $200-million PREDICT program, which aimed to predict when a virus might spill over from animals into human populations and cause an outbreak of disease. Ebright was one of the first leading scientists to ask whether SARS-CoV-2 might have escaped from a lab—Shi Zhengli's lab.

David Evans
Professor in the Department of Medical Microbiology and Immunology at the University of Alberta, advisor to the Public Health Agency of Canada and the WHO's subcommittee on variola virus research. He is an expert on smallpox and rescued its close cousin, horsepox, from its genome sequence alone. His publication of that experiment raised fears a terrorist could follow the same recipe to make smallpox.

Ron A.M. Fouchier
A virologist with a special interest in avian influenza virus. He is deputy head of the viroscience department at Erasmus Medical Center in the Netherlands, where his 2011 experiment, which made avian flu virus capable of infecting ferrets, caused an international debate over the safety and value of gain-of-function experiments.

Gao Fu (George F. Gao)
A virologist educated in China, the US, Canada, and the UK.

He has been director of China's Center for Disease Control and Prevention since 2017, as well as being a professor at the Institute of Microbiology, Chinese Academy of Scientists, and president of the Chinese Society of Biotechnology, while also holding various other academic posts in China. Gao is a nexus through which civilian and military research flows together in China. He was a collaborator with Xiangguo Qiu on experiments done at the National Microbiology Laboratory in Winnipeg.

Gary Kobinger
Canadian virologist and vaccine developer. He is a professor at Université Laval, the University of Manitoba, and the University of Pennsylvania. He is the co-developer with Xiangguo Qiu of a monoclonal antibody cocktail for fighting Ebola, for which the two of them received the Governor General's Innovation Award. He was Xiangguo Qiu's boss at the National Microbiology Laboratory in Winnipeg.

Marion Koopmans
Head of the viroscience department at Erasmus Medical Center in the Netherlands. She works closely with Ron Fouchier and was one of ten scientists okayed by China to be on the WHO-convened joint study team to investigate the origin of SARS-CoV-2.

Qiu Xiangguo (Xiangguo Qiu)
Chinese medical doctor, immunologist, Ebola expert, and winner of the Governor General's Innovation Award with Gary Kobinger. Born in China and educated in China and the US and at the University of Manitoba, she became an Ebola expert while employed by the National Microbiology

Laboratory in Winnipeg. She brought in "students" from China to study with her at the U of M and the NML until her security clearance was withdrawn in July 2019. As head of the NML's pathogens and vaccines department, she worked closely with the Wuhan Institute of Virology, George Gao, and various members of the People's Liberation Army, including Major General Chen Wei, its leading bioweapons and Ebola expert.

Shi Zhengli (Zhengli Shi)
Born in China and educated in China and France. She is a coronavirus expert who has gathered viruses from bats by trapping them in their habitats all over China since 2004, earning the nickname Bat Woman. She is director of the Center for Emerging Infectious Diseases at the Wuhan Institute of Virology. Her lab is at the centre of concern that SARS-CoV-2 leaked into the population of Wuhan and the world after gain-of-function experiments done by her. She is also the editor of the journal *Virologica Sinica*.

Wang Linfa (Linfa Wang)
Born in China, educated in China and the US. He is a Chinese and Australian citizen and was instrumental in describing the genomes of two new, lethal bat-borne viruses, Hendra and Nipah. He is a professor at East Normal University in Shanghai and also professor and director of the Emerging Infectious Diseases program at Duke-NUS Medical School in Singapore. He put the Wuhan Institute of Virology's scientists (Hu Zhihong and Shi Zhengli) together with scientists from Australia's BSL-4 in Geelong, and Peter Daszak of EcoHealth Alliance, to investigate bats as reservoirs of coronaviruses.

Zhihong Hu (Hu Zhihong)
Born in China and educated in China and the Netherlands. She is an expert on invertebrates and molecular biology and former director of the Wuhan Institute of Virology. She was responsible for the creation of the Wuhan Institute of Virology's program to sample bat colonies for viruses of interest. She co-authored a breakthrough paper showing that horseshoe bats are the probable reservoir of SARS.

1. The Past Is Prelude

WE LAID MY mother to rest in the small Jewish cemetery just north of our hometown, Saskatoon, Saskatchewan, early in November 2019. We had no idea that a virus multiplying in human lungs on the other side of the world was about to change everything, everywhere; that millions would sicken and die, especially the old and vulnerable; that in the wealthy parts of the world, people of colour, especially South Asian immigrants, would be that virus's prime victims[1] while the 1 percent would largely be spared and the 0.1 percent would make out like bandits; that we would stand on porches to bang pots in appreciation of those on the health care front lines; that a prime minister and two presidents would insist this was just another flu, nothing to worry about, though they knew better, and then would themselves be brought low; that mad stories based on wild theories would roil and fester about where this disaster came from, and why.

Weirdly, though none of this had yet come to pass, on that November day I was thinking about a pandemic one hundred years before—the great Spanish flu pandemic of

1918–19—and how it shaped my mother's life, and mine too.

We scrambled out of our cars to follow the coffin to the grave, heads down and shoulders hunched against a bitter wind. The grass crunched as we walked over it, half-frozen yet still bright green, pretending summer might still remain. The cold, seeping up from the ground and invading our shoes, told a more truthful story. That night, winter would smother everything with thick snow that clotted eyebrows and clung to coats, that mounded on sidewalks and shimmied across the roads when the wind kicked up.

Only the cemetery's wrought iron entry gates looked familiar. My father's father, an ornamental iron worker from Ekaterinoslav, Russia, who arrived in 1903 to homestead in what was then called the Northwest Territories, made them himself. When he created them, this place of burial was far from town amid a vast grassland. Now, it was surrounded by all manner of construction machines ready to rip and rend the greatest agricultural soil anywhere to make way for Saskatoon's growth. All this change was disorienting. I couldn't get my bearings though I'd been to this cemetery many times. I couldn't remember what our father's headstone looked like, never mind where it was, and where our mother's fresh-dug grave must lie.

The rabbi, who knew where he was going, positioned himself at the right headstone, a double that had been erected to mark the graves of both our parents. Our father died first. Dr. Sam Landa, or plain Sammy when our mother was yelling for him, was our family hero. He was the kind of doctor who came very close to killing himself from overwork, the kind of doctor who kept a telephone on the dining room table in case the hospital or a patient might call, who left the house before we woke to do the surgeries he was scheduled to perform. (Except when we went to the lake on holiday. Then he left before we woke to get to the golf course.) He was the kind of doctor who piled us into the back of the car on summer evenings for family outings but would stop along the way to check on children suffering from the polio epidemic of the 1950s.

My mother had lived to 102, a beautiful woman and hardy almost to the end, yet emotionally fragile. Hardiness and beauty were her birthrights; the fragility was the dark gift of that pandemic which, when it finally disappeared after a multi-year killing spree among the young and strong around the globe, seemed to be immediately forgotten, though it left 50 million dead. We'd celebrated her birthday just two weeks before—her white hair done perfectly in a French roll, her lipstick a match for her scarlet silk blouse, her face lighting up with laughter as her great-grandchildren chased each other around the furniture. Her own mother had not been so lucky. She had been killed by the Spanish flu at the age of 24, exactly one day before our mother's first birthday.

Her family lived then in the little village of Southey, in southwestern Saskatchewan, not far from Regina. I have a picture of my mother from those days, a cute little girl with a Buster Brown haircut, sitting on a wooden chair with one leg tucked under her. There is the unpainted siding of a store to her right, the prairie flat as a board behind her. Her father, a handsome man also named Sam, operated that store. His parents had their own homestead a few miles away. His eldest brother, known to some of us as Bad Beril, had abandoned his wife and three children and absconded to America, while his eldest sister, her daughters, and her clever husband lived in the nearby village of Earl Grey. Southey and Earl Grey were commercial hubs for a small group of Jewish colonists from eastern Europe who had come to Canada to homestead courtesy of loans from Baron de Hirsch. They called themselves the Lipton Colony. My father's parents were part of it too, but they came from Russia, while my mother's parents were from Romania. All of them thought that pioneer life on the Canadian prairies, living in sod huts with dirt floors through brutal winters and blazing summers while struggling to turn grassland into wheat fields, was much better than dodging pogroms in the cities of eastern Europe. Or, worse, getting drafted into some king's army.

That was the public story most of the colonists shared—escape from armies and pogroms—but in my mother's family there was another one, a secret one, about why her grandparents, then in their fifties, fled the bright lights of Romania for a dirt-floored, sod-roofed house in the middle of nowhere. According to a memoir written 90 years later by my mother's first cousin, Hope Richman,[2] they had owned a coffee and tobacco shop in a town near Bucharest until Moshe, my great-grandfather, was arrested for selling black market tobacco. Though all the other merchants in town were doing the same, he was the only one dragged off to jail. His wife, great-grandmother Leah, had a friend, a certain Madame Lupescu according to Hope, who happened to be both Jewish and the mistress of King Carol I of Romania. When Moshe was arrested, Leah climbed into a horse and buggy at 4 a.m. and drove hell-bent for leather to Bucharest to plead for help from Madame Lupescu. (Did I mention that Leah was an accomplished midwife who delivered many of the babies born to the Lipton colonists?)

Madame Lupescu sent a note to the king, who arrived at her house immediately and wrote a note for Leah to give to the chief of police, a get-out-of-jail-free card. The king also arranged for a customer to buy the shop, after which my great-grandfather was given a position as the supervisor of an estate.

When this secret story of our origin was first shared with me, years before Hope published her memoir, I had already been a journalist for more than a decade. I knew well that origin stories twist and evolve as they pass from one mind to another, from one generation to another. Details drop off, imagined embellishments replace them, until the whole bears little resemblance to what really happened. I knew right away that this one had to be wrong. It was way too smooth, way too improbable, and a generation removed from the experience of the aunt who recounted it for me. So, I did what journalists do: I fact-checked it.

That Leah was a midwife is true. There may have been a tobacco shop in her family, because she was a smoker from an early age. There *was* a woman named Lupescu who lived in Bucharest and was the famous mistress of King Carol. But that's where invention buried fact. Madame Lupescu wasn't Jewish: her mother and father had been Jewish once, but both had converted. Madame Lupescu was a Catholic. More to the point, her boyfriend was King Carol II, not King Carol I. Magda Lupescu was born in 1899, which would have made her four years old when my Romanian family escaped to Canada. These facts be damned, Aunt Hope believed it, and so did my great-aunt Beckie, the youngest daughter of Moshe and Leah. Aunt Beckie believed it right to the end, and was so ashamed of it she refused to share it with a historian interested in the origin stories of Jewish homesteaders on the Canadian prairies. Even when her own daughter asked her to tell it, assuring her it was safe, Great-Aunt Beckie clamped her mouth shut in a tight line and turned her head away rather than spill these beans.

ORIGIN STORIES HAVE great power. They can become downright dangerous when someone shoves them in a file labelled *Secret*. There was another origin story in my family, one kept secret from my mother for many years. It was not about Madame Lupescu and my great-grandfather's brush with the law; it concerned the life and death of my mother's biological mother—specifically, that she was killed by the Spanish flu pandemic on October 24, 1918.

Hope, who was seven at the time, described it many years later in her memoir. In the fall of 1918, she wrote, everyone in her immediate family was sick with the flu, "but not seriously." There was no doctor available. As another distant relative, Sol Sinclair, explained in his own memoir, the closest doctor lived more than 20 miles away at Fort Qu'Appelle and only came through the Lipton Colony later that winter on a horse-drawn cutter.[3]

It fell to Moshe and Leah to come to Earl Grey to take care of Hope's family. By then, Hope's best friend, Cathy, had died; Hope had attended her funeral. There was a 19-year-old boy named Don whom Hope had a crush on. He worked in the bank. He used to come by her house every day. But one day he failed to appear. Hope went to look for him at the bank and the manager told her he was ill and over at the town hall, but he'd be fine, she should go home. Instead, she ran to the town hall, which was serving as the local hospital, and managed to sneak past a nurse.

The place was filled with beds, and her friend Don was in the second-to-last one. He waved to her and told her to go home; he said her dad would be upset if he knew she was there. Don was dead the next day. That night there was a phone call after she was in bed, never a good sign. She heard her mother crying. "My beautiful aunt Rose," as Hope called my mother's mother, had just died too.

A year or two later, my grandfather married my step-grandmother, a woman from North Dakota with a spectacular sense of humour who, according to one version of the family story, egged my grandfather on to some kind of business that got her the nickname "the Typhoon." They had three children together. The death of Rose, my mother's mother, was known to everyone in the extended family. Yet my mother's grandparents, her uncles and aunts, her cousins, her father and stepmother, kept it from her. No one told her that her stepmother was not her mother, that her own mother had died from the flu. She only found out the truth when, in the depths of the dust bowl Depression, the family moved north to Prince Albert, because it was green, my grandfather always said. Someone who was not a family member said something to my mother that made her ask questions and demand answers. Finally, they told her. The knowledge that everyone she loved had kept this secret from her for 14 years, that everybody she loved knew who her real mother was and what had happened to her—everybody *but* her—and never told her, rocked her for the rest of her life.

No doubt they were trying to spare her pain; instead, they gave her a lifetime of anxiety. Hearing her describe what this discovery meant to her, how it threw her, how she had to struggle to right herself—may be when my abhorrence of secrets began. It is probably why, when I fell by accident into journalism, which entails dragging secrets into the light, it felt like a calling.

IT IS POSSIBLE that the very day we said goodbye to our mother in Saskatoon, deep beneath our freezing feet, down through the centre of the earth to the city of Wuhan, Hubei, China, a new and dangerous virus found its way up someone's nose. Viruses are neither alive nor dead and measure in nanometres, yet as virologist Ron Fouchier explained when asked why he studies them, tiny though they are, they can bring down elephants. This virus had the capacity to lock on to a receptor found on the outside of several kinds of human cells. When it got close to that receptor, a dance ensued between the cell and the interloper. A spike of protein on the virus's surface divided in two, thanks to a protein-splitting enzyme called a furin on the outside of the human cell, helping virus and cell become one. The information system that directs the creation of the virus, a string of RNA (ribonucleic acid) wrapped in a bit of fat, took over the cell's copying machinery. New viruses burst forth and moved into cells next door, to do the same thing again and again and again. Some of the descendants burrowed deep into the host's lungs, some found their way through the vascular system to the heart, some lodged in the kidneys, some fogged the brain. Some were borne by a sneeze or cough to another person, and so on.

Some hosts never noticed the virus's presence; their immune system killed it off without fuss. Others didn't do so well. They got a fever. Their bodies ached. Their heads pounded. Their lungs filled with what looked like ground glass when viewed on an X-ray film. Their blood clotted in the wrong places. They couldn't get their breath. Eventually, they made their way to hospitals. Pneumonia was what it looked like, but a pneumonia caused by what? It did not respond to the usual treatments for pneumonias caused by bacteria or fungi. Some people died.

Over in northern Italy, only a little later, it was the same story. A cough, a sneeze, aches, difficulty breathing, a visit to a hospital. Perhaps coincidentally, Italy was the first European country to sign on to China's Belt and Road initiative, a massive network of new infrastructure built with the help of China-supplied loans and meant to shift the focus of world commerce from America to China. It was the same story in Iran—a sneeze, a cough, a visit to the doctor in the ancient city of Qom where Chinese companies were building a high-speed rail system as part of the same global project. Viruses cannot move on their own; their hosts must expel them close enough to another victim that they can be breathed in and so keep on reproducing. Business meetings are ideal virus–host exchange events. A sneeze, a cough, a handshake followed by a hand moving to a mouth, can do the trick.

Wuhan is China's Chicago, though settlement there dates back 3,500 years. It was formed by the merger of three cities, Hankou, Hanyang, and Wuchang, in 1949. Now it is a metropolis sprawling on both sides of the great Yangtze where it meets the Han River. Wuchang was the capital of the Wu dynasty (220–280 CE), while Hankou was a centre of trade, the site of concessions granted by China's beleaguered government to European invaders in the nineteenth century. More than 11 million people live in Wuhan now. It is a leader in chemical industries, has several universities, many laboratories. Some of its factories used to make fentanyl for shipment to places like the Downtown Eastside of Vancouver, where it wreaked havoc. Wuhan is both ancient and hyper-modern, with many lakes and forests, broad streets and high-rise towers, tiny tumbledown structures, narrow lanes, and ancient shrines. It is also China's largest centre of research into the type of virus that by that November day in 2019 was finding its way into people's noses, a coronavirus distantly related to two others—SARS and MERS—that also have the capacity to sicken and kill.

Wuhan, like Chicago, is a major port. It has direct flights to just about everywhere (except Canada). It is home to China's first Biosafety Level 4 (BSL-4) laboratory for the study of deadly pathogens such as Ebola and Marburg. The BSL-4 lab is located at the Wuhan Institute of Virology (WIV), which also holds the largest collection of coronaviruses in the world, numbering in the thousands. These viruses, or RNA fragments from them, which can be assembled into full viral sequences, have been laboriously collected by the WIV's leading coronavirus researchers, Shi Zhengli and Peng Zhou, and their students. They have been found in the anuses, feces, and blood of bats living in caves and mines all over China. There are other laboratories in Wuhan that also study coronaviruses and send researchers and students to the field to trap bats. Chinese military medical researchers do the same work, trapping bats, then isolating or assembling the viruses they carry to see what they can be made to do. Bats harbour many dangerous viruses, from coronaviruses such as SARS and MERS to Ebola, Nipah, and Hendra.

It took three years for the 1918 flu virus to spread around the world and burn itself out. It took scientists 85 years to isolate that virus, determine its sequence, and "rescue" it—that is, re-create it from its genome sequence alone. It took less than two months for SARS-CoV-2 to spread around the planet, the deadliest pandemic since the one that killed my grandmother. It took Chinese scientists just a few days to determine the genome sequence of the virus, and just one week for scientists in Switzerland to make it in the lab from its published sequence.

Things have speeded up since 1918. As the road signs say: Speed Kills.

I TOLD YOU my mother's story to explain why, when SARS-CoV-2 began to burn like a bog fire through the hospitals of Wuhan, China, and dire stories about its origin and allegations of a cover-up spread with it, it caught my full attention. As the internet filled with chatter about a student at the WIV who might have been patient zero but had disappeared, of an ophthalmologist punished for giving his colleagues a heads-up about something SARS-like making the rounds, of publications retracted, of refusals to admit human-to-human transmission or the true number of the afflicted, it was as if my dead mother fairly screamed at me:

Secrets: dig!

From the beginning, I wondered about the virus's origin. The newspapers and television newscasts insisted that all serious scientists agreed it could only have spread to humans from what they called a natural reservoir—a bat, a pangolin, a snake—likely held in a cage in one of Wuhan's wet markets, where wild animals are killed for the culinary delight of connoisseurs and those who believe wild food is the healthiest. The Huanan Seafood Market was fingered almost immediately as the place where this disease began; it was shut down and sterilized by officials on January 1, 2020. But by the end of January, as the first cases of what was soon called COVID-19 appeared in Toronto and Vancouver and New York and Washington State, that story did not feel right to me, if only because scientists rarely agree on anything, and this wet market consensus seemed to have formed awfully quickly. By the middle of March 2020, after the WHO (World Health Organization) finally got around to declaring a pandemic, though its presence had been obvious for six weeks, it reminded me of the tale of Madame Lupescu—way too smooth, and likely to be wrong.

I know when a story has me in its grip: I start ripping articles out of the papers, followed by net searches and downloads. Eventually there is a big pile of papers on my desk. I'd been doing that since 2016 with stories about China, especially about Canada's relationship with China. Soon there were two large piles on my desk, the China articles on one side, those about the virus on the other. At a certain point I realized they were the same story, that I couldn't understand one without the other.

There was one little problem about digging into the origin of the virus. From March 16, 2020,[4] the prime minister asked Canadians to stop all non-essential international travel, and two days later the borders were shut. Canadians stranded abroad had to be flown home, Parliament was closed, the provinces declared lockdowns, and scientists and officials left their labs and offices to work from home. How could I investigate the origin of this virus if I couldn't get on planes and go knock on doors? I couldn't even get people on the phone. Calling most offices just sent me to voice-mail hell. If I'd held my breath waiting for calls to be returned, I'd have been dead many times over. Email was no better. I finally got coronavirus expert Ralph S. Baric of the University of North Carolina on the phone when he happened to be in his lab and picked it up reflexively. He had done a lot of fascinating work with the leading coronavirus expert at the WIV in China, Shi Zhengli, courtesy of money from the US National Institutes of Health (NIH) and the United States Agency for International Development (USAID). The two of them had joined the sequence for the spike from one coronavirus strain to the backbone from another and ended up with a chimera that can infect human cells and humanized mice and for which there are no treatments or vaccines. (Humanized mice have been genetically modified so they can serve as models for human disease.) In other words, they made a virus with pandemic potential.

Baric was suspicious of me. "What email? I never saw an email," he said when I told him I'd sent one to him explaining who I was and asking for an interview.

"Well, maybe I sent it to the wrong address," I said, though I'd picked the address off one of his published papers. "Is there an address you prefer?"

"If you sent it, you already know the address," he said, as if to suggest I was lying, there'd been no email at all.

"Okay, I'll send it again," I said. "But can we please have a conversation?"

He said he was reluctant. He was not inclined to do so.

"Why?"

"I don't know your take," he said.

That's how politicians talk, I said to myself as I hung up. Why would my "take" matter to a scientist? What's going on here?

AFTER A FEW more truncated conversations and prickly email exchanges, I decided that if I was really going to investigate the origin of this pandemic—and I couldn't seem to stop myself—I would have to do my best to mimic what a journalist named I.F. Stone did long ago. For years, he published a weekly out of Washington, DC, without the benefit of Bob Woodward-style interviews with presidents or off-the-record drinks with political staffers and lobbyists. Instead, he read public documents and wrote about what he found. And what he found was often explosive.

So these field notes have been mainly drawn from months of reading: newspapers; magazines; online publications; articles uploaded to what are known as preprint sites; articles in peer-reviewed journals; chains of emails; the blacked-out responses to access to information requests. In the end, I did manage to do some crucial interviews, but I had to do them on the phone. These voice-to-voice contacts were rare triumphs of journalistic finagling. It's hard to tell human stories when you can't look a person in the eye and build trust. Yet the phone is better than nothing. And it is definitely better than Zoom. On Zoom calls you get distracted by images, especially your own. On the phone you can focus on the other person's emotions, their hesitations. A shift in breathing may suggest something held back that you need to probe. On the phone, you can make better judgments about the value of the information conveyed.

Were important secrets revealed? Oh yes, as you will see. I followed money trails and traced networks of scientists as complex and self-serving as the most intricate spiderweb. I learned that the enterprise of biological science, which has adapted at warp speed to a globalized world, is now done in a way that my father would not have recognized. No scientist with hot news—and there is no news hotter than something on the origin of SARS-CoV-2—is content to wait for a peer-reviewed journal to publish their work. Now, preprint articles get uploaded to servers long before a journal's peer reviewers get around to saying whether the work is filled with error or a masterwork for the ages. Preprint articles are to peer-reviewed science what self-publishing is to literature: the editors of peer-reviewed journals have lost control of the gates to scientific fame and fortune.

In the hunt for the origin of the virus, practicing scientists in Canada, Switzerland, India, the United States, and France teamed with bankers and anarchists and anonymous folks in private Twitter groups to churn out fascinating work uploaded to sites such as bioRxiv or ResearchGate. This turned out to be a good thing from the point of view of getting to the truth, because most research into the possible lab origin of the pandemic had been shoved by editors of the leading peer-reviewed journals to the science equivalent of off-off Broadway. Peer-reviewed journals are supposed to be fair umpires of scientific discourse. Yet they not only failed to publish reliable information about the closest relatives of SARS-CoV-2, which could point to where it came from, they also failed to insist that their authors list their obvious competing interests and failed to publish articles that cried out to be read. Some—*Nature*, *The Lancet*, *EMI*—turned themselves into platforms for the science equivalent of propaganda. The story of the search for the origin of SARS-CoV-2 makes clear that scientific publishers, and some scientists who should know better, allowed themselves to be corrupted by their desire to stay on the good side of the country that will soon be the largest science funder and publisher in the world—China.

But I get ahead of myself.

Investigating the origin of SARS-CoV-2 helped me understand who we should closely question about how this pandemic spread from one side of the globe to the other, sickening and killing millions. And holding people, institutions, and nations to account for that is something we must do, or the tragedy of the pandemic of 2020–21 will be repeated over, and over, and over again.

2. SARS: The Prequel

JANUARY 3, 2020. It was one of those mornings when I had time to read everything in the papers, even a very small note that appeared in the *Globe and Mail*, a pickup from Reuters. It said a serious pneumonia was infecting people in China. Health officials there were investigating 27 cases. The first thing that popped into my head was, "Shit, here we go again." The symptoms sounded like SARS (Severe Acute Respiratory Syndrome), a pandemic that occurred in 2002–3. Was this another SARS outbreak? A bigger one?

Now, in the 20/20 hindsight of the SARS-CoV-2 pandemic, the original SARS seems almost insignificant (except to the families of those who died). SARS was slow to infect and therefore to spread, and too lethal, doomed to die out in humans because it killed too many of its hosts. But it scared the crap out of Canadian public health officials when it arrived in Toronto in the winter of 2002–3. Anyone who lived in the city then, especially if they worked in health care, kept watch for its return, for the sequel.

Toronto's SARS Case A became sick after his mother, who'd recently come home from Kowloon, China, became ill and died on March 3, 2003. He had taken care of her. On March 7, he felt so sick himself he went to the emergency room of a Toronto hospital. At that point, the SARS virus had been circulating in China since November 2002, but the government of China had done an excellent job of suppressing the facts about it, insisting it was just a chlamydia outbreak, treatable with antibiotics, no big deal. Apparently, no one in that emergency room in Toronto believed SARS might reach so far. The patient was not isolated. Those who examined him did not drape themselves in personal protective equipment (PPE), such as N95 masks. And the virus spread from him to them, from hospital worker to hospital worker, from hospital worker to patient, and from hospital to hospital. The WHO issued a travel advisory telling travellers to stay away from Toronto.

Months later, it was officially determined that the virus that caused SARS was a previously unknown coronavirus.[5] These nanometre-sized bundles of fun are basically just the genetic instructions for their own replication. An outer envelope protects a long strand of about 30,000 nucleotides of RNA that encode six genes and several open reading frames—systems that organize the translation of sequence information into proteins via nucleotide triplets that signal when to stop transcribing. Coronaviruses are decorated, as virologists put it, with protein spikes reminiscent of a crown, thus the name. These spikes hook onto molecular receptors on their victims' cells for which they have a chemical affinity, creating entry bridges through which their RNA flows. Once inside the cell, the viral RNA is copied and the host's cellular organelles manufacture the amino acids the RNA sequence specifies into proteins. These proteins self-assemble into new viruses—newly made RNA sequences tucked inside newly made envelopes surmounted by protein spikes. The new viruses—called virions—make an explosive exit and then do the same thing in the cells next door. They're like Aliens with Terminator attributes.

Viruses probably date back to the earliest days of life on this planet, when organic molecules began to self-assemble and combine to form single-celled critters. They are to biology what Silicon Valley is to industry: what they infect, they change, and as they infect, they are changed themselves. As they move from one cell to another, they sometimes pick up genetic bits from their hosts and sometimes leave bits of their own genetic information behind. The human genome is full of viral sequences that are leftovers from past infections passed down through the generations. Some sequences do nothing. Some jump around in the genome as DNA is copied during cell division. Finding their way to new positions may wake them up, renewing their capacity to call up amino acids to form new proteins. They reshape us as they enter our cells, and as they are copied, mistakes—mutations—reshape them. A mutation may turn out to be useful, helping them become more efficient at entry, helping them adapt to new hosts.

The virome—the totality of all the world's viruses—is uncharted territory. It is not known how many viruses lurk in the world; since they're always changing, what's known today may be unknown tomorrow. Different kinds of viruses infect—maybe *predate* is the better word—every kind of living thing. Those that infect mammals pass from one victim to another through sneezes, coughs, touch, and feces. RNA sequences of the SARS-CoV-2 virus have been detected in sewage systems from Italy to Spain to Ontario.[6]

They are characterized by the way their genetic information is arranged. Some, like SARS-CoV-2, have only a single strand of RNA, which makes them prone to rapid mutation. As their RNA is copied by a cell, mistakes happen and go uncorrected. Double-stranded DNA viruses have several systems, including polymerases and enzymes, which recognize errors as they occur and cut out the mistakes.

Viruses are named according to the taxonomic system that categorizes everything alive. Assigning a name involves comparisons with known viral genome sequences and forms to determine where the new virus fits among viral genuses, species, subspecies, strains, and clades. Though viruses are not alive, they aren't dead either. They evolve and change through the pressure of natural selection just as living things do. If they find their way into the right host, they exhibit lifelike intent—the intent to multiply. They can move from one species of animal to another if both have the same type of cellular entry receptor that the virus can exploit. When viruses move from animals to humans, scientists give them a special name—zoonoses—which is ridiculous because humans are animals too.

Coronaviruses were discovered in 1965 in human patients who appeared to have ordinary colds. Since then, four types of coronavirus have been characterized: alpha, beta, delta, and gamma. SARS is a beta coronavirus. So is SARS-CoV-2. They don't mutate as fast as RNA viruses without an envelope, but when they're copied, random mistakes do happen, which are passed on to their virions. The right mutation may help them evade host immune systems, and if it does, virions with that mutation will do well while virions without it may be extinguished. If two different viruses happen to find themselves in the same cell at the same time, they sometimes swap genetic material in a process called reassortment. Mutation, reassortment, and recombination allow viruses to change quickly, which is how they adapt to and evade their hosts' immune systems. Host immune systems must learn to recognize them as they change to fight them off. Shi Zhengli, the leading expert on coronaviruses at the Wuhan Institute of Virology, accused of letting SARS-CoV-2 leak from her lab, has called this interaction an arms race between virus and host.[7]

Until the SARS beta coronavirus came along, alpha and beta coronaviruses were thought to be easily defeated by human immune systems, just like ordinary colds. SARS was a shock: it caused a terrible pneumonia for which no treatments or vaccines were available. SARS took 44 Canadian lives, mainly those of health care workers exposed to it in hospitals. Worldwide, SARS infected 8,098 people in 26 countries and killed 774—a 9 percent mortality rate—before it disappeared.[8]

The SARS virus, like SARS-CoV-2, exploits the ACE2[9] receptor. These are enzymes found on the outer surface of many kinds of cells in the human body—in the vascular and digestive systems, in the lungs, kidneys, gall bladder, and airway. They are also found on the cells of other mammals, including mink, ferrets, cats, rats, even whales. In the mini-drama of the SARS pandemic, and in the coronavirus outbreaks that followed, bats are believed to have been the mammal that spread the virus to humans, either directly or through an intermediary animal, such as camels in the case of MERS (Middle East Respiratory Syndrome). In a groundbreaking article published in *Science* in 2005, bats in China were found to harbour viruses with genomes similar to SARS. Shi Zhengli and several other people who feature in the SARS-CoV-2 origin story—such as Linfa Wang, a scientist in Singapore, and Peter Daszak, who directs the American charity that sent US government money to Shi Zhengli's lab—co-authored that article.

At sundown, bats emerge from their roosts in caves, rock shelters, mines, trees, attics, barns. Mini Draculas on the wing, they hunt insects or perform sexual services for plants as they feed on their fruit. They are the most widespread land mammal and are found everywhere except polar regions. There are about 1,400 species, some of which migrate long distances, like birds, while others stick closer to home. Some hibernate in winter but wake from time to time to get water. Many different viruses circulate in their systems at the same time, making them recombination factories from which new viruses with new attributes will spring. They carry viruses that inflict the most lethal diseases on humans, such as Ebola, Marburg, Hendra, and Nipah. Yet these viruses do not make the bats ill. An explanation for why tiny bats can coexist with such dangerous pathogens has been famously offered by Linfa Wang.[10] Wang believes that because bats are the only mammals that fly, they have fast metabolisms operating at high temperatures, and those high temperatures keep viruses at bay.

THE SPANISH FLU pandemic that killed my grandmother was caused by an avian influenza virus. It is now believed that it jumped from wild migrating birds to humans (and pigs) in Haskell County, Kansas, in February 1918. At that time, about 50,000 US Army recruits willing to fight in World War I were gathering at nearby Camp Funston (later called Fort Riley). A lot of them got sick, very sick, with something that seemed like the flu. Some died, which mystified the military doctors.

Since when do fit young men die of flu? The virus travelled with the soldiers to training camps farther east, then onto ships and across the Atlantic to the trenches of Europe. Eventually it made its way right around the globe to China. It returned to North America as World War I came to an end. It spread to civilians, like my grandmother, who lived along the rail lines that brought the soldiers home. In major US cities like Philadelphia, hospitals and funeral homes were soon overwhelmed. The sick died alone in their beds and in the streets. Pictures of the period show people wearing cotton masks. It was in the effort to figure out what caused this public health disaster, and how to stop it, that medicine in the US was reinvented from a lowly art to evidence-based science according to John M. Barry, who describes the pandemic and its aftermath so vividly in *The Great Influenza.*[11]

Virologists have been able to synthesize functional viruses from their genome sequences alone since the turn of the millennium. Some have done their work while employed by the military, or with the aid of military research grants. That does not necessarily mean they are making bioweapons. It does mean they are pursuing inquiries that the military considers to be consequential, to guard either the health of troops or the security of their nations. In every war, disease kills more troops than bullets or bombs; military medicine has a long history.

The first synthetic virus was made by Jeronimo Cello working with Eckard Wimmer at the University of Stony Brook in the US with funding from DARPA—the US Army's Defense Advanced Research Projects Agency. In 2002, they synthesized polio from oligonucleotides (short runs of RNA or DNA molecules) they ordered from a commercial company. They published their method in *Science*[12] to a great outcry; after all, the rest of the world had been trying to rid itself of polio, not hand bad guys the recipe for how to make it.

In 2005, a researcher at the US Centers for Disease Control, Terrence Tumpey, synthesized the 1918 flu virus that killed my grandmother. Synthesizing an old virus that seems to have disappeared from circulation, or a new one, is characterized in virology-speak as a rescue. Synthesizing the original form of a virus that killed 50 million people and calling it a rescue is like calling an election loss a landslide win and expecting others to believe it.

To synthesize the flu virus, someone first had to figure out its whole genome sequence. It was determined bit by bit, gene by gene, and published in a series of papers between 1997 and 2005 by Jeffery Taubenberger and Ann Reid. Taubenberger was then with the US Armed Forces Institute of Pathology, as was Reid. They plucked bits of the genome sequence out of samples army pathologists had taken from the lungs of soldiers who died during the pandemic and stored for future scientists to study. The samples were extremely small and preserved in paraffin and only got Taubenberger and Reid partway to a full sequence. A larger sample came from the lungs of a woman who died of the flu in 1918 in the Alaskan village of Brevig. Almost everyone else in that village died of it too. In 1951, Dr. Johan Hultin, an American pathologist, tried to get samples of live virus by digging up the remains of people buried there; he'd hoped burials in permafrost would preserve viable flu. He failed. But he succeeded 50 years later when he tried a second time after reading one of Taubenberger's papers.[13] He dug into a mass grave with the help of two teenagers and came upon a woman's body—a very large woman's body. It turned out that her lungs did have live virus thanks to the insulation provided by her significant body fat. He sent the samples to Taubenberger, who finished the sequence. It took Terrence Tumpey a few long, hard weeks in the summer of 2005 to synthesize the virus, working alone in a BSL-3 lab with extra precautions.[14] Centers for Disease Control (CDC) officials spent a lot of time figuring out which precautions should be taken. In their view, the risk of causing a pandemic, if the virus escaped the lab, was very low. People would have immunities, they reasoned. Others were not so sure.

By contrast, a group at the University of Bern, unaffiliated with any military or government agency, synthesized SARS-CoV-2 right after its genome sequence was published. By then it was easy to order up small segments of the necessary base pairs from a company that makes RNA and DNA to order. They made viable SARS-CoV-2 virus in one week flat. They published their method on a preprint server called bioRxiv on February 21, 2020.[15] Unlike Tumpey and his colleagues at the CDC, these researchers could not be sure that a lab escape would cause no problem. No one in Switzerland had yet tested positive for SARS-CoV-2. No one was immune.

IT TOOK ABOUT 18 months for Linfa Wang, Zhihong Hu, Shi Zhengli, and Peter Daszak, along with colleagues from the Wuhan Institute of Virology and Australia's BSL-4 at Geelong, to demonstrate that the SARS epidemic probably began when a coronavirus jumped from horseshoe bats in southern China to people.[16] For some time, most scientists believed that this jump was indirect. The hypothesis was that a bat infected a civet cat, then an infected civet cat was brought into a wet market in southern China as a wild food, and from there, in a spray of fluids or feces, the virus spread to humans. Later work suggested it might have been a direct jump. That publication in *Science* in 2005 was a big deal for the Chinese scientists who worked on it. They were not well-known in the West. It was soon followed by many more papers done by a multinational network of virologists working on bat-borne viruses, especially coronaviruses.

The next bat-harboured coronavirus to cause havoc was MERS (Middle East Respiratory Syndrome). In 2012, MERS jumped from bats to camels to people, spreading to 27 countries, infecting at least 2,249 people, and killing 35 percent of them. It spread less quickly than SARS but was more lethal.

In 2016, an alpha coronavirus called SADS jumped from bats in southern China to piglets, devastating China's pork industry. Viruses that infect pigs can sometimes infect humans. By then, Ebola, another bat-borne horror, had broken out in various countries in west Africa. Again.

The multinational network of virologists was well funded, in part through grants from US agencies, including the National Institute of Allergy and Infectious Diseases, USAID, and the Department of Defense, but also from the European Union and China. They were supposed to be working out how to predict a virus's emergence before it caused disease, figuring out how to snuff it out either with vaccines or with treatments. They did this work by hauling bats and samples from bats, including their feces, back to their labs, trying to isolate viruses or at least assemble their genome sequences, and manipulating isolated or synthesized viruses to see if new iterations might infect humans.

They argued that the purpose of this work was to create vaccines in advance that would protect us from whatever nature might spit out through reassortment, recombination or mutation. USAID called the program it devised to support this work PREDICT. It spent $207 million on it over ten years.

After that first *Science* article, Shi Zhengli became known throughout China as Bat Woman. Until 2004, Shi had studied viruses that infect shrimp. But in the aftermath of SARS, and at the behest of the then director of the WIV, Zhihong Hu, she became a coronavirus expert. Climbing through caves and mines in southern China in search of bats to sample, she and her colleagues eventually managed to isolate a few SARS-like viruses from bats found in a cave in Yunnan, a province about 1,000 kilometres southwest of Wuhan. She worked on them in a BSL-2 lab, in which work is done in a containment cabinet by people wearing PPE, such as masks and eye shields and gloves. BSL-2 labs are used for handling microbes that might infect humans but should not cause severe disease. A hood over the work cabinet sucks all fine particles into a filter and duct system and away from the researcher and the lab. Autoclaves are available to sterilize equipment. BSL-2 labs are found in most university biology departments and hospitals. Before SARS, coronaviruses were not considered to be dangerous pathogens like Ebola, a Risk Group 4 pathogen requiring the highest level of containment.

The Wuhan Institute of Virology where Shi has worked since 1990 publishes a peer-reviewed journal called *Virologica Sinica*. It wasn't much of a journal until it stopped publishing in Mandarin in 1998 and switched to English, the global language of science. After that, scientists based in other countries, including Canada, joined its editorial board. One of those Canadian associates is Xiangguo Qiu, who worked at the National Microbiology Laboratory (NML) in Winnipeg, where she specialized in Ebola. Opened in 1999, the NML has Canada's only BSL-4 labs (one for human diseases, the other for animals), where the most dangerous pathogens are studied.

For many years, the NML in Winnipeg was led by Dr. Frank Plummer, who also taught at the University of Manitoba, and created a relationship with the University of Nairobi, Kenya, to find out why certain women in the sex trade were immune to HIV, the virus that causes AIDS. When the MERS virus first appeared in a patient in Egypt in 2012, a doctor there sent a sample to virologist Ron Fouchier at Erasmus Medical Center in the Netherlands. Fouchier had become famous the year before for making avian flu virus more infectious through a method known as passaging, in a type of experiment called gain-of-function. By injecting the virus into a ferret, which the virus would not normally infect, then extracting serum from the ferret, then injecting that serum into the next ferret and repeating such "passages" ten times, the virus had mutated sufficiently to infect ferrets by direct contact, and then without any contact at all. Fouchier in turn conferred with Plummer, the NML's director.

I point to these cross-the-globe connections to demonstrate that in the aftermath of the SARS pandemic, virology became an intensely globalized science. Researchers in various countries began to work together with the aid of grant money that also flowed across borders. While virology aims, among other things, to protect humans from viral disease, it is also important to national security. Study of deadly pathogens has a dual use: the eradication of suffering by learning how a pathogen is spread, how it infects, and how it might be neutralized or vaccinated against; but also, using the same tools, the creation of militarily useful pathogens, making viruses more infective and lethal, figuring out how to distribute them and how to protect against them in the context of war.

The use of pathogens and chemicals in warfare goes back to the Hittites in 1450 BCE, who apparently found a way to spread tularemia to their enemies in western Anatolia.[17] Poisonous gases were repeatedly used against troops in World War I. The Nazis used Zyklon B to murder millions of Jews and other prisoners. The Japanese, before and during World War II, tested various pathogens on Chinese prisoners. This work was led by Unit 731, a group of military doctors based near Harbin, China. They exposed their Chinese prisoners, both military and civilian, to a long list of infectious diseases of military interest—such as plague—letting the diseases run their natural course, even vivisecting prisoners to see what the pathogens were doing as the prisoners sickened and died. They referred to the prisoners as logs, and they treated them worse than logs. They also worked on delivery systems that could carry pathogens to the enemy while protecting their own troops from infection. The techniques the Japanese developed were transferred to the US military during its postwar occupation of Japan. The US military let the men who'd conducted those experiments escape war crimes punishment in exchange for their information.[18] During the Korean War, the US was accused by China of deploying pathogens on Chinese and North Korean troops with delivery systems perfected by those Japanese military doctors and engineers.[19] After that war, many countries signed the Biological and Toxin Weapons Convention, which forbids offensive—but not defensive—research on pathogens. China did not sign until 1984. It has long been suspected by the US State Department of doing offensive research on biological and chemical weapons.[20] Russia is also a signatory. After the fall of the Soviet Union, Russia reiterated its commitment to the treaty and said it had eliminated its biological weapons. But that hasn't stopped it from using Novichok, a nerve agent developed in the former Soviet Union, to try to assassinate those considered threats to Vladimir Putin.

Because of the dual-use potential of the scientific study of dangerous pathogens, security clearances are required for personnel working at the NML in Winnipeg. Xiangguo Qiu and her husband Keding Cheng, both born and educated in China, had such clearances. Xiangguo Qiu worked mainly on bat-borne pathogens. In 2018, she shared a Governor General's Innovation Award with Gary Kobinger, formerly her boss at the NML, now at Laval, for a cocktail of monoclonal antibodies they devised in association with an American company to treat Ebola. But on July 5, 2019, Xiangguo Qiu and her husband, as well as her students from China, were escorted from the NML by the RCMP. Their security clearances had been revoked, their work at the NML therefore suspended, and their adjunct status at the University of Manitoba terminated.

Winnipeg is a smallish town; word gets around. On July 14, 2019, the story of their suspension and removal made the CBC news. The rumour was that it had something to do with smuggling pathogens to China. Staffers at the NML complained to a reporter that Xiangguo Qiu had spent a lot of time training people to work in China's new BSL-4 lab at the WIV. But official spokespeople for the NML insisted the reason they'd been escorted out had nothing to do with anything untoward; it was just a possible "policy breach," no risk to the Canadian public,[21] so move right along, nothing to see here....

Thanks to an access to information request by Karen Pauls of the CBC, Canadians learned in the fall of 2019 that Xiangguo Qiu had not *smuggled* anything to China but had arranged with all necessary approvals the shipment of various strains of Ebola, Hendra, and Nipah viruses to the WIV, possibly without a material transfer agreement that should have set out how they could and could not be used. Material transfer agreements are routinely signed between institutions sharing samples for scientific study. They clearly establish ownership so that the recipient cannot patent and profit by what the sender has developed.

I had read those stories with interest. They were in the China pile on my desk. Then I came upon a preprint article by two Chinese scientists who alleged that SARS's sequel—SARS-CoV-2—might have leaked from one of two labs studying bat coronaviruses in Wuhan. They mentioned the Wuhan CDC lab and Shi Zhengli's lab at the WIV where Xiangguo Qiu was said to have spent time training staff. That's when I put the China clippings and the SARS-CoV-2 clippings together in one big box.

BACK TO SARS. Through the late winter and spring of 2003, like the late winter, spring, and fall of 2020, Canadians became too familiar with the faces of infectious disease experts who appeared on television night after night explaining what was going on. They seemed to know what they were talking about, yet SARS's second wave caught them by surprise. After SARS was finally snuffed out, a commission of inquiry led to major changes in the way Canadian hospitals deal with patients who arrive in emergency departments with respiratory ailments. Isolation cubicles were set up. Triage nurses' instructions were changed. Much more care was taken in infection control throughout hospitals. Stockpiles of PPE were stored in strategic locations by the federal government and the province of Ontario because there had not been enough PPE available when it was needed.

Of all the countries outside China that suffered from SARS, Canada, specifically Toronto, took one of the harder hits.[22] The blame fell on China.

The Chinese government's public health officials had fumbled the ball. Its political leadership had made things worse by suppressing the fact that from late 2002 they knew a severe infectious pneumonia was circulating in Chinese hospitals that did not respond to normal treatments and killed people. Though experts at two institutions, including China's Academy of Medical Military Sciences, had identified the cause of the illness[23] as a coronavirus, public health officials at the China Center for Disease Control and Prevention insisted that people were getting sick from chlamydia, a bacterium treated with antibiotics. This was the approved line adopted by the Communist Party leadership for at least three months. China failed to inform the WHO, failed to warn other nations, and failed to ask for help. Eventually, pulmonologist and Party member Zhong Nanshan took his life in his hands and bucked the party line in public. He insisted that SARS was very serious, that it was not caused by chlamydia but by a virus, that it was not treatable with antibiotics, but that it responded to cortisone, an anti-inflammatory drug.

An Italian doctor, Carlo Urbani, working for the WHO in Vietnam, Laos, and Cambodia, got a sample from an American patient in Hanoi late in February 2003 and isolated the virus. He died of SARS himself a few weeks later. One reference strain of SARS—Urbani—is named after him. Another—the Tor2 strain—is named for virus isolated from a patient in Toronto. When the SARS genome was published in *Science* that May, it became utterly clear that the Chinese Communist Party's political line had prevailed over good science for far too long, permitting the virus to spread and kill.

Even after the WHO was allowed to send an international team of experts to China to investigate, uncooperative Chinese officials went so far as to load SARS-infected health workers into ambulances and drive them around to keep the investigators from interviewing them. When rumours circulated that SARS cases in China's military hospitals were not being reported, and WHO investigators tried to follow up, officials denied them access. Eventually, cooler heads prevailed. Scapegoats were found and duly punished. The minister of health was fired. A public relations campaign began. Leading virologists and epidemiologists from Australia, the US, the UK, etc., were invited to China to give advice on strengthening its public health service. Zhong Nanshan, the whistle-blower, was touted as a national hero.

When it was all over, the WHO drafted new International Health Regulations that were adopted by its member states to prevent any future prevarications regarding infectious diseases circulating within their borders. These rules make clear that if a SARS-like disease occurs in a signatory's territory, it must notify the WHO immediately. They also make clear that borders are not to be shut unnecessarily to contain an outbreak, in order to avoid stigmatization and the kind of economic losses that China and Canada suffered over SARS. Canada signed on. China signed on. George F. Gao, who had studied at Harvard and Oxford and briefly at the University of Calgary as one of the first generation of Chinese students allowed to study in the West after the overthrow of the Gang of Four,[24] was enticed to come home and help reorganize public health.

For a time, China appeared to be co-operating with the rest of the world. For a time, it *was* co-operating. China determined that it needed the highest containment laboratory, a BSL-4, so deadly pathogens could be safely studied. It had only two BSL-3 labs. Both were operated by the People's Liberation Army (PLB). China entered into an agreement with France for French engineers and technicians to design and build a BSL-4 at the WIV. (Chinese contractors ended up building it and the French architects refused to sign off on what they'd done.) France entered this relationship on the understanding that the world's scientists would be able to study in the WIV's BSL-4 when it was complete, and that French scientists would be welcome. French military and intelligence officials had warned repeatedly against helping China,[25] worried that China would use a BSL-4 for military purposes. But the politicians overruled their fears and even sent China three mobile BSL-3 labs. Later, China announced that it intended to build a total of five BSL-4s, two for its military.[26]

Leading Western scientists were invited to China to give lectures and to take up major teaching positions, with labs and living expenses and large salaries provided. If they came, they were honoured and given prizes as friends of China. Thousands of aspiring Chinese biology researchers had already headed to universities in Canada, the US, the UK, France, Germany, and the Netherlands to study and to build this burgeoning international house of biological science. Many became citizens in the countries where they studied, building careers, stepping into important posts, linking East to West. Like the radical change in medical education and research in the US after the 1918 Spanish flu pandemic, China, after SARS, remade itself into a champion of global biological science. But the funding mainly flowed one way, from the West to China, even though by 2019 the Chinese economy was almost as large as the American. In 2016, a year before the WIV's BSL-4 was finished, China asked France for "dozens" of the airtight containment suits that must be worn by researchers working in BSL-4s. This time France's dual-use commission said no, fearing that China was trying to outfit military BSL-4s at France's expense. But France still gave five million euros for research to be jointly conducted in the new Wuhan BSL-4 by scientists from both countries working together. Apparently, that didn't work out well either. Only one French researcher was allowed to work in the WIV's BSL-4.[27]

After 2013, with the installation of Xi Jinping as president of China, and also as general secretary of the Communist Party and then as chairman of the Central Military Commission, China grew less co-operative. While insisting China supported globalization, Xi's idea of globalization turned out to be like Xi's version of socialism, an odd ideology rooted in ethnic nationalism (the Han are 90 percent of China's ethnic mix; minorities including

Tibetans, Uyghurs, and Mongolians are supposed to assimilate) expressed through military expansion into neighbouring territories, all wrapped in a cult of personality—Xi's. Xi's Socialism with Chinese characteristics means billionaires may flourish with the help of the state so long as they obey the Party, but they may find themselves in jail or disappeared altogether if they pose a threat to the leader. On the grounds of stamping out corruption,[28] Xi's officials purged and prosecuted thousands inside and outside China.[29] They also began the practice of hostage diplomacy, in which foreign nationals working in China were snatched and held in appalling conditions (lights on 24/7, solitary confinement) until China got what it wanted from their governments. In December 2018, Canadians Michael Kovrig and Michael Spavor were grabbed and jailed and eventually charged with espionage after Meng Wanzhou, the daughter of a favoured billionaire, the founder of telecom giant Huawei, was arrested at Vancouver International Airport. The US government had made an extradition request for her.

GLOBALIZATION WITH CHINESE characteristics was also expressed through the Belt and Road initiative, designed to place China at the centre of world commerce as it had been long, long ago. Its projects include high-speed rail lines and mega ports to link China to places such as Nigeria, Greece, the Congo, Iran, Italy, and Australia. More than a million Chinese entrepreneurs, armed with loans from state-owned banks, fanned out to strategic places, especially in Africa. The Chinese military set up a new PLA naval base in Djibouti. It flexed its military muscles in all directions, from the Himalayas to the South China Sea. In the Himalayas, on the border with India, its troops skirmished with Indian troops, both sides using sticks and fists. It squatted on disputed land on the border with Bhutan. To the outrage of neighbours, China set up military bases on man-made islands in the strategically vital South China Sea, claiming as its traditional waters areas already claimed by others such as the Philippines. It set about bringing democracy activists to heel in Hong Kong while "re-educating" Uyghurs in Xinjiang in prison camps to stamp out the possibility of terrorist attacks by Islamist radicals. Eventually, word trickled out to human rights groups about what else was going on, besides instruction in Mandarin and Party dogma, behind the camps' walls topped with electrified wire and surveillance cameras. They charged that China was also using Uyghurs as forced labour, forcing Uyghur women to have IUDs inserted, or sterilizing them, tearing down mosques, indulging in torture. The Canadian House of Commons Subcommittee on International Human Rights would eventually declare China's actions against Uyghurs to be a form of genocide.[30]

After the election of Donald Trump as president of the United States, Xi's Make China Great Again strategy ran head-on into Donald Trump's US version. Trump raged about China's alleged bad economic behaviours—cheap labour and lax environmental controls that lured US corporations to manufacture in China instead of at home; currency manipulation; widespread theft of intellectual property; alleged cyber and industrial spying; attempts to corner markets in strategic commodities; attempts by China's leading communications technology companies, especially Huawei, to become lead suppliers to the West's 5G telecommunications equipment market. If Huawei equipment was used in these new networks, the US argued, Huawei could spy. If the government of China told Huawei to spy, it would have no choice but to comply, because China had passed two laws requiring such compliance.[31] Canada is the only country in the Five Eyes intelligence-sharing network (the US, Australia, New Zealand, Canada, and the UK) still considering using Huawei's 5G equipment, though US lawmakers warned that Canada could be cut off from US intelligence if it chose to do so.[32] Huawei's 5G products have already been banned in the other Five Eyes nations. Canada has studied the matter for so long that Canadian telecom companies bought 5G equipment elsewhere. The US imposed wave after wave of tariffs designed to strip Chinese companies of advantage in US markets. China imposed counter tariffs, especially on American farm products, hoping to damage President Trump's re-election prospects. This was called a trade war and sometimes a new Cold War.

Eventually, the US and China settled into trade talks just as the US House of Representatives voted to impeach Trump and SARS-CoV-2 began to circulate widely in Wuhan. And then it was SARS revisited: déjà vu all over again.

3. Political Science 1.0

ON JANUARY 8, 2020, another news story about the mysterious pneumonia circulating in Wuhan appeared in my *Globe and Mail*. The headline said it was caused by a virus from the same family as SARS and MERS, though no other details were given. It said the case count in Wuhan had climbed from 27 to 59. If the first story on January 3 had seemed ominous, this one reverberated in my head like a drum roll. The numbers had doubled in only five days, the mark of an epidemic disease. A top official in China's Ministry of Transport explained that disinfection would be done in major transportation hubs because at least 440 million trips were about to be made on trains, and 79 million on planes, as people gathered with their families for Lunar New Year, China's most important annual holiday. That suggested someone in authority was worried, but apparently not worried enough to tell people to stay home. Another official was quoted as saying this virus seemed to cause severe illness in some but "does not appear to pass easily from person to person."[33]

How else, I found myself thinking, could the case count double in five days unless it *was* passing easily from person to person?

On January 14, a news story appeared in *Science*, one of the world's top scientific journals. Any scientist whose article is published by *Science* or *Nature* counts it as a career high. Both also carry news items written by their own staff journalists. The *Science* story was about the SARS-like disease circulating in Wuhan, China. It quoted Chinese authorities who claimed that only one person out of 42 had died, so this new virus was not as lethal as SARS. Again, the story said that "no evidence suggests the virus easily passes between humans."[34]

There is an old mantra I learned when doing a book on archaeological science: absence of evidence is not the same as evidence of absence. Now, whenever someone makes an argument based on no evidence, I pull out the magic word: *yet*. And how exactly had 59 cases shrunk to 42?

"It's a limited outbreak," an official from China's Center for Disease Control and Prevention told *Science*. "If no new patients appear in the next week, it might be over."[35]

The *Science* story also suggested that Chinese researchers were being more open than they had been about SARS: they had just shared with the world the full genome sequence of the as yet unnamed virus so it could be compared with other known sequences of SARS-like coronaviruses, and so test kits could be manufactured to permit accurate diagnosis. Yet in immediate contradiction of this supposed openness, the report noted that it wasn't officials from China's Center for Disease Control and Prevention who'd released the genome sequence first; it was two academic researchers.

On January 10, Australian virologist E.C. (Eddie) Holmes, of the University of Sydney and China's Fudan University, and Zhang Yong-Zhen, of Fudan University and the Center for Disease Control and Prevention in Beijing, had uploaded the full sequence to a blog called Virological.org. The blog is run by Andrew Rambaut, a virologist at the University of Edinburgh. Blogs published out of Edinburgh are not subject to Chinese internet censorship. The very next day, Chinese groups investigating the disease under the direction of the National Health Commission of China posted three more genome sequences to GISAID, a database normally used for sharing flu virus sequences.[36] It was as if China's officials had intended to hold off on publication of the genomes but had been forced to act by the academics.

By 2020, suppression of facts about an epidemic in China was much harder to manage than it had been in 2002.[37] Western journalists prowl Chinese social media; China's WeChat has over a billion monthly users. Weibo, like Twitter, is also heavily used. On these sites, rumours about a SARS-like pneumonia in Wuhan had been circulating since late December. That's why, by January 3, a story about an infectious disease in Wuhan had made it to my morning newspaper. The social media rumour mill knew better than *Science* that these four posted genomes were not the only ones sequenced—or at least partially sequenced. Three commercial labs had sequenced samples that were sent to them for identification by doctors in Wuhan hospitals in late December. The first company, Vision Medicals, noted that the sequence they found was similar but not identical to SARS and reported that to authorities. The biggest commercial sequencing company in China, BGI, which has interesting ties to China's military, also sequenced samples and noted the similarities to SARS. The third, Capital BioMedlab, shared its results with the Wuhan hospital where an ophthalmologist named Dr. Li Wenliang worked. Dr. Li saw the report. It said this virus looked like SARS. On the evening of December 30, he warned a few colleagues in a small online chat group that SARS might be circulating again and to take precautions. Somebody shared that warning, which sparked a social media uproar. Dr. Li and seven others were arrested by police and charged with publishing untrue material. He was made to sign a confession. On Chinese social media, Dr. Li was soon dubbed The Whistleblower, even as censors tried to scrub his name and any references to him from the net.[38]

By early February, Dr. Li had contracted SARS-CoV-2 and died, his status elevated from whistle-blower to truth-telling martyr. He became a figure that the Cyber Administration of China, created by Xi in 2014, reporting directly to the Central Committee of the Communist Party, was hard pressed to contend with. A video circulated of his mother sadly reminiscing about her son. Li's Weibo account was flooded with comments. Orders went out to the Cyber Administration's hundreds of thousands of paid trolls. These people are teachers, low-level government workers, and students who are paid 25 dollars per post longer than 400 characters, 40 cents for flagging posts for deletion, and one cent per share according to one leaked document.[39] Trolls were told to watch out for posts with images of candles, or no image at all, just blackness, the better to play down the "incident." But they didn't scrub them all: I found a YouTube video of people offering flowers and bowing in front of Dr. Li's image.

Under the terms of the International Health Regulations set out by the WHO in the aftermath of SARS, Chinese officials should have alerted the WHO to this SARS-like problem back in December, as soon as they were informed that something SARS-like was making people ill. Instead, before any officials had said anything in public about the virus, the commercial labs involved were ordered by the authorities to destroy or turn in to unnamed designated testing institutions any remaining samples they had.[40]

Professor Zhang Yong-Zhen at Fudan University was sent samples to test after the commercial labs. He received the packages, packed in dry ice, on January 3. Two days later, his lab, working with E.C. Holmes, had the full sequence.[41] Two days after that, while Zhang was on a plane and Holmes was in Australia, they conferred about what to do—post it to Rambaut's blog or hold it? Zhang decided to post it. They also submitted an article to *Nature* co-authored with scientists from a Wuhan hospital and Tian Jun-Hua from the Wuhan Center for Disease Control and Prevention. Tian Jun-Hua, like Shi Zhengli, had been crawling in and out of bat caves, sampling them for viruses for years.

Zhang and Holmes and colleagues had determined the virus's sequence from a fluid sample taken from a sick person who had been admitted to Wuhan's Central Hospital on December 26. The man worked at the Huanan Seafood Market. The genome sequence was of an unknown coronavirus. They labelled it Wuhan 1. They noted that this new coronavirus made use of the human ACE2 receptor and showed no sign of any recombination event in its immediate history. Recombination happens when two strains of a virus, while in the same host cell, swap genetic material as they are being replicated. No recent recombination event suggested that this virus had not undergone any sudden big change. When they compared the sequence to other known coronaviruses, such as SARS and MERS, and to other published coronavirus sequences taken from bats, they found the closest genomic relative had been extracted from a bat sampled in a city on China's eastern seaboard more than a thousand kilometres from Wuhan. Although their paper was submitted to *Nature* on January 7, it was not accepted until January 28 and not published online until February 3.[42] Every day counts in a pandemic. The longer you wait, the harder it is to contain. I figured Holmes and Zhang must have been extremely concerned and that's why they uploaded the sequence onto a blog first instead of waiting for *Nature* to publish. I wrote to Holmes to ask for an interview, but he did not respond to my email.

ON JANUARY 15, one day after that *Science* story appeared, Donald Trump signed the trade pact with China.[43] One day after that, January 16, a Thursday, his impeachment trial began in the Senate (he was acquitted on February 5). The following Monday, January 20, President Xi Jinping finally issued a public statement about this new infection, admitting it was spreading person-to-person. By then, Chinese officials had known for at least six days that they were facing a pandemic, according to internal documents later obtained by the Associated Press. Yet 40,000 people had been allowed to gather for a civic banquet in Wuhan, the Communist Party had held two important meetings in the city, and many people had been allowed to travel out of Wuhan and China either on business or for the upcoming holiday.[44] By that point, according to a CNBC report, 3,000 people were likely infected.[45]

That day, Monday, January 20, Nathan VanderKlippe,

the *Globe and Mail*'s Asia correspondent, published a story under the headline: SARS-LIKE VIRUS HAS BEEN TRANSMITTED FROM PERSON TO PERSON, CHINESE HEALTH AUTHORITIES SAY. The third paragraph put the official case count at 224. Only six days before, it had been 42. Or was it 59? VanderKlippe quoted the hero of the SARS debacle, Zhong Nanshan, who had been appointed on December 31 to lead the expert investigatory teams sent by China's National Health Commission to find out what was going on in Wuhan. Zhong said he was not concerned. "I believe the outbreak will not cause the same impact on society and the economy as SARS did 17 years ago."

That seemed like slippery talk to me. It would turn out to be a Möbius strip of a statement, on one side true, on the other false, truth meeting falsity at the twist point. It was certainly true that the impact on society would not be the *same* as SARS. The total number made sick worldwide by SARS had numbered in the thousands and the dead in the hundreds. SARS-CoV-2 would kill millions and sicken hundreds of millions. The economic impact would be beyond calculation.

Yet the *Globe* story made clear that Canadian officials were pleased to buy this no-big-deal line China was selling. The Public Health Agency of Canada assured Canadians thinking of visiting Wuhan, which would be brutally locked down only three days later, that the risk was assessed as low. And, said VanderKlippe: "The agency also told the *Ottawa Citizen* that the risk is low in Canada because there are no direct flights from Wuhan."[46]

What the hell, I muttered as I clipped the story. Since when does a virus stop being infectious when somebody changes planes?

THE SAME DAY Xi Jinping made his announcement, several groups of China's scientists submitted five articles on the new coronavirus to three of the world's leading peer-reviewed scientific journals, *The Lancet*, *New England Journal of Medicine*, and *Nature*, as well as to a lesser journal edited in China but with editors in the US called *Emerging Microbes and Infections* (EMI). Clearly, these submissions had been coordinated—how else could people working in different institutions in different parts of the country submit their articles on the same day? This suggested they had been reviewed first by a higher authority and that higher authority wanted to control the story of the outbreak. Most were published online on or after January 24, the day after the lockdown of Wuhan, except for an article from Shi Zhengli's lab. It was sent to *Nature* on the 20th but appeared first on the preprint site bioRxiv on January 23. *Nature* published it online, along with Holmes and Zhang's much earlier submission, on February 3. That these papers were all published so quickly suggests peer review was scant, which sometimes happens at major journals when timely articles are submitted by well-respected authors. And what could be timelier than articles from Chinese experts about the Wuhan problem?

They seemed to describe objectively the nature of the virus, its genome sequence, which SARS-like coronaviruses it was most closely related to, the chief symptoms of the disease, which treatments were proving useful and which were not, with some hints about where the virus came from. Submission dates are shown on both peer-reviewed publications and most preprint sites. That five papers (but not Zhang and Holmes's paper) had the same submission date should have made alarm bells ring in editorial offices that this was not ordinary science, but political science.

George F. Gao (Gao Fu) was a co-author on two of the papers. By then Gao had become the director of China's Center for Disease Control and Prevention. He is a man with many friends in the West and is thought to be a friend to the West. Yet the head of the US Centers for Disease Control (CDC), Dr. Robert Redfield, had been trying to find out from Gao since January 3 what was going on with these pneumonia cases in Wuhan. As with most other cases of Westerners trying to get information from their contacts in China, Gao had told Redfield very little, which didn't help Redfield's reputation with the White House.[47]

Three of the articles suggested the outbreak had begun at the Huanan Seafood Market in central Wuhan, a market that had been shut down and sterilized on January 1. But one article in *The Lancet* with 29 co-authors (the first author, Chaolin Huang, is at Jin Yin-tan Hospital, Wuhan, and the last author, Bin Cao, has four appointments in different institutions) contradicted those assumptions. It made clear that the earliest known case, a man whose symptoms first appeared on December 1, had had no contact with that wet market at all. That same paper also showed that person-to-person transmission was known as of December. Titled "Clinical features of patients infected with 2019 novel coronavirus in Wuhan, China," the *Lancet* article said that by January 2, 41 patients had been identified by PCR (polymerase chain reaction) test as having been infected with the novel coronavirus. Of those 41 patients, only 27 had been exposed to the Huanan Seafood Market and one family cluster had been found, signifying person-to-person transmission. Sixteen people—or 32 percent—had been admitted to the ICU, and six had died. These were horrible numbers, especially considering only half the patients had underlying diseases and their median age was 49. All showed pneumonia in lung X-rays. Some had headaches and diarrhea, some had acute cardiac injuries and kidney problems. The first patient to die was a man who had had continuous exposure to that market. He was admitted to hospital after seven days of fever and cough. "Five days after illness onset, his wife, a 53-year-old woman who had no known history of exposure to the market, also presented with pneumonia and was hospitalized in the isolation ward." The patients studied had been admitted to hospital between December 16 and January 2. As the earliest patient's symptoms had appeared on December 1, the first known infection had begun in November. That patient's infection with the new virus had been confirmed retrospectively by a lab test specific to the new virus. More to the point, "no epidemiological link was found between that first known patient and later cases," meaning the Huanan Seafood Market was certainly *not* the point of origin.

More disturbing still, the authors reported that by January 22, 2020:

> 835 laboratory-confirmed 2019-nCoV [translation: novel coronavirus] infections were reported in China, with 25 fatal cases. Reports have been released of exported cases in many provinces in China, and in other countries; some health-care workers have also been infected in Wuhan. Taken together, evidence so far indicates human transmission for 2019-nCoV. We are concerned that 2019-nCoV could have acquired the ability for efficient human transmission. Airborne precautions, such as a fit-tested N95 respirator,

and other personal protective equipment are strongly recommended.[48]

This was a blunt warning: a very nasty virus much more infectious than SARS was dancing across the landscape. Oddly, though the information about the first patient could have provided a vital clue to the origin of the virus, the authors said nothing about where he lived or worked. Only his gender was referred to and the fact that no one in his family had exhibited symptoms. Yet these authors *were* interested in the point of origin, referring to an earlier paper on which Shi Zhengli was the last author and therefore the senior leader on the project. In 2013, Shi Zhengli's group at the WIV had reported in *Nature* that they found two SARS-like viruses circulating in bats in Yunnan that made use of the ACE2 receptor.[49] That meant these viruses could likely jump straight to people.

Virologists reading the *Lancet* paper would have known that the genome of one of the viruses isolated by Shi's group and described in that 2013 article had been used in very controversial experiments published in 2015 in a collaboration between Shi's lab and virologist Ralph Baric's group at the University of North Carolina. Baric's group took the gene for the spike from one of the isolates and combined it with the genes of another coronavirus that infects mice. From this recombined sequence they synthesized a new virus that easily infected human cells and humanized mice. Their experiment also showed that no known antiviral treatment worked against such a virus and there is no vaccine against it. Though the virus was not naturally created in the wild, their assumption, explained in the paper, was that nature might well create such a virus, and if it did, the virus would have pandemic potential. Why do such a study in the first place? To try to figure out what nature might do before it does it and prepare vaccines in advance.

This paper was published four years after Ron Fouchier, at Erasmus Medical Center in the Netherlands, and Yoshihiro Kawaoka, at the University of Wisconsin in the US, did gain-of-function experiments on avian flu virus. It was an avian flu virus that jumped to humans and unleashed the 1918 pandemic. Both Fouchier and Kawaoka genetically modified avian flu viruses to see if they could make them more infectious. Using the passaging method, Fouchier succeeded in making it spread by airborne transmission between ferrets.

Fouchier's and Kawaoka's papers describing their methods and results were submitted to major journals—*Science* and *Nature*. The editors of both journals were afraid to publish for fear of teaching terrorists how to cause a pandemic. The government of the Netherlands had given Fouchier permission to do the experiment but later argued that publishing it amounted to exporting a known pathogen, something that required government permission. Fouchier's group had a Dutch grant and a European Commission grant, and included co-authors who were advisors to and shareholders of pharmaceutical companies. Kawaoka claimed relationships through speaking engagements with a long list of pharmaceutical companies. The National Institutes of Health (NIH) in the US had also funded both scientists. When *Nature* and *Science* raised a fuss, the NIH told people doing similar work to stop those experiments until further notice. Ralph Baric had to put his other gain-of-function experiments on hold while the NIH held a series of meetings and conferences to try to figure out whether the risks of this kind of work were worth taking and whether the work should be published. Scientists outside the US voluntarily stopped doing similar work for 60 days.[50]

By 2012, Fouchier's and Kawaoka's works were published[51] after the WHO and US National Science Advisory Board on Biosecurity decided they should be. *Science* then published an article in 2013 by a Chinese group led by Hualan Chen at Harbin Veterinary Research Institute (which will house China's second BSL-4). Chen's group had combined genetic sequences from two viruses, an avian flu virus and a swine flu virus that had caused a small pandemic in 2009, arguing they might do the same on their own in nature. They found the hybrid could infect guinea pigs by airborne transmission.[52]

Rules and oversight for flu experiments were then formulated by some countries, requiring that such experiments only be done in enhanced BSL-3 labs, though Canada required they be done in BSL-4 conditions. Everything quieted down. But beginning in 2015, stories in *USA Today*[53] revealed that even ultra-secure BSL-4 labs at the US CDC and at the NIH had "a trio of biosafety failures involving variola virus (the causative agent of smallpox), *Bacillus anthracis* (the bacterium that causes anthrax), and avian influenza."[54] No one died, no one got sick, but if leaks could happen from the most secure labs in the US—and there are 15 BSL-4s in the US, handling the most dangerous pathogens, and about 42 worldwide—what might be leaking from university labs, which are mostly BSL-2s? The *USA Today* stories showed there are at least 200 BSL-3 labs in the US alone, and public reports on accidents do not provide the names of labs where accidents occur or list what they're working on. "Even the federal government doesn't know where they all are, the Government Accountability Office has warned for years," one story said.[55] This set off another round of investigations and debates on whether gain-of-function experiments should be permitted, this time led by the White House. But by 2017, gain-of-function experiments were once again being funded by the US government under rules considered by some to be less than rigorous.

A significant number of scientists had complained since 2014 that these experiments are plain reckless. They created an organization, the Cambridge Working Group, to protest and to inform the public.[56] To those in the know, the reference in the *Lancet* article to Shi's 2013 isolation of bat viruses brought Shi and Baric's dangerous 2015 gain-of-function experiment to mind. Some wondered: Had Shi been doing similar gain-of-function experiments in her own lab at the WIV? Had this new virus been whomped up there—and escaped?[57]

THE SECOND ARTICLE submitted January 20 and published online on January 24 in *The Lancet* didn't appear to accept as proven fact the first article's contention that the virus was efficiently transmitted between people. Titled "a novel coronavirus outbreak of global health concern," its four authors included George F. Gao. Despite the skyrocketing number of infections, despite known transmission of cases among hospital staff, these authors seemed unconvinced about human-to-human transmission, though if shown, they averred, it would obviously be worrisome. "The virus might further spread to other places during this festival period and cause epidemics, especially if it has acquired the ability to efficiently transmit from person to person."[58]

This article was also unclear about the role of the Huanan Seafood Market, suggesting it was the likely point of origin, though the first article in the same journal had taken that off the table. "Exposure history to the Huanan Seafood Wholesale Market served as an important clue at the early stage, yet its value has decreased as more secondary and tertiary cases have appeared." The authors warned of possible epidemics due to the 15 million trips usually made to Wuhan for the Lunar New Year festival. That's why Wuhan had been ordered shut down the day before. They also offered that speed of information sharing would be very important, which was an odd thing to put in writing after Gao failed to give information to Redfield, head of the US CDC, who had been asking for it since January 3. "Rapid information disclosure is a top priority for disease control and prevention," they said.[59] And yet 14 days had gone by since the first publication of the genome on a blog, and still more days had transpired since medical staff in Wuhan became ill after treating patients, demonstrating efficient human-to-human transmission. According to the first article, an epidemic, possibly a pandemic, was already well under way. According to this one, not yet.

George F. Gao was also a co-author on the third paper submitted the same day as his *Lancet* article to the *New England Journal of Medicine*. It too was published online on January 24. It was titled "A Novel Coronavirus from Patients with Pneumonia in China, 2019." It explained that local Wuhan health officials had first reported clusters of pneumonia cases beginning on December 21, 2019. Yet it wasn't until December 31, five days after samples were extracted from three patients and a novel coronavirus was identified, and one day after Dr. Li's social media post to colleagues warning of SARS circulating in Wuhan, that China's Center for Disease Control and Prevention dispatched a "rapid response team" to work with Hubei province and Wuhan local health authorities to find out what was going on. Once again, the Huanan Seafood Market was fingered as the source of the outbreak. The article explained that tests run on samples washed from the lungs of patients were negative for known pathogens, so the researchers used a polymerase chain reaction to probe for an RNA sequence typical of a well-conserved region common to all coronaviruses. They did not spell out why they were looking for coronaviruses, but it was certainly because commercial labs had already told Wuhan authorities a new SARS-like coronavirus was infecting patients. According to this paper, three of the four samples tested were positive for a coronavirus. Patient one was a retailer at the seafood market, patient three visited there. The article made no mention of the other patient's relationship to the market, only that patient two died.

From these samples, the paper said, genomes were successfully sequenced of a beta coronavirus, lineage B. The authors isolated the virus, grew it on human epithelial airway cells, put the cells under the microscope, and saw protein crowns—spikes—on the virions. Three full sequences were then uploaded to GISAID. The authors said that when compared with other known coronaviruses, the new virus's genome was most like two others whose sequences had been recovered from bats and published previously. One was labelled SL-CovZC45 and the other ZXC21. These were the same closest relatives that Holmes and Zhang had pointed to in their paper, submitted on January 7 but not yet accepted by *Nature*. Neither paper gave the complete affiliations of the scientists who'd pulled those sequences from those bats.

Please remember: as of January 24, one day before Lunar New Year, when this article was published online, two bat-derived viruses called ZC45 and ZXC21, whose sequences had been previously *published* (as opposed to known), were found to be the closest relatives of this novel virus, soon to be named SARS-CoV-2.

THE FOURTH PAPER by Shi Zhengli and her group at the WIV's Laboratory of Special Pathogens and Biosafety was submitted on January 20 to *Nature* and accepted on January 29. Zhang and Holmes's paper on the same subject had been accepted the day before, though submitted 13 days earlier. The lead author of Shi's paper was Peng Zhou, who has worked closely with Shi for years. Shi was the last author. With the title of the article, the group set out to form a narrative about the origin of the virus. It was called "A pneumonia outbreak associated with a new coronavirus of probable bat origin." They cited their own previous work that had reported many SARS-like coronavirus sequences found in bats or their feces in China. The co-authors stated that the "epidemic" started on December 12 in Wuhan, and that samples from seven very ill patients in hospital had been sent to the WIV to be examined.[60]

"We first used pan-CoV PCR primers [meaning primers that locate RNA sequences common to all coronaviruses] to test these samples given that the outbreak occurred in winter and in a market—the same environment as SARS infections." Six of the patients were described as sellers or deliverymen at the Huanan Seafood Market. The paper said nothing about the seventh. Five of the samples were positive for coronavirus.

The paper does not say when Shi's lab received those

samples, but in an interview she later gave to *Scientific American*, she said she was called by her boss on the evening of December 30, while at a conference in Shanghai, and told to come back to the lab urgently. Shi and her group soon isolated the novel coronavirus from five of the patient samples sent to the WIV. They found that it entered human cells through the ACE2 receptor, just as SARS had done. Seven days later they had the sequence of one genome, and then got several more, all of which were uploaded to GISAID, labelled as WIV02, 03, etc. Their analysis showed these sequences were 99.9 percent identical to each other. They did not spell out what that meant, but virologists would have found it interesting that barely any mutations had taken place since the virus started circulating in Wuhan. That pointed to a recent date of origin and indicated that this virus was very well adapted to humans from the start. But how could that be, for a virus of "probable bat origin"? Jumping from bats to an intermediary animal and then to humans would have required a lot of mutations. Jumping straight from bats to humans would also require significant mutations as the virus adapted itself to the human immune system and got better at evading it. This suggested an origin of a very different kind, especially as these authors noted that the virus might be becoming even more efficient at infection. They pointed out that there had been "2,794 laboratory-confirmed infections including 80 deaths as of 26 January 2020" and 33 infections had been confirmed in ten other countries. In other words, in the four days since the *Lancet* article's case tally, the number of laboratory-confirmed infections had tripled and the number of deaths had doubled. The virus was getting into human cells with great alacrity and hadn't had to change itself much to do so.

When they compared these sequences with other SARS-like viruses, they found that this new coronavirus was only 79.6 percent identical to the original SARS but 96.2 percent "identical at the whole-genome level to a bat coronavirus." The most similar sequence was labelled RaTG13, and its accession number was given, meaning the lab had published this sequence on a public site. Shi's group offered no information about RaTG13 other than to say it had been "previously detected in *Rhinolophus affinis* [horseshoe bats] from Yunnan province."

Yunnan is more than 900 kilometres southwest of Wuhan. Horseshoe bats in southern climate zones don't migrate more than about 180 kilometres. If RaTG13 was the closest relative of this new virus, how had the new virus made its way to Wuhan from Yunnan? This question was neither raised nor answered by the article, though other authors, including Shi, would soon argue there must have been an intermediary animal, for example a pangolin. Perhaps a pangolin had been infected by a bat in Yunnan and that pangolin had been brought to the Huanan market. There, in the hot mess of slaughter, the virus it carried might have jumped to a human. Articles published later described a coronavirus, almost as close to SARS-CoV-2 as RaTG13, that was discovered in pangolins smuggled into China. The hypothesis that a pangolin was the intermediary between Yunnan bats and humans in Wuhan even made it to the *New Yorker*.[61]

However, it would soon be learned that no pangolins had been sold in the Huanan market before it was shut down. And the pangolin-to-human-in-a-wet-market argument made little sense if the first known victim had no relationship to any wet market. Other researchers would soon propose a different explanation as to how the virus found its way to Wuhan. Two Wuhan scientists would suggest that perhaps researchers at the WIV or at the Wuhan CDC had picked it up when they were taking samples in Yunnan bat caves and brought it back to Wuhan.[62]

After examining the sequence of SARS-CoV-2 carefully, Shi's group declared (as had Holmes and Zhang's) that no recent recombination event had occurred to create it, although they thought there might have been one further back in time. Shi's group said that though the new virus was much like RaTG13, both were so distant from other known coronavirus sequences that the two "form a distinct lineage from other SARSr-CoVs [SARS-related coronaviruses]." The spike protein of the new coronavirus, the part that permits it to enter a host, was only 75 percent like most other known SARS-like spike sequences, but it was 93.3 percent like the spike sequence of RaTG13. Both were longer than other known spike sequences and contained certain interesting insertions. "The close phylogenetic relationship to RaTG13 provides evidence that 2019-nCoV [the new coronavirus] may have originated in bats," they concluded.[63]

The article ended with a push for more surveillance of natural reservoirs of SARS-like viruses—bats. "Most importantly, strict regulations against the domestication and consumption of wildlife should be implemented." That was the line USAID's PREDICT program had been pushing for years. Yet although PREDICT had long been a funder of Shi's work, no mention of PREDICT was made at the end of this article. The only funders acknowledged were Chinese granting agencies.

And why does what Shi said in *Nature* (and also on the bioRxiv site where the article was first posted on January 23) about the closest known relative to SARS-CoV-2 matter? Because on January 20, 2020, the same day that Shi and her colleagues at the WIV submitted their article to *Nature* describing the closest known relative as a sequence called RaTG13, Shi and two colleagues *not* associated with the WIV submitted another article, the fifth submitted that day. It said something quite different about the closest known sequence to SARS-CoV-2. *Emerging Microbes and Infections* (*EMI*) accepted this second article on January 21, the day after its submission and two days before Shi's article submitted to *Nature* appeared on the bioRxiv preprint site. The *Emerging Microbes* article was published online on January 31.[64]

Shi's co-authors on this second paper in *EMI* were Shibo Jiang and Lanying Du, both at the Kimball Research Institute at the New York Blood Center. Shibo Jiang also has an appointment at the Key Laboratory of Medical Molecular Virology at Fudan University in Shanghai and is an editorial board member of *EMI*. This article was called "An emerging coronavirus causing pneumonia outbreak in Wuhan, China: calling for developing therapeutic and prophylactic strategies." It compared the sequence of the new virus to other known coronavirus sequences, just as Shi's group had done in the article in *Nature*. But here, no mention was made of the sequence known as RaTG13. In this paper, the new virus's closest relatives were said to be the same two identified by Zhang and Holmes in their article in *Nature*, and in the *New England Journal of Medicine* article co-authored by George Gao. These sequences had been pulled from horseshoe bats in Zhoushan City, about 1,000 kilometres southeast of Wuhan and about 2,000 kilometres from Yunnan. Though Shi's *Nature* article, submitted the same day, placed these sequences in a different clade from the new virus and its closest relative, RaTG13, the second paper on which she is last author, submitted to EMI, asserted that the new virus is closest to two sequences pulled from horseshoe bats in Zhoushan City "between 2015 and 2017, which can infect suckling rats and cause disease," known as ZXC21 and ZC45.[65]

I looked up the paper that described those two sequences. Of the 15 co-authors listed on it, 12 are military scientists affiliated with the People's Liberation Army's Nanjing Command.

To repeat: In her *Nature* article submitted on January 20 and uploaded to a preprint site on January 23, Shi's group found the new virus to be 96.2 percent identical to RaTG13, a sequence recovered from Yunnan, about 900 kilometres southwest of Wuhan. In a second article on which Shi is also last author, and also submitted on January 20, RaTG13 is not mentioned. Instead, the closest sequences are deemed to be the ones Shi's *Nature* article described as being in a different clade from SARS-CoV-2. The *Nature* article appeared four days after the *EMI* article, so there was time to fix the *Nature* article if it was incorrect, or to fix the *EMI* article if that was incorrect. One of them is incorrect. Yet no correction or explanation has been published.

It would become clear, later, that Shi's group had not published the full sequence known as RaTG13 until the *Nature* article, though the sample from which it came had been in her lab since 2013, she had sequenced and published a portion of it in 2016, and she had assembled the full sequence in 2018.

That would cause quite a ruckus.

4. On Trust

TRUST IS FUNDAMENTAL to all healthy societies. We trust that the politicians we elect will put our nation's interests before their own. We trust that when we go to a restaurant, we won't be poisoned, that when we say hello to a neighbour, he'll say hello back, not cry "Die, heretic" as he hauls out a knife. In every healthy society, trust will be violated some of the time, but not all the time. Trust is its own reward.

Science, like good government, is also based on trust, but as Ronald Reagan once said about a nuclear arms treaty, it's trust but verify. Scientists must be able to rely on the published findings of their peers. No one has the time to redo every experiment to be sure the methods and results were accurately reported and properly interpreted. Scientific publishing is pockmarked with citations of others' works that are being relied upon, but anyone can ask to see the raw data or samples that were used to form those published papers. Most publications make a lot of this underlying material available online. At least one of the authors will list an email address or phone number so colleagues can ask questions and get answers.

In a pandemic, the need to trust both the science that describes it and the governments that decide what to do about it climbs with the body count. But as I soon learned, in the early days of the SARS-CoV-2 pandemic, trust in government was not rewarded. Trust in scientists' assertions about the virus's origin was not rewarded either.

AN IMPORTANT ARTICLE from outside Mainland China appeared online in *The Lancet* on January 31, 2020. It was written by Joseph Wu, Kathy Leung, and Gabriel M. Leung, researchers at the WHO Collaborating Centre for Infectious Disease at the Li Ka Shing Faculty of Medicine, University of Hong Kong. In their paper, titled "Nowcasting and forecasting the potential domestic and international spread of the 2019-nCoV outbreak originating in Wuhan, China: a modelling study," they estimated the potential domestic and international spread of the virus to that point. By their reckoning "75,815 individuals have been infected in Wuhan as of January 25, 2020."[66] That number was 35 times greater than the number reported by Shi's group as confirmed by tests, but in an epidemic only a fraction of the population will be tested.[67] These modellers used statistical methods and models to get a better fix on what the real rate of infection and number of cases could be.

They calculated the R number (the number of people one infected person can infect) to be 2.68. That meant new infections were growing exponentially "in multiple major cities of China with a lag time behind the Wuhan outbreak of about 1–2 weeks." They argued that since the virus had already spread in China, large cities with transport links to China could also become outbreak epicentres "unless substantial public health interventions at both the population and personal levels are implemented *immediately*." [Italics mine.]

Self-sustaining outbreaks in major cities around the globe would otherwise be inevitable because of the substantial export of pre-symptomatic cases. The lockdown in the greater Wuhan area would not stop this international spread; it had been put in place way too late. They wrote:

> Our simulation suggested that wholesale quarantine of population movement in Greater Wuhan would have had a negligible effect on the forward trajectories of the epidemic because multiple major Chinese cities had already been seeded with more than dozens of infections each.... As such, given the substantial volume of case importation from Wuhan, local epidemics are probably already growing exponentially in multiple major Chinese cites. Given that Beijing, Shanghai, Guangzhou and Shenzhen together accounted for more than 50% of all outbound international air travel in mainland China, other countries would likely be at risk of experiencing 2019-nCoV epidemics during the first half of 2020.

Their recommendation? Get going on a vaccine. And:

> Above all, for health protection within China and internationally, especially those locations with the closest travel links with major Chinese ports, preparedness plans should be readied for deployment at short notice, including securing supply chains of pharmaceuticals, personal protective equipment, hospital supplies and the necessary human resources to deal with the consequences of a global outbreak of this magnitude.[68]

China, the main manufacturer of personal protective equipment for the world, was way ahead of these researchers. It had already sent agents *in early January* to buy up in foreign markets all masks, gowns, and gloves they could lay their hands on in places as far away as Nigeria.[69] The resulting shortages faced by everybody else crippled attempts to protect health care workers from the virus, which amplified its spread.

The same day this *Lancet* article appeared online, WHO director-general Dr. Tedros Adhanom Ghebreyesus declared a public health emergency of international concern (acronym: PHEIC), which is not the same as a pandemic. It is a lesser terror. There is a well-understood definition of *pandemic*. It is: "an epidemic occurring worldwide, or over a very wide area, crossing international boundaries and usually affecting a large number of people."[70] The *Lancet* article made clear that the situation was well beyond a health emergency of international concern; a pandemic was already in progress. By January 31, the virus had spread to Sichuan, Yunnan, Shanghai, Guangxi, and Shandong in China, as well as to Hong Kong and Thailand, with numbers doubling almost daily. Europe and the Middle East, Canada and the United States all had cases by then too. Yet Ghebreyesus didn't get around to declaring a pandemic until March 11, almost six weeks later. The US House Foreign Affairs Committee's minority staff report would later explain that before declaring a public health emergency of concern, which half his emergency committee had voted for on January 22, the director-general first travelled to China and met with the political leadership. The House Foreign Affairs minority staff report concluded the reasons for delay were political.[71]

ON JANUARY 15, a 35-year-old man who had travelled home from Wuhan to Snohomish County, Washington, fell sick with a cough and fever and four days later went to an emergency room. The first Canadian case appeared in Toronto's Sunnybrook Hospital emergency ward just days after that. This man too had just travelled home from Wuhan. Soon his wife was sick as well. In Vancouver, the sixth case presented a nasty surprise: health workers had been told to watch for sick people who said they'd been in Wuhan. But this woman had not travelled from Wuhan, but from Iran, one of the main links on China's Belt and Road chain. Later it became known that Iranian doctors had been trying to draw their government's attention to this pneumonia since late December, to no avail. The first man to die in Iran was a businessman from Qom who had been in China.[72] Qom was the centre of the Iranian infections. China Railway Engineering Corp. is building a high-speed rail line through Qom and, according to one story, Iranian officials had been afraid to say anything about the virus for fear of losing business.

In other words, by January 31, competent epidemiologists and intelligence officials who read scientific journals, and the rest of us who saw newscasts showing Chinese workers racing to erect two huge field hospitals in Wuhan, knew that a pandemic was loose, no matter what the WHO called it. Some believed it could be the big one that scientists such as Shi Zhengli, Ralph Baric, and Peter Daszak had been warning about for years.

According to Bob Woodward's book *Rage*, President Trump was told on January 28 by his national security advisor, Robert O'Brien, that "this will be the biggest national security threat you face in your presidency." By then, O'Brien's deputy, Matthew Pottinger, had been on the phone for days with contacts in China and Hong Kong. Pottinger had been a reporter in China for the *Wall Street Journal* for seven years. He spoke fluent Mandarin and had learned to distrust official statements from Chinese authorities while covering SARS. According to Woodward, Pottinger had been told by one expert that this was no SARS 2003, he should think "influenza pandemic 1918."[73] Another source told Pottinger that 50 percent of the spread was from people who showed no symptoms. Pottinger had even been warned by people in the know that China had a "sinister goal. China's not going to be the only one to suffer from this."

On January 31, the same day that modelling article came out in *The Lancet*, the same day the WHO declared a public health emergency of international concern, the US closed its borders to travellers from China (unless they had American passports or green cards).[74] The US did this in spite of International Health Regulations that call on member states to refrain from unilaterally shutting their borders to keep out disease. On February 6, the day after Trump's first impeachment trial ended in acquittal, President Xi himself told Trump the virus was airborne.[75] Trump explained to Woodward the next day that this meant it was more deadly than the flu.

Yet governments, including the US and Canadian governments, said something quite different in public. Until the pandemic was declared by the WHO on March 11, though the science that officials insisted they relied upon had said "Get ready now!" six weeks before, they kept saying the risk was low.

What were the consequences of six weeks of "the risk is low"?

A paper was published in *Science* in December 2020 by epidemiologists at the Broad Institute, which is affiliated with Harvard and MIT. The co-authors traced a specific strain of the virus that first appeared at the end of February 2020 in Boston at a conference organized by a pharmaceutical company called Biogen. The strain had a particular mutation that made it possible to trace its movements across the US and around the globe. It was first discovered in two patients in France. It is believed to have been imported just once to the United States. How it got to the Biogen conference is not known, but people from Italy, Switzerland, and Germany attended. By the end of October 2020, that one strain was shown to have infected approximately 245,000 Americans, perhaps as many as 300,000, and to have spread to Slovakia, Sweden, and Australia.[76] In other words: In those six weeks, many people became sick who might otherwise not have been infected. Many people died who might have lived.

The government of Canada blindly followed the WHO's recommendation to keep international borders open. When asked why Canada was not following the US lead, Theresa Tam, the chief public health officer of Canada and an advisor to the WHO, referred to the International Health Regulations and said there could be consequences for such actions. Testifying before the Health Committee of the House of Commons on January 29, 2020, Tam had said: "Canada's risk is much, much lower than that of many countries. It's going to be rare, but we are expecting cases... It's going to be rare but we are going to have some."[77] The minister of health, Patty Hajdu, was asked after the US shut its doors to visitors from China why Canada was not doing the same. She too muttered about the WHO and consequences, and offered that "here in Canada we have very different processes in place than in the United States. For example, we do not need to call a public health emergency here because we already have the structures, the systems and the authorities to spend appropriate dollars necessary to respond, treat and maintain our public health systems." She also scolded Opposition MPs regarding the spread of "misinformation and fear across Canadian society."[78] As if real evidence had not been published by leading journals on pre-symptomatic individuals being able to infect others; as if the risk to Canada, where cases had already appeared, remained "low"; as if she'd never heard of the precautionary principle. Poor judgment was displayed by provincial leaders as well. The premier of Quebec did not suggest that spring break travel might be a bad idea, and so Quebec was inundated with infections as people returned from holidays in France and Florida. The premier of Ontario, just before the declaration of a pandemic, told people to go where they liked for spring break and enjoy.

On April 30, 2020, it was revealed that the federal government had let its stockpile of gloves, gowns, and N95 masks (put in place after the SARS crisis) degrade.[79] Items thrown out when they were past their best-before date had not been replaced. The federal government was scrambling for supplies on international markets just as many other countries were doing the same. Though Ontario had borne the brunt of the SARS crisis and had instituted a stockpile of its own, Ontario had also failed to keep replenishing it. Front-line health workers, especially in long-term care facilities, were unprotected in the critical first phase of the pandemic, further enabling SARS-CoV-2 to spread.

What about masks? Chief Public Health Officer Tam

insisted that masks were not only unnecessary for ordinary Canadians but might increase risk as the untrained would not know how to remove them safely. She maintained this absurd argument long after other countries had demonstrated that mask wearing dampened the spread of the virus. The WHO offered the same recommendation that masks should not be used by the general public until June 5, 2020.[80]

By July, any remaining trust in the foresight and preparedness of the government of Canada was reduced to rubble when it was revealed in the *Globe and Mail* that the Public Health Agency of Canada (PHAC) had sidelined its well-regarded epidemic intelligence unit upon which the WHO also depended. The GPHIN (Global Public Health Intelligence Network) had been put together in the 1990s to scour media reports, travel bulletins, websites, etc., to give warning of infectious disease outbreaks in countries whose governments might not be forthcoming with timely warnings.[81] Public health functions in Health Canada had been moved to a new agency—PHAC—by the Harper government. Bureaucrats from other parts of government—such as Treasury Board—had been brought in to run it, bureaucrats with no training in epidemiology or medicine. Eventually, GPHIN's scientific staff were told they could no longer pass warnings directly to any departments that might have an interest, but only to their bosses, who would decide whether or not to issue warnings up the line. Warnings soon fell to zero, and unit scientists were tasked with tracking domestic issues as opposed to outbreaks brewing in the far reaches of the globe. Bureaucrats without science backgrounds even began to treat attendance at scientific meetings abroad as perks they were entitled to. GPHIN issued its last warning before the pandemic on May 19, 2019.

Only when this story hit the *Globe and Mail*, thanks to reporter Grant Robertson,[82] were there any consequences for poor PHAC decisions. The president of the agency and the vice-president left their roles.[83] The minister of health was forced to order an "independent" review (which the minister did not make public until after the House of Commons rose for the summer, on July 12, 2021[84]). The auditor general eventually issued a scathing report on the government's lack of pandemic preparedness, especially PHAC's deficiencies, pointing out, among so many other failings, that it did not raise its risk assessment from low until the day after a pandemic was declared by the WHO.[85] In response, PHAC blamed a faulty algorithm.

There was no trust left to lose in the government of China. It had withheld vital information during SARS, and it was doing so again. According to *Science*, it was a *Wall Street Journal* story in early January on the pneumonia in Wuhan that made China change course from trying to suppress all information about it to shaping the narrative.[86] Most troubling of all: China withheld or had ordered destroyed early samples of the virus and had at first refused permission for outside experts from the WHO to enter Wuhan.[87] Which, of course, raised the question: What was China trying to hide?

TRUST IN SCIENCE is based on two things: that what is published represents full disclosure of methods and results, and that political agendas or interests play no role in shaping what authors report or fail to report. But political agendas and interests were clearly shaping what China's scientists reported or failed to report, as well as when they reported. China's own official account of events, published in June 2020, would make that clear:

> Having forged the idea that the world is a global community of shared future, and believing that it must act as a responsible member, China has fought shoulder to shoulder with the rest of the world. In an open, transparent, and responsible manner and in accordance with the law, China gave timely notification to the international community of the onset of a new coronavirus, and shared without reserve its experience in containing the spread of the virus and treating the infected....General Secretary Xi Jinping has taken personal command, planned the response, overseen the general situation and acted decisively, pointing the way forward in the fight against the epidemic. This has bolstered the Chinese people's confidence and rallied their strength. Under the leadership of the CPC, the whole nation has followed the general principle of "remaining confident, coming together in solidarity, adopting a science-based approach, and taking targeted measures," and waged an all-out people's war on the virus....
>
> As of 24:00 May 31, 2020 a cumulative total of 83,017 confirmed cases had been reported on the Chinese mainland, 78,307 infected had been cured and discharged from hospital, and 4,634 people had died. This demonstrates a cure rate of 94.3 percent and a fatality rate of 5.6 percent.[88]

This official statement regarding timely notification and sharing was untrue. The case numbers, allegedly the product of science, were highly suspect. They barely changed after that publication in June 2020. On December 14, 2020, I checked graphs on the WHO website, a dashboard showing the number of SARS-CoV-2 infections and deaths in various countries around the world. Most displayed a rise-fall-rise pattern over the course of the year. But the line for China rose steeply at first then dropped to the bottom of the graph and stayed there.[89] China has a population of 1.4 billion. As of December 14, 2020, the total number of infections in China was said to be 94,300. The number of deaths was said to be 4,754.[90] Canada has a population of 37.59 million. As of the same day, Canada had recorded a total case count of 468,475 and 13,553 deaths.[91]

The numbers of the afflicted posted by China, especially the numbers of the dead, have been disputed since they were first made public. Early on, Radio Free Asia, a news operation, ran a simple fact check. It reported that there are seven funeral homes with crematoria in the three-city megalopolis known as Wuhan. These funeral homes have the capacity to do about 2,500 cremations a day. People interviewed claimed these crematoria were working at capacity for at least 17 days in February 2020, which adds up to a much higher number of deaths than 2,500, the official number at that time. It comes out to 42,500. Yet in June 2021, more than a year later, according to the WHO, the number of deaths in the whole of China had only risen to just over 5,000.

JUST BECAUSE GOVERNMENTS dissemble doesn't make facts go poof; they make themselves known eventually, and when they do, people consider future government statements on that subject with skepticism or worse. A cover-up or downplaying of danger may be the obverse of the boy who cried wolf but will have the same effect. When officials later try to mobilize citizens to face the crisis, they may not be listened to. And there is another result of the death of public trust: alternative facts rise up and take over the podium.

I can claim no immunity to the intellectual toxins that flow from the death of trust—such as paranoia, such as instant skepticism, often unwarranted, of official pronouncements. I have developed a lifelong habit of searching for patterns in what might well be random events, and if I find them, I try to extract meaning from them, warranted or not. My husband Stephen was a great antidote to this form of brain poison. *That's not a pattern, that's a two-point curve*, he'd shout when I brought him some terrific nugget I'd carved out of nothing. He meant: two points on a graph don't tell you anything. Early on, long before the upsurge of cases became terrifying, before the old and sick and their caregivers in long-term care became the favoured hosts of the virus, before arguments broke out on buses and in the streets about whether to wear a mask, I began to lean toward a lab leak as the origin of the pandemic.

At first, I was just fascinated by how fast SARS-CoV-2 was spreading. I said, right out loud, that SARS-CoV-2 is smarter than SARS.

"Whaddya mean *smart?*" barked an old friend, a guy who is usually contrary and therefore good to argue with. "It's just a virus."

"It transmits easily between people and doesn't kill too many," I said. "If a virus kills off too many hosts, it dies out. So it's smart."

What troubled me, what made me think there might be something unnatural afoot, was that it appeared to have been born smart. Usually, when a virus first jumps from one species to another, it doesn't do well. As a leading

Canadian virologist, Dr. David Evans, put it to me in an email, "it's like a bull in a china shop." As Ron Fouchier's experiments had shown, it takes many passages for a virus to adapt itself to a new host's immune system; it takes time to develop the mutations that will be good at evading that immune system. Yet this virus appeared to have settled into human tissues as if to the manor born.

A group of Chinese and American scientists who published a paper at the end of February 2020 in the *Journal of Medical Virology* made that quite clear. They called their article "Evolutionary history, potential intermediate animal host, and cross-species analyses of SARS-CoV-2." The first author, Xingguang Li, works at Wuhan University's Engineering Research Center of Viral Vector. The last authors, the senior ones who did the analysis, included Brian Foley of Los Alamos National Laboratory and Antoine Chaillon of the University of California, San Diego. They dismissed the argument some were making that SARS-CoV-2 probably appeared first in pangolins, who then infected humans. They pointed out that the viral sequences found in pangolins were not as close as RaTG13 to SARS-CoV-2, but more to the point, that SARS-CoV-2 has a unique peptide not found in the pangolin version of the virus, and that peptide had an impact on host range and ease of transmission. By comparing 70 genome sequences collected from the afflicted in 12 countries across the world (most from China) between December 24, 2019, and February 3, 2020, using several different methods to calculate rates of mutation (called "clocks"), they worked back to the most recent common ancestor of the sequences. What the various methods all showed was "an explosive, star-like evolution of SARS-CoV-2, and recent and rapid human-to-human transmission." They pinpointed the origin of SARS-CoV-2 to November 24, 2019.[92]

Trevor Bedford of the Fred Hutchinson Cancer Research Center in Seattle, Washington, had come to the same conclusion. At the end of January 2020, he told Jon Cohen of *Science:* "one of the biggest takeaway messages [from the viral sequences] is that there was a single introduction into humans and then human-to human-spread." In other words, the virus must have been human-adapted from the start, but where and when could that have happened?[93] There were new strains, yes, and Bedford was putting them up on a website called nextstrain.org as they came in, but the variations among them were minimal, just a nucleotide here, another there, at the rate of about two mutations per month. That low mutation rate continued for many more months.[94]

As I read these papers and stories, I began to think of SARS-CoV-2 as a viral Einstein. Not only did it transmit easily between people from the start, but it often did so not only pre-symptomatically but also asymptomatically, meaning many who are infected feel perfectly well yet can infect others. Since a virus's sole purpose is to replicate, asymptomatic infection is the viral equivalence of extreme brilliance. This capacity was demonstrated by Dr. Andrea Crisanti, then an infectious disease expert at Imperial College London. He was on sabbatical at the University of Padua, where he is now a professor, when the virus hit northern Italy like a tsunami. As northern Italian hospitals were overwhelmed, Crisanti decided to test everybody in the small town of Vo' (against the wishes of Italian public health authorities) to see how many had been infected. The tests showed that 3 percent of the village had the virus and more than 75 percent of those who tested positive showed no symptoms at all (verifying Matthew Pottinger's information from China and what the Canadian minister of health had brushed off as a "very small study from Germany"). Crisanti's advice to Canadians, which he gave through the CBC and through Eric Reguly, who wrote about him in the *Globe and Mail*, was: "test widely, catch all possible cases early, and isolate them to prevent the virus from spreading like wildfire."[95]

In late spring, an article in the journal PNAS detailed exactly how clever SARS-CoV-2 is. The paper came from the lab of Fang Li of the Department of Veterinary and Biomedical Sciences at the University of Minnesota. Li had the kind of background that was becoming familiar to me, a pattern I noticed and tucked away for future reference. Born and brought up in China, he had earned his BSc from Beijing University but then went to the US in 1996 to do a PhD at Yale. From there, he went to Harvard Medical School for post-docs, and from Harvard to his own lab at the University of Minnesota.

His group had already published in *Nature* important facts about the molecular structure of SARS-CoV-2. Using X-ray crystallography, they had shown that its receptor binding domain—the part of the spike that attaches to the ACE2 receptor in humans—made it significantly more effective at binding to human cells than SARS. They had also shown that the sequence called RaTG13 would also use the ACE2 receptor and that its receptor binding motif (which is within the domain) was similar to that of SARS-CoV-2. "RaTG13...contains a similar four-residue motif in the ACE2-binding ridge, supporting the notion that SARS-CoV-2 may have evolved from RaTG13, or a RaTG13-related bat coronavirus," they wrote. They concluded both could directly jump from bats to humans without an intermediary.[96]

In the second paper published in PNAS in May, Li's group reported on chemical inserts at SARS-CoV-2's receptor binding domain's cleavage site, the place where it splits into two segments, one of which merges with the host cell. They reported that on host cells, a furin—an endopeptidase that will cut an amino acid chain at a particular point—pre-activates the virus's spike, efficiently starting the merger. They contrasted the behaviour of SARS's receptor binding domain to SARS-CoV-2's. In SARS, they said, the molecules of the receptor binding domain "stand up," which makes them more discernible to the host's immune system. In SARS-CoV-2, they lie down flat, which makes the virus harder for the host's immune system to spot. But lying down should have made SARS-CoV-2 less infective, yet the addition of those inserts permitted the virus to enter three different kinds of human cells while also staying hidden, the viral equivalence of genius. "The hidden RBD [receptor binding domain] can evade immune surveillance, potentially leading to insufficient immune responses and longer recovery time," they wrote.[97] So: not only was SARS-CoV-2 a possible descendant of RaTG13, whose origin was not known, it was also much more efficient at entry than SARS and particularly good at fooling the human immune system.

These papers really made me wonder: Could SARS-CoV-2 have been developed from RaTG13 in a lab?

This idea clearly occurred to others too, some of whom had no compunction about leaping directly from question to allegation. Because Li's group had also previously shown exactly how the molecular structures of SARS and MERS align with human receptors, Li was accused of creating the blueprint for SARS-CoV-2. Hong Kong-based Lude Media called him a key figure "behind the CCP [Chinese Communist Party] virus."[98] Lude Media is an anti-CCP lobby group aimed at turning China into a federal state, started by Steve Bannon and Guo Wengui, a Chinese billionaire in exile in the US.

Social media, preprint platforms, and everyday gossip were by then replete with many crazy theories about SARS-CoV-2, both its origin and its nature. There was the 5G network conspiracy theory: COVID-19 is caused by 5G networks, so the cell towers need to come down. There was the it's-all-a-hoax-to-bring-down-the-economy-and-President-Trump theory, often heard at Trump rallies, though Trump knew better and would get very sick from SARS-CoV-2 himself. There was the it's-just-like-the-flu-that-kills-a-few-old-people-every-year-never-mind-party-on theory. I heard that one from a neighbour who I thought was an intelligent fellow, yet nothing in the way of evidence, such as SARS-CoV-2's climbing death rate, could move him from his conviction. I ran into another neighbour who said her husband sits on a corporate board along with several people from China who have medical backgrounds, and one of them said at a board meeting that the virus was made in a lab. When I asked to interview that person, the husband, through the wife, said no. And of course, there was the Bill Gates conspiracy theory. Gates had been warning about pandemics for years, pointing to the need to predict their emergence and prepare vaccines in advance. The Bill Gates theory is that SARS-CoV-2 was just cover so he could get us all to line up for vaccines that will slipstream chips into our arms, the better to control us all.[99]

Perhaps not surprisingly, since China had moved on January 20 to control the narrative, China-friendly origin theories also appeared almost immediately in Chinese media. *Global Times*, an unofficial voice of the CCP, reported that some were saying the US was responsible for bringing the virus to China. The World Military Games had been held in Wuhan in October 2019. American troops had participated. Apparently, they'd stayed in a hotel close to the Huanan Seafood Market. Therefore, they had let it loose, maybe on purpose.[100] A Chinese foreign ministry spokesman actually made that allegation from a public platform, Twitter.[101] But when a leading academic in Spain mentioned at the tail end of a paper on disease tracking with waste water that he'd found bits of RNA from the virus in a Barcelona sample collected in March 2019,[102] Chinese officials switched theories immediately to Spain as the place of origin.

On January 29, noting all the theories circulating on the origin of SARS-CoV-2, the *Washington Post*'s Adam Taylor wrote what was titled a debunk piece.[103] The week before, the *Daily Mail* in London had asserted that there was a possible link between the virus and the "Wuhan National Biosafety Laboratory," which had been the subject "of safety concerns" articulated in an article in *Nature* in 2017. Taylor had followed up with Tim Trevan, the biosafety expert quoted in that *Nature* article, about the Wuhan Institute of Virology's new BSL-4 lab, which was then about to open. Trevan told Taylor he hadn't suggested to *Nature* that anything was wrong in Wuhan, just that it's hard to predict how complex systems may fail, and BSL-4 labs are complex systems. However, when I checked the original *Nature* story, I found Trevan had said something different.

According to *Nature*'s reporter, David Cyranoski, Trevan had said "an open culture is important to keeping BSL-4 labs safe, and he questions how easy this will be in China, where society emphasizes hierarchy." Cyranoski had also quoted Richard Ebright, a professor of chemical biology at Rutgers and a well-known critic of gain-of-function experiments, on China's plan to open five to seven BSL-4 labs by 2025. Ebright told Cyranoski that building that many would give rise to suspicions in other governments that their purpose was biowarfare, because these kinds of facilities are all "dual use."[104]

Trevan told the *Washington Post* that he didn't think bioweapons were being researched at the WIV because nobody thinks bioweapons work well (they have a nasty habit of infecting those who release them). But another expert seemed to think bioweapon research could be going on in China, just not at the WIV. Why not there? Such work would be covert, Milton Leitenberg, a chemical weapons expert at the University of Maryland, told Taylor, and the WIV's BSL-4 has close connections with the Galveston National Laboratory, the only privately owned BSL-4 in the US. Leitenberg told Taylor that he had talked about the bioweapon possibility in emails with others around the world, but "no one had found convincing evidence to support that theory."

Taylor then turned to Elsa Kania, an adjunct senior fellow at the Center for New American Security, to find out what she thought about whether this pandemic could have resulted from China working on bioweapons. The Center is a very well-connected Washington think tank. It is bipartisan and gets donations from anybody who is anybody, including investment bank Goldman Sachs; military contractors Northrop Grumman, Lockheed Martin, and Booz

Allen Hamilton; Palantir Technologies, a cybersecurity/technology company with CIA connections; and foreign governments like Canada through the Department of National Defence.[105] What Kania (an informal advisor to the Biden election campaign while finishing her PhD at Harvard) told Taylor about China's biowarfare research was downright unnerving. "While Chinese officials had expressed public interest in the potential weaponization of biotechnology, a coronavirus would not be a useful weapon," she told Taylor. "Hypothetically," Kania said, "a bioweapon would be designed to be highly targeted in its effects, whereas since its outbreak the coronavirus is already on track to become widespread in China and worldwide." In other words, while China might well be building bioweapons, SARS-CoV-2 would not be what it would build.

I wasn't so sure her logic fit the facts. If the government of China was to be believed, it had contained the virus promptly. By contrast, its most serious competitor, the US, would endure a terrible onslaught of sickness, death, and economic injury. If SARS-CoV-2 was a weapon aimed at wreaking havoc on competitors' economies rather than on their militaries, it was doing a hell of a job.

Kania was the go-to person on China and bioweapons because in August 2019 she had co-authored with Wilson Vorndick a short piece called "Weaponizing Biotech: How China's Military Is Preparing for a 'New Domain of Warfare.'"[106] I looked that paper up. It was not a speculative article: she and her co-author quoted from writings on this subject by China's own PLA leadership. She referred to a book published in 2010 called *War for Biological Dominance* by Professor Guo Jiwei of the Third Military Medical University. He wrote on the impact of biology "on future warfare." In 2015, according to Kania and Vorndick, the then president of China's Academy of Military Medical Science, He Fuchu, argued that "biotechnology will become the new 'strategic commanding heights' of national defense, from biomaterials to 'brain control' weapons." In 2017, Zhan Shibo, a retired general, wrote: "'Modern biotechnology development is gradually showing strong signs characteristic of an offensive capability,' including the possibility that 'specific ethnic genetic attacks' could be employed." The PLA's National Defense University, the authors said, uses a textbook with a new section on "biology as a domain of military struggle, similarly mentioning the potential for new kinds of biological warfare to include 'specific ethnic genetic attacks.'"

After reading that article, I found another by Monika Chansoria that appeared in *Japan Forward*, published by the Japan Institute of International Affairs in March 2020. It too listed indicators of China's interest in biological warfare, both offensive and defensive. It pointed to possible sites of research and manufacture, such as a facility near Lop Nor in Xinjiang where there had been an outbreak of hemorrhagic fever in the 1980s, as well as biological products factories located in Wuchang, which is part of the megalopolis of Wuhan, and in Kunming, the capital of Yunnan province.[107]

Corey Pfluke, an analyst with US defence contractor Lockheed Martin, put dots together in a similar way in a piece also published in March 2020 in Air University's *Wild Blue Yonder* (located at Maxwell Air Force Base, Alabama). She pointed out that biowarfare research work in China would be no surprise. "The British used contaminated blankets, and the Japanese used flea bombs." [The Japanese put fleas carrying plague in little ceramic vessels and dropped them from planes on Chinese troops. The Chinese later accused the Americans of using the same filthy trick during the Korean War.[108]] Like Kania, Pfluke pointed out that China "has shown interest in biological weapons, biodefense, and even genetic weapons—a new subfield of biological weapons."

Back in 2005, according to Pfluke, the US State Department identified two facilities in China with "links" to an offensive biological weapons program: China's Ministry of Defense's Academy of Military Medical Sciences Institute of Microbiology and Epidemiology in Beijing, and the Lanzhou Institute of Biological Products. In addition, "it is estimated that there are at least 50 other laboratories and hospitals being used as biological weapons research facilities." China has the capability of delivering biological weapons in aerosol form, and yet, according to Pfluke, because of the way China suffered under Japanese occupation—the use of biological weapons on Chinese soldiers, Unit 731's horrific experiments on prisoners—she thought it unlikely they had an interest in using them. Nevertheless, she too brought up the genetic weapon idea, essentially the use of the CRISPR-cas9 gene editing technique for nefarious purposes.

She cited James Clapper, the former director of US National Intelligence, who had argued as far back as 2016 that gene editing had to be added to the list of major threats. CRISPR makes moving genes around, or altering them, or deleting them, almost simple by comparison with earlier techniques, which is why Emmanuelle Charpentier and Jennifer Doudna, who invented CRISPR, won the Nobel Prize in Chemistry in 2020.

> CRISPR could be used to make genetically engineered killer mosquitos, plagues that target and wipeout specific crops, and possibly that can snip people's DNA...it might be possible to use CRISPR to alter diseases to target entire races by focusing the disease on a certain genetic trait. In this way, China could, hypothetically, build a disease that targets the Japanese and release it, without worrying about it infecting China's own people. This may sound like a science fiction movie plot, but it is no longer inconceivable. Not only can genes be edited, but China is already successfully doing it.[109]

At first, I thought these comments about genetic weapons were ridiculous. I couldn't imagine how anyone could tweak a disease such that it would only trouble a particular ethnic group. But then I came across a fascinating article on the preprint site bioRxiv. This site is managed by Cold Spring Harbor Laboratory, on Long Island, New York. At the turn of the last century, the Cold Spring Harbor Laboratory was funded by the rich and clever (like Alexander Graham Bell) to advance the pseudo-science of eugenics. Invented by Darwin's cousin Francis Galton, eugenics involved spurious arguments built on meaningless measurements purporting to show that the best and brightest were being outbred by the stupid and dangerous, whose breeding should be discouraged, even stopped by law. After eugenics was debunked, Cold Spring Harbor fell on hard times until Barbara McClintock did groundbreaking work there on the genetics of corn—establishing long before the advent of molecular genetics that chromosomes break and re-form, and when they do, changes happen. Later still, it became a famous centre for postgraduate courses and conferences on leading-edge work in molecular genetics. James D. Watson of DNA fame presided there for many years and used it as a platform to launch the Human Genome Project. Then it created a degree-granting school named after Watson. But there must be something in the water at Cold Spring Harbor, something lingering from eugenics days. Watson was shown the door after making eugenic/racist comments that could not be ignored, remarks he apologized for in 2007 but made again in 2019, leading CSHL to sever all remaining connections to him.[110]

The article that caught my eye on its bioRxiv site was by Svante Pääbo and Hugo Zeberg. Pääbo is a leading researcher at the Max Planck Institute for Evolutionary Anthropology in Leipzig. Zeberg works there too, but also in the Department of Neuroscience at the Karolinska Institute in Stockholm. I knew Pääbo's work from a previous book I'd done that in part described the use of genomic science to trace the first people to inhabit the Americas.[111] Pääbo's group at the Max Planck was the first to extract a complete genome sequence from ancient Neandertal remains and compare that sequence with the modern human genome. His group demonstrated that modern humans must have had children with Neandertals about 50,000 years ago when modern humans moved into Europe, because some modern humans still carry a few Neandertal genes. Those findings overturned prior theories (his own included) that modern humans had wiped out Neandertals without any "gene flow" (sex that produces offspring) between them.[112]

Pääbo and Zeberg relied on work by others who had identified a gene cluster on human chromosome 3, which predicts a high risk of respiratory failure from SARS-CoV-2. They showed that "the risk is conferred by a genomic segment of about 50 kb [50,000 base pairs] that is inherited from Neandertals." They then looked at which human populations carry this Neandertal variant. They found it occurs at a frequency of 30 percent in South Asian populations (people from Bangladesh, India, Pakistan) and 8 percent in Europeans, and is not seen at all in African populations.[113]

Their article was eventually published by *Nature*. The *Nature* peer-reviewed version revised these frequency estimates significantly. In *Nature*, Pääbo and Zeberg said that the Neandertal anomaly can be found in 50 percent of South Asians and 16 percent of Europeans. The lowest frequency outside Africa was in East Asian populations—people from Mainland China, Hong Kong, Taiwan, Japan, Macau, Mongolia, North and South Korea. "The Neandertal variant may thus be a substantial contributor to COVID-19 risk in certain populations,"[114] they concluded.

That paper, in the context of Kania's article and Clapper's comments on genetic weapons, formed a three-point curve. The more I thought about it, the more it seemed to mean something. East Asian countries like Japan and South Korea and Taiwan had done well at staving off the worst of SARS-CoV-2. So had most African countries—to everyone's surprise—at least until a new and more infectious version of the virus appeared in South Africa. South Asia, and in particular India, had done fairly well in the first wave, though not as well as China. But in the spring of 2021, its second wave would turn into a catastrophe.

I called the contrary friend and told him about Pääbo's article, about Pfluke's and Kania's arguments. "What the hell do you make of that?" I demanded.

"What do *you* make of it?" he countered.

"I don't know," I said.

EARLY IN FEBRUARY 2020, an interesting article appeared on a preprint server, only to be rapidly taken down. It was called "The possible origins of 2019-nCoV coronavirus," by Botao Xiao and Lei Xiao.[115]

Botao Xiao is with the Joint International Research Laboratory of Synthetic Biology and Medicine, School of Biology and Biological Engineering, South China University of Technology in Guangzhou. He is also associated with the School of Physics, Huazhong University of Science and Technology in Wuhan. Lei Xiao is with the Tain You Hospital at Wuhan University of Science and Technology.

Arguing that it is crucial to know where the virus arose, the authors considered the Huanan Seafood Market origin idea and found it wanting. SARS-CoV-2 was only found "in 33 out of 585 samples collected in the market after the outbreak," they said. They mentioned the two papers that had just appeared in *Nature* comparing SARS-CoV-2 to other bat-carried coronaviruses. They considered it very unlikely bats had flown into the market, and "according to municipal reports and the testimonies of 31 residents and 28 visitors, the bat was never a food source in the city, and no bat was traded in the market." No intermediate host had been identified, either.

However, the authors noted, there are two facilities in Wuhan that study bats and try to extract viruses from them. The first, the Wuhan Center for Disease Control and Prevention, has "specialized in pathogens collection and identification. In one of their studies, 155 bats including *Rhinolophus affinis* [a horseshoe bat] were captured in

Hubei province (Wuhan is its capital), and another 450 bats were captured in Zhejiang province." They pointed out that one scientist at the center, Tian Jun-Hua, had been the subject of a documentary showing him crawling into bat caves unprotected by a haz-mat suit. He explained to the filmmakers that he had once been forced to quarantine for 14 days after being attacked by bats, and another time after bats peed on him.

I recognized that name: Tian Jun-Hua was one of the co-authors on the Zhang and Holmes paper in *Nature* that described the SARS-CoV-2 genome sequence. Tian had also captured a live tick on a bat that became the subject of another paper co-authored with E.C. Holmes.

Surgery is performed in this lab, said these authors, which also houses caged animals as well as samples extracted from bats in the wild, so there could have been a waste handling problem. Therefore, they wondered if this virus might have escaped from Tian's lab, which is only about 280 metres from the Huanan Seafood Market and the "Union Hospital where the first group of doctors were infected during this epidemic."

The other possibility they pointed to was that the virus escaped from the second lab, which is 13 kilometres from the market—the Wuhan Institute of Virology. The authors explained that the "principle investigator participated in a project which generated a chimeric virus" and reported the potential for human emergence. Perhaps, they said, SARS-CoV-2 might have leaked from that laboratory. The principal investigator was not named, but they meant Shi Zhengli: they footnoted the paper describing the chimera experiments done at Ralph Baric's lab and published in 2015.

"In summary, somebody was entangled with the evolution of 2019-nCoV coronavirus," they wrote. "In addition to origins of natural recombination and intermediate host, the killer coronavirus probably originated from a laboratory in Wuhan. Safety level may need to be reinforced in high risk biohazardous laboratories. Regulations may be taken to relocate these laboratories far away from city center and other densely populated places."[116]

The authors had quickly pulled this piece from the preprint site. The *Wall Street Journal* wanted to know why. Botao Xiao explained that he had taken it down because it was based on others' published writings, not on hard evidence. This was a strange statement, as most scientific papers have long lists of citations—other people's writings—and opinion pieces and essays also appear in scientific publications. Though it was taken down, it did not disappear. Nothing on the internet disappears. People found it. Spread it. Added to it. The WIV became the focus of lab origin theories.

On April 1, a sinister variation on this Wuhan lab-leak theme began to circulate. An American filmmaker/actor named Mathew Tye (who documented his life in China on a YouTube channel called laowhy86 and is known as C-Milk)[117] uploaded to YouTube a documentary claiming to have the answer to the origin of the virus. His piece concerned a young woman who allegedly answered an ad for a position at the WIV just before the pandemic broke out, a girl who he thought was probably patient zero, and who he claimed had disappeared.

I found Tye's documentary while going down an internet rabbit hole. I can no longer recall what I was chasing, but I found Kiwiblog, whose author, a blogger, pollster, and political activist in New Zealand named David Farrar, cited as his authority on the Tye documentary an article by Jim Geraghty, a writer on politics for the right-of-centre *National Review* in the US. Geraghty's article included screenshots of job postings that had appeared on the WIV's website in November and December 2019. In one, Peng Zhou, a collaborator of Shi Zhengli's, advertised for a post-doc who would work with him on bats' apparent immunity to SARS-like viruses and Ebola viruses; he referenced a significant number of viruses found in bats and rodents. The posting was dated November 18, 2019, just six days before Li's group had calculated that the virus exploded into the human population from a single entry. The other job opening was for a scientist to "research the relationship between the coronavirus and bats to work with Dr. Zhengli." The screenshot in Geraghty's story showed the job was posted on December 24, 2019.

The implication drawn by Kiwiblog is that the person who took the first job had died of the virus and the second job posting was to find someone to replace her, though there were no facts to support that.

> A new job posting at the WIV is phrased in such a way that states in essence "we've discovered a new and terrible virus and we need to hire people to come and help deal with it"...Huang Yan Ling was the original researcher and early contractor of the virus and all her biographical material and photos were removed from the WIV website. No one has seen her despite government assurances that she is alive and well with no proof. There has been a massive search for her on the Chinese internet with most articles getting quickly scrubbed by Chinese government censors.[118]

The questions on social media about this young woman grew so intense that on February 15, according to Kiwiblog, the WIV "denie[d] that Huang Yanling was patient zero addressing 'fake information' as to her whereabouts and yet she has disappeared from public view."

Then, according to Kiwiblog:

> Zhen Shuji, a Hong Kong correspondent from the French public-radio service *Radio France Internationale*, reported: "when a reporter from Beijing News of the Mainland asked the institute [WIV] for rumors about patient zero, the institute first denied that there was a researcher Huang Yanling, but after learning that the name of the person on the Internet did exist, acknowledged that the person had worked at the firm but has now left the office and is unaccounted for."

Geraghty, who described the arguments about the possible origin of the virus, said something quite different about the missing girl. His piece was more fact-based, but just as incendiary. He pointed out that according to the spokespeople at the WIV, Huang Yangling had been a post-doc there until 2015, after which she left Hubei. She had published no research articles since 2015. Information about her had been scrubbed from the website, but so had information about two of her fellow students who were at the WIV at the same time. But where was she? Dead? Or alive? Geraghty asserted that if the Chinese government knew how to find her, she'd have been brought forward immediately.[119]

Though these facts were very thin, the story did not die.

That is because concern about what might have been going on at the WIV was not confined to blogs or YouTube documentarians or preprint sites. In fact, suspicion had been expressed first by members of the Western scientific community, who know what virologists get up to. On January 31, as the article from the two Chinese researchers appeared on the bioRxiv site pointing to the possibility of a lab accident at the WIV, Jon Cohen, a staff reporter for *Science*, published a piece called "Mining coronavirus genomes for clues to the outbreak's origins."[120] He interviewed Trevor Bedford, who was putting all the published SARS-CoV-2 genome sequences on the nextstrain.org site. As Bedford saw it, the difference between Shi Zhengli's RaTG13 sequence and SARS-CoV-2 is about 1,100 nucleotides, which is significant in a genome only 29,000 nucleotides long. It would take a significant amount of time for so many mutations to accumulate, so SARS-CoV-2 must have been circulating for a long time, in some animals, somewhere.

"One of the biggest takeaway messages [from the viral sequences] is that there was a single introduction into humans and then human-to-human spread," he told Cohen. SARS-CoV-2 had acquired only seven nucleotide changes since the pandemic burst on the scene, a very slow rate of change. He calculated that for SARS-CoV-2 to have evolved from the closest relative, RaTG13, while circulating in some other animal species would have taken at least 25 years.

However, Bedford's calculation was disputed by Richard Ebright, who argued that the mutation rate may have changed if the virus passed through another host before infecting humans. Then Ebright added a shocking suggestion: that the data were "consistent with entry into the human population as either a natural accident or a lab accident."

Ebright is not some basement-dwelling conspiracy theorist speculating wildly after enjoying too many edibles. Richard Ebright is a serious scientist. He is the Board of Governors Professor of Chemistry and Chemical Biology at Rutgers University. So, Cohen tried to get comment from Shi Zhengli. She did not respond. Her papers list an email address, which I tried too, to no avail. As an alternative, he sought out Peter Daszak of New York-based EcoHealth Alliance, the long-time collaborator—and funder—of Shi's work on bats. Daszak pooh-poohed Ebright's suggestion.

"Every time there's an emerging disease, a new virus, the same story comes out: This is a spillover or the release of an agent or a bioengineered virus. It's just a shame...." Daszak explained to Cohen that in the course of the work he and Shi had done together over the past eight years, they had pulled samples from over 10,000 bats and 2,000 other species. The closest one to SARS-CoV-2, RaTG13, was "fished out of a bat fecal sample they collected in 2013 from a cave in Mojiang in Yunnan province." He expected that after they sampled across all of China's southern provinces, they would find other samples closer to SARS-CoV-2.[121]

Which was an odd thing to say. His collaborator, Shi, had surveyed bats in various provinces for that first article they co-authored on SARS's origin, published in *Science* in 2005. The article reported that the highest number of bats carrying SARS-like coronaviruses had been sampled in Hubei, where Wuhan is located. Why hadn't he said that Shi might find a closer relative than RaTG13 in Hubei?[122]

Daszak was likely thinking harder about his future than about findings from the past. It was normal for the internet to be awash in rumours and theories. No one important would pay attention to that. But it was another thing entirely for *Science* to ask about a possible lab accident at the WIV. Unless this notion was squashed like a bug by a heel, it could take hold and spread like the pandemic itself. And then what? For years the National Institutes of Health (NIH) and its National Institute of Allergy and Infectious Diseases (NIAID) and USAID had been funding Shi's work at the WIV through *his* organization.[123] If serious people believed this pandemic might have come from a virus collected by or manipulated in Shi's WIV lab, if the blame for this catastrophe fell on her, it would also fall on him, on EcoHealth Alliance, on the NIH and USAID. The China virus would become the American virus.

Only two days later, Daszak was drafting a public statement for *The Lancet* and also a letter for the National Academies of Sciences, Engineering, and Medicine to send on to the White House to counter the lab-leak theory. He conferred on both with Ralph Baric, Linfa Wang, one of his board members, Rita Colwell, a former director of the National Science Foundation, as well as other people with aligned interests (the need to do work in China, the need for funds from the NIH).[124] At the same time, another group with similar interests that he was also in touch with organized a letter to appear in *Nature Medicine*.[125] Daszak's protest statement was signed by a group of 27 scientists.[126] But the statement and the letter quashed nothing: they provoked the opposite response.

5. Political Science 2.0

THE *NATURE MEDICINE* letter appeared first on February 9 on the blog Virological.org. It was titled "The proximal origin of SARS-CoV-2." *Proximal* means approximate. The letter was pockmarked with similar weasel words, such as "possible," "plausible," and "likely." Yet for months after its publication on that website and then online in *Nature Medicine*, this letter was pointed to again and again as definitive proof that SARS-CoV-2 could only have had a natural, not a lab, origin.

The co-authors listed what was then known about SARS-CoV-2's genome sequence and how it binds to the human ACE2 receptor. They pointed out that its binding affinity is better than anything computer modelling programs suggested. In effect, they argued that nature is smarter than computer models, so nature must have made the virus because no lab, guided by modelling, would have made it this way. "Thus, the high-affinity binding of the SARS-CoV-2 spike protein to human ACE2 is most likely the result of natural selection on a human or human-like ACE2 that permits another optimal binding solution to arise," they wrote. "This is strong evidence that SARS-CoV-2 is not the product of purposeful manipulation."[127]

That's not evidence at all, I muttered to myself. That's speculation.

Then they argued the genome sequence showed, "irrefutably," that another SARS-type "backbone" had not been used to make this virus. In other words, it was not a chimera such as had been made by Baric's and Shi's groups in 2015. Baric's group had taken the spike sequence from a virus isolated by Shi and attached it to a SARS-related sequence whose original spike sequence had been removed.[128] To their surprise, they managed to grow a functional virus in culture and infect human cells with it, and then humanized mice. But the authors of this *Nature Medicine* paper mainly ignored the fact that making a chimera is not the only way to make a virus more lethal or more infectious. The easier method is letting nature do it for you by passaging, which leaves no tracks of unnatural insertions or deletions in the genome sequence. Fouchier made an avian flu virus infect normally immune ferrets by injecting it into them, one after the other, until the virus changed enough to thrive in them. And yet for most of this article, these authors did not discuss passaging; they insisted instead that the virus was most likely the result of natural selection in an animal followed by a transfer to humans, or "natural selection in humans following zoonotic transfer."

They suggested that since the unusual amino acid inserts in the virus's receptor binding domain had been found in other beta coronaviruses, natural selection was a probable explanation for their existence in SARS-CoV-2. But natural selection where? While they thought a pangolin might have carried a viral ancestor of SARS-CoV-2, they found it more likely that an animal carrying a virus with both the amino acid inserts and the peculiar SARS-CoV-2 spike would "probably have to have a high population density (to allow natural selection to proceed efficiently) and an ACE2-encoding gene that is similar to the human ortholog." To put that in English: pangolins didn't fit the bill. They don't hang out in crowds. These co-authors found it more probable that this bit of natural selection had happened in human beings without being noticed, which flew in the face of Terence Bedford's insight that this virus had entered humans once and exploded. "Once acquired," they wrote, "these adaptations would enable the pandemic to take off and produce a sufficiently large cluster of cases to trigger the surveillance system that detected it."

The co-authors loosely pinpointed "emergence of the virus in late November 2019 to early December 2019." Their evidence for this date range was supplied by one of the group, who had posted his own estimate without peer review to his Virological.org website. There were other papers, peer-reviewed papers, they could have relied on for an estimate.[129]

Only toward the end did they finally admit that a lab leak was even possible, because lab leaks have happened before, or that passaging had been done for years with SARS-like viruses in BSL-2 labs. They mentioned "documented instances of laboratory escapes of SARS-CoV." But they did not specify instances, such as the fact that SARS had escaped several times, twice from a leading CDC laboratory in Beijing, resulting in a death. "In theory, it is possible that SARS-CoV-2 acquired RBD [receptor binding domain] mutations during adaptation to passage in cell culture, as has been observed in studies of SARS-CoV," they wrote. However, they insisted they did not find this plausible. They believed the better explanation was that they were acquired through recombination or mutation. Why? If the strange furin cleavage site had been created through cell passage in a lab, there would have to have been a precursor virus much closer to SARS-CoV-2 than RaTG13, "which has not been described." Even then, they wrote, more passages in cell culture, or in animals with ACE2 receptors similar to humans', would be needed to arrive at SARS-CoV-2 and "such work has also not previously been described." These features would suggest "the involvement of an immune system."

Exactly! I shouted. The creation in a lab of the curious features of SARS-CoV-2 could have resulted from a virus adapting itself to humans through multiple passages in human cells or mice with humanized immune systems. Just because such work had not been described did not mean it hadn't been done. Military researchers don't always publish their finest work.

To be certain of the date of origin, they argued, studies of banked samples from patients in Chinese hospitals with unusual pneumonias should be tested, a sensible suggestion if those samples had been held in the US, the UK, Canada, or Australia. But they were held in China.

On March 3, two weeks before this letter appeared in *Nature Medicine*, China's State Council issued a secret notice. "During the period of the epidemic prevention and control, all universities, research institutions, medical institutions, enterprises and their staff shall not publish information on scientific research related to epidemic prevention and control without approval." The notice set out guidelines for getting that approval, including aligning work with social needs and propaganda requirements, and warned that any who failed to comply would be held "accountable."[130] Only a week after this letter first appeared on Virological.org, and long before *Nature* published the letter online, China's Center for Disease Control and Prevention passed along even more stringent orders to all its offices. "No one can under their own name or in the name of their research team provide other institutions and individuals with information related to the COVID-19 epidemic on their own, including data, biological specimens, pathogens, culture, etc.... Anyone who violates the above regulations shall be dealt with severely in accordance with discipline, laws and regulations."[131] As the Associated Press put it when it acquired these notices, China had "centralized all COVID-19 publication under a special task force... applying to all universities, companies and medical and research institutions. The order said communication and publication of research had to be orchestrated like 'a game of chess' under instructions from Xi, and propaganda and public opinion teams were to 'guide publication.'"[132]

At least one of the co-authors would have been aware of these orders. Nevertheless, they arrived at this conclusion: "Although the evidence shows that SARS-CoV-2 is not a purposefully manipulated virus, it is currently impossible to prove or disprove the other theories of its origin described here. However, since we observed all notable SARS-CoV-2 features, including the optimized RBD and polybasic cleavage site, in related coronaviruses in nature, we do not believe that any type of laboratory-based scenario is plausible."[133]

Yet the evidence they cited did not show that SARS-CoV-2 "is not purposefully manipulated." It only showed one kind of manipulation had not been done. Passaging in a lab is as purposeful as sequence swapping, it just doesn't leave tracks. And *plausible* is a word that applies to arguments, not to short pieces relaying factual evidence, i.e. a *Nature* letter. So why had *Nature Medicine* published it?

Because of the track records of its co-authors, all highly respected scientists.

Unfortunately, some were also interested parties whose work depended on entry to China, or who had close relationships with the NIH/NIAID or with other funders of Shi's experiments. Yet only one author declared competing interests. The co-authors were: Kristian Andersen, E.C. Holmes, Andrew Rambaut, W. Ian Lipkin, and Robert F. Garry. E.C. Holmes is not just a professor at the University of Sydney, he also has an appointment with Fudan University in China. He should have known about the notices issued warning China's scientists not to publish anything or share any samples without approval from the government of China. Along with Zhang, his colleague at Fudan, he published the SARS-CoV-2 genome on January 10 on the Virological.org website. He has published numerous papers with Chinese scholars over the years and needs access to China to do his work. Andrew Rambaut is a virologist at the University of Edinburgh and administers the Virological.org website that posted the SARS-CoV-2 sequence. W. Ian Lipkin is the John Snow Professor of Epidemiology at the Mailman School of Public Health at Columbia University. According to his bio on the school's website, Lipkin helped China during the SARS crisis and holds honorary positions in Shanghai and Beijing. He is also director of the NIH/NIAID Center for Research in Diagnostics and Discovery and advises the WHO. The Mailman School at Columbia got about $1 million a year for several years from USAID's

PREDICT program through Peter Daszak's EcoHealth Alliance. Kristian Andersen is a professor at the Scripps Research Institute in California and, as I would later learn, at the time this letter was written, was an applicant to NIAID, as was Daszak's EcoHealth Alliance, for grants from a new $82-million NIAID program. Centers for Research in Emerging Infectious Diseases—CREID—is a little like the $200-million PREDICT program that ended (with an extension) in October 2019, to the chagrin of Daszak and PREDICT founder Dennis Carroll.[134] Robert F. Garry is a leading virologist at Tulane and part owner of a private company called Zalgen Labs, which creates vaccines and is a subcontractor on a grant won from the NIH for an antibody cocktail to treat Lassa fever. Garry was the only one who declared any competing interest on the paper.[135]

I thought it "likely" that this article got started when Peter Daszak mentioned to Lipkin, whom he knew well, that something needed to be done to demolish the lab-leak theory before it took off. Lipkin had once been a research fellow at Scripps, where Andersen is now a professor. He has an institutional interest in the continuation of funding from NIAID for viral science in China. I didn't think Daszak would have approached Holmes, Rambaut, and Andersen himself. They had publicly disdained the PREDICT program's goal of trying to predict pandemics by surveying for emerging viruses in animals, and Daszak had played a large part in the PREDICT program. Holmes, Rambaut, and Andersen had argued in *Nature* in 2018 that there is an unknown number of viral species in the world undergoing constant evolutionary change, so no one could predict, even with the aid of the best artificial intelligence system, which one might start a pandemic. They said that making false claims about predictive capacity and building up stockpiles of vaccines for diseases that might never emerge would just break down trust. They had also been dismissive of a larger project being promoted by Daszak, Dennis Carroll, and the Gates Foundation to discover the genome sequence of every virus in the world that might infect humans. This multi-billion-dollar Viral Genome Project, according to Andersen, Rambaut, and Holmes, was plain "arrogant."[136]

On the other hand, it was possible Daszak *had* appealed to Andersen, who then brought in the others. NIAID's CREID program planned to fund ten groups to survey for emerging diseases in various hotspots around the world. Months after this letter appeared, NIAID announced that both Andersen's and EcoHealth Alliance's applications had succeeded. Months after that, US Right to Know published Ralph Baric's emails, obtained under a freedom of information application, which showed that Andersen and Daszak were in touch by email by February 1.

Three of the five authors of the *Nature Medicine* letter shared an interest in getting more money for viral surveillance out of the US government's NIH/NIAID, as did Daszak. NIH/NIAID had a strong interest in avoiding blame over something done with NIAID money in Shi's lab that might have kicked off the pandemic, as did Daszak. Two of the co-authors had a deep interest in continuing to do research in China on bat-borne viruses. Daszak's name did not appear on this paper, but it might as well have.

THE *LANCET* STATEMENT was posted online on February 18, just a few days after the "proximal origin" article first appeared on Virological.org. The *Lancet* statement was titled: "Statement in support of the scientists, public health professionals, and medical professionals of China combatting COVID-19":

> ...we have watched as the scientists, public health professionals, and medical professionals of China, in particular, have worked diligently and effectively to rapidly identify the pathogen behind this outbreak, put in place significant measures to reduce its impact and share their results transparently with the global health community.

According to the authors, this transparency was being threatened by spurious rumours.

> The rapid, open, and transparent sharing of data on this outbreak is now being threatened by rumours and misinformation about its origins. We stand together to strongly condemn conspiracy theories suggesting that COVID-19 does not have a natural origin. Scientists from multiple countries have published and analysed genomes of the causative agent...and they overwhelmingly conclude that this coronavirus originated in wildlife.

This last claim about the scientific consensus on the virus's origin cited nine articles, including the proximal origin letter that had just appeared on the Virological.org website. None of the nine articles, not even the letter, were sufficiently definitive to support the claim that "scientists... overwhelmingly conclude that this coronavirus originated in wildlife."

The *Lancet* statement closed with:

> We invite others to join us in supporting the scientists, public health professionals, and medical professionals of Wuhan and across China. Stand with our colleagues on the frontline![137]

What was more interesting than what the statement said was who signed. Of the 27 signatories, six had relationships with the Wuhan Institute of Virology, the main target of these "rumours," through EcoHealth Alliance. Though it spoke of the value of transparency, these relationships were not declared as competing interests. Dennis Carroll, the organizer of the PREDICT program at USAID, from which he had only recently retired, was identified only as a fellow with the Scowcroft Institute of International Affairs in Texas. Peter Daszak was identified as the president of EcoHealth Alliance, but the *Lancet* readers might not know that his organization had funded Shi Zhengli's work in China, and that her lab was a target of a "conspiracy" theory. William Karesh was not identified as the executive vice president of EcoHealth Alliance, but only with the World Organization for Animal Health. Jonna Mazet was identified with University of California, Davis, but not as a partner/director of PREDICT, which funded Shi. Rita Colwell was not identified as a board member of EcoHealth Alliance. Hume Field was not identified as an employee of EcoHealth Alliance.[138]

US Right to Know would later publish emails obtained through access to information requests written between Daszak, Rita Colwell, and Baric as this statement was being drafted.[139] The emails make clear that this statement was Daszak's baby.

These publications did not endear their co-authors to

scientists asking reasonable questions about the origin of SARS-CoV-2. But they were successful for a time in the main goal: tarring anyone who might suggest this virus could have leaked from a lab as a conspiracy theorist and therefore untouchable by major media. The point was to shape the story. No science journalist would question a paper from W. Ian Lipkin or E.C. Holmes published in one of the world's top peer-reviewed journals. No daily journalist would have time to search out the relationships and interests they shared.

It took time for alternate views to find their way into the public sphere. Karl and Dan Sirotkin would be the first to publish a peer-reviewed article explaining why a lab leak, as a result of gain-of-function/serial passage experiments, was not only plausible but probable.[140] They explained how serial passage in a lab gooses natural evolution, they pointed to the high affinity of SARS-CoV-2 to the human ACE2 receptor and the unusual nature of the amino acid inserts that worked so well to enhance its merger with human host cells. They called for a proper investigation and an examination of the risks versus the rewards of gain-of-function experiments.

Soon, a loose group of scientists, annoyed at being told their questions were illegitimate, began to form. They even got themselves a name: DRASTIC (see chapter 15). Their number grew as they roamed through the WIV's online archives and databases, wrote papers, put their findings out on Twitter and on preprint sites. Some of their papers eventually appeared in peer-reviewed journals, which made major media take a second look. Their findings began to shape a new origin story.

ON MARCH 26, nine days after the proximal origin letter appeared in *Nature Medicine*, the director of the NIH, Francis Collins, joined the smack-down-the-lab-leak campaign. On his blog, he described the *Nature Medicine* letter as the lab leak's refutation, though he weirdly conjoined it to a bioweapon theory. Collins wrote:

> Some folks are even making the outrageous claims that the new coronavirus causing the pandemic was engineered in a lab and deliberately released to make people sick. A new study debunks such claims.

"The reassuring findings," Collins continued, "are the result of genomic analyses conducted by an international research team, partly supported by NIH." He mentioned by name only Kristian Andersen of the Scripps and Robert Garry of Tulane. Collins said "their analysis showed that the backbone of the new coronavirus's genome most closely resembles that of a bat coronavirus discovered after the COVID-19 pandemic began." This was plain wrong. The "backbone" reference in the *Nature Medicine* letter was to a lab-made chimera, not to its resemblance to a bat-borne coronavirus. The bat sequence Collins referred to—RaTG13—had been pulled out of a fecal sample in 2013, not recently, though it was not published until after the pandemic began. Collins reiterated the letter's argument that the strongest evidence against a lab origin was that computer models failed to predict SARS-CoV-2's high binding affinity. He did not refer to passaging at all.

However, on March 27, 2020—as *Newsweek* would report in April—the US Defense Intelligence Agency "updated its assessment of the origin of the novel coronavirus to reflect that it may have been accidentally released from an infectious diseases lab."[141] The report ruled out that the disease was genetically engineered "or released intentionally as a biological weapon." It seemed to DIA analysts "very unlikely" that researchers would release "such a dangerous virus, especially within China, without possessing a known and effective vaccine." The Agency partly based its opinion on a publication "in early February [by] China's Academy for Military Medical Sciences [which] concluded that it was impossible for them to scientifically determine whether the Covid-19 outbreak was caused naturally or accidentally from a laboratory incident." But the DIA suggested the release could have been the result of unsafe laboratory practices.

Clearly, the DIA had not asked itself how the Chinese military could have formed that opinion so quickly. Chinese military virologists were not sent into Wuhan until January 26. At first, the People's Liberation Army's lead virologist and bioweapons expert, Major General Chen Wei, worked in a mobile lab in the city, trying to devise treatments for the sick and a protective spray for health workers, but then she and her team moved into the WIV's BSL-4, where they worked flat out on a vaccine for SARS-CoV-2. Together with CanSino Biologics—a company with Canadian connections using old technology developed at the National Research Council—they had that vaccine in Phase 1 trials by March 16, only six weeks after Major General Chen Wei arrived in Wuhan. She and her team wouldn't have had time to thoroughly investigate anything by early February, let alone issue a report the US Defense Intelligence Agency should rely on.[142]

Newsweek's story on the DIA's change of view did a thorough job of going through the unresolved questions, including a critique of the *Nature Medicine* letter. It mentioned that the WIV had been doing gain-of-function experiments for five years and that there had been lab accidents in China. It pointed out that officials from the US embassy in Beijing had visited the WIV BSL-4 lab and been told by WIV staff that they lacked sufficient trained technicians. *Newsweek* quoted James Le Duc, director of the Galveston National Laboratory, which houses a BSL-4 and helped train people in Wuhan, who insisted that the WIV lab was very well run and statements about a lab leak were a "malicious move to purposefully mislead the people."

But the reporters didn't leave it there. They pointed out that gain-of-function experiments were being conducted in many laboratories around the world and reminded readers that there had been many dangerous lab accidents at well-run facilities in the US. They quoted Harvard epidemiologist Marc Lipsitch, who had written a critical article in *Nature* on gain-of-function work, stating that it "entails a unique risk that a laboratory accident could spark a pandemic, killing millions." The scientists in the group he helped found, the Cambridge Working Group, had been keeping tabs on lab accidents and concluded they occur at the rate of two per week in the US. *Newsweek* noted that, after a pause, the NIH had decided in favour of funding gain-of-function experiments again, to be conducted in BSL-3s only, so that "by the time the current pandemic hit, animal-passage experiments had become commonplace."

Newsweek raised PREDICT's role in funding the WIV, and the dubious value of surveying animals (as opposed to people) for emerging viruses was questioned too. Once again, Richard Ebright had his say: "The PREDICT program has produced no results—absolutely no results—that are of use for preventing or combating outbreaks. There's no information from that project that will contribute in any way, shape or form to addressing the outbreak at hand. The research does not provide information that's useful for developing antiviral drugs. It does not provide information that's useful for developing vaccines."

The same day the *Newsweek* story ran, *Politico* published one of its own: EcoHealth Alliance's just-renewed NIAID grant of $3.7 million, which would have helped fund some very risky experiments to be done by Shi Zhengli, had been cut off. All money remaining in the grant account had been frozen.[143]

6. The Canadian Connection

IN MARCH 2010, when Richard Fadden was director of the Canadian Security Intelligence Service (CSIS), he gave a speech to the Royal Canadian Military Institute which was later televised by the CBC. He said that China was trying to exercise undue influence on Canadians in high places to shape Canadian decision-making. He referred to some municipal politicians and cabinet ministers in two provinces but did not name names, though the suggestion was that these officials were of Chinese origin. He warned about Chinese-government-backed Confucius Institutes that were offering boards of educations in Canada and around the world free programs involving instruction in Mandarin and Chinese history (no mention of Tiananmen Square). He said these institutes were funded by China and run out of Chinese embassies and consulates whose employees also organized demonstrations against Canadian policies having to do with China. He was raising a red flag about what spy novelists call agents of influence, suggesting that China was trying to make Canadian policy favourable to its own interests in this overt/covert fashion. He had told his minister and deputy minister what he intended to say, and he got permission to say it. He did not apologize when apologies were demanded. Many considered Fadden's speech an assault on Canadian multiculturalism and on the dignity of Chinese-Canadians.[144] Fadden and the government were rebuked by the House of Commons Standing Committee on Public Safety and National Security. CSIS was told to apologize for defaming people and to stick to its knitting—defending the security of Canada.

That's when I first ripped an article out of the papers on Canada-China relations. But I got religious about it in 2016, after the 2015 federal election that brought a Liberal government to power with Prime Minister Justin Trudeau at its helm. After the campaign was over, there were private fundraising dinners organized by people who appeared to want things from the government, cash-for-access dinners they were called. In November, the *Globe* published a story about a dinner that had taken place the previous May at the home of Benson Wong involving billionaires of Chinese descent who had been in Canada for some time. But the *Globe* also named other attendees who were active in the government of China. They did not pay for dinner: only Canadians may make political contributions. After the dinner, a businessman who attended got the green light to open a Schedule 1 bank, while another attendee, Zang Bin, described as an advisor to the government of China, donated, with a partner, $1 million for the creation of a statue to honour the prime minister's father, Pierre Elliott Trudeau, and to support the Faculty of Law at l'Université de Montréal.[145]

By then, a certain uneasiness about Canada's relationship to China had grown right alongside a balloon of optimism about the great opportunities China's huge market offers Canadian businesses. President Xi had come to power and enunciated a new direction, starting with an anti-corruption campaign, which might be a good thing, who knew? The *Globe and Mail*'s Nathan VanderKlippe had not yet produced front-page stories on the re-education/prison camps for Uyghurs in Xinjiang and the dispersal of Uyghur workers across China as forced labour. Nor was it understood that huge sums of money were flowing out of China and into casinos and real estate in Toronto and Vancouver, the Vancouver houses sometimes purchased with cash by people who listed their occupations as student or housewife. There had been several well-publicized investments and takeover attempts of Canadian energy, mining, and tech companies by Chinese state-owned companies, some with national-security implications.[146] Still, in 2016, few Canadians grasped that every Chinese institution has a Communist Party director guiding its path, that every China-based company, large or small, is expected to follow the party line. In the fall of 2016, many believed that welcoming China into the globalized business tent would inevitably result in China becoming a liberal democracy. And what could be bad about that?

On a day when I had nothing better to do, I looked up the annual financial reports for Prime Minister Justin Trudeau's Papineau riding association. This may seem like a strange way to fill one's time, but I'm a journalist and we do things like that. I did this before I heard of China's United Front Work Department—a political-influence operation with thousands of employees whose job is to shape opinion and policy in countries of interest to China—such as Australia, such as Canada—and whose tactics include everything from creating friendship organizations, to giving prizes for good works, to bribery and intimidation of citizens of Chinese descent, or students from China sent to such countries to study. The United Front Work Department reports directly to the Central Committee of the Communist Party of China.[147]

Papineau is a working-class riding in Montreal. As the 2015 election day drew near, and it seemed likely that Justin Trudeau might bring the Liberals back to government from third-party status, the Papineau Liberal riding association's annual report to Elections Canada showed that it had received a lot of money from donors with Chinese surnames whose addresses were listed as Richmond Hill, Vancouver, Richmond, Thornhill, Markham, all of which are distant from Papineau and have their own Liberal riding associations to contribute to. Papineau reported total contributions of $237,054 for 2015. Of 213 contributions, 122 came from persons with Chinese surnames who lived outside of Quebec, totalling $170,800. Many of those contributions came weeks before or several weeks after the election. The previous year, 2014, total contributions were $40,787.61. Out of 37 donations that year, 25 came from persons with Chinese surnames from outside Quebec. In 2016, donations again came mainly from persons with Chinese surnames from outside Quebec. But in 2018, and again in 2019, another election year, total contributions to Papineau fell to a little over $11,000 each year and few persons with Chinese surnames from outside the province were among the donors. By then, Canada-China relations were in the dumper. When I checked former prime minister Stephen Harper's riding association for similar donations in 2015, it appeared that donors to Calgary Southwest were mostly local. In the early years of Harper's time in power, he had been vocal in his disdain for China's form of government and the way it deals with minorities. But as prime minister, Justin Trudeau's father, Pierre Trudeau, had led Canada to officially recognize China in 1970, opening the door to US recognition a few years later, and to China taking its rightful place on the world stage. Perhaps China had high expectations of the son.[148]

As I went through the returns I was thinking: Is something going on here? If so, what?

Early in his first term, Prime Minister Trudeau and his ambassador to China, John McCallum, a well-respected former Liberal cabinet minister and a Sinophile, went in hot pursuit of a free-trade deal with China. Some of the items Canada wanted to negotiate had to do with the environment and labour issues plus the Chinese state's propensity for turning a blind eye to the theft of intellectual property, something American companies doing business in China had complained of for years. China, for its part, was interested in an extradition treaty with Canada, the better to extend its reach in the Fox Hunt, Xi Jinping's project to haul back to China for punishment miscreants who'd skipped out with a lot of money. At first, the trade talks seemed to go well. This may be why my husband's company, which had patented in many countries, including China, a new shape for the leading edges of the blades of large industrial fans, was suddenly contacted by a Chinese company that had been manufacturing them for some time without a license. My husband had chased that company to no avail. Now, out of the blue, the company suddenly asked for a license, and even wanted to pay to make up for past bad behaviour. To say the least, that is unusual business practice.

Soon, we realized we could predict the licensee's behaviour according to what was happening between the governments as portrayed on the front pages of the newspapers. When things were going well, things went well with the licensee. When things went wrong, radio silence. The talks fell apart.

After Huawei's chief financial officer, Meng Wanzhou, was arrested in December 2018 at Vancouver International Airport, due to an extradition request from the US, the relationship between the governments collapsed. First came the detention of the two Canadians, Michael Spavor and Michael Kovrig, who had been working in China for years. Spavor had a business arranging tours to North Korea. Kovrig, a Canadian diplomat on leave, was working with The International Crisis Group. They were locked up in jails, lights on 24/7, charged with nothing. The link between their arrests and Meng's was made clear. This behaviour by the government of China became known as hostage diplomacy, then coercive diplomacy, though it is not diplomacy at all, it's thuggery. The arrests were followed by economic punishment, the spurning of Canadian products by Chinese officials on spurious grounds, which caused significant harm to the Canadian economy. Ms. Meng's appearances at court were at times augmented by small groups of protestors who decried her treatment—being forced to wear an ankle bracelet to make sure she did not flee from her comfortable life in one of her two lovely Vancouver properties. As the case moved along, the Chinese ambassador to Canada, Lu Shaye, kept ratcheting up his rhetoric, insisting the extradition proceedings had nothing to do with the law, only with politics, that Canada was dancing to a Trumpian tune. In the end, he was as vituperative as a ghosted lover, though he was clearly expressing the views of his government: he was rewarded with a posting to France. The new Chinese ambassador to Canada, Cong Peiwu, appeared to threaten the safety of 300,000 Canadian citizens living in Hong Kong. The city had been in a state of political turmoil after China moved to end its one-country-two-systems relationship with it, bringing Hong Kong more directly, and in draconian fashion, under Beijing's control. If Canada dared accept Hong Kongers as political refugees, the ambassador warned—refugees created by China as it stamped hard on Hong Kong's democracy movement—who knows what might happen to those Canadian citizens?[149]

By then the licensee had long since stopped responding to emails.

IT WAS ONE of those things you find while googling for something else. The pandemic's first wave had passed its peak in Toronto. The story was labelled "opinion" by the *Gold Star Daily*, a paper located in Mindanao, a politically restive group of islands in the Philippines. The author, Perry Diaz, writes a regular column for a chain aimed at Filipino readers wherever they may be. The chain is headquartered in San Diego, California. Diaz, who's based in Sacramento, began by musing about whether SARS-CoV-2 might have escaped from a lab. Then he referred to an *Epoch Times* article of January 31, 2020, by a freelancer named J.R. Nyquist. Normally, I wouldn't have spent five seconds considering material published by *Epoch Times*, but my eye was caught by the word "secret."[150]

The *Epoch Times* is published by the Epoch Times Media Group, which is associated with Falun Gong, an organization the government of China has banned as a cult (no doubt because it hopes to overthrow the Communist Party of China, but also because it operates like one). In addition to teaching relaxation/meditation exercises and a philosophy somewhat akin to Taoism, Falun Gong's leader, Li Honghzi, preaches that all Communists and their supporters should/will go to hell. He has similar views about certain forms of sexuality and abortion. Falun Gong was tolerated by the Communist Party of China until it gained too many adherents. In 1999, the Party decided Falun Gong had to go. Falun Gong members can be seen protesting silently in front of Chinese consulates and embassies around the world, pressing into the hands of passersby pamphlets bearing lurid claims about how China harvests the organs of Falun Gong prisoners. According to Wikipedia, and an article by NBC News, Falun Gong's headquarters are in a 400-acre compound called Dragon Springs in New York State, where members of the heavily advertised Shen Yun dance troop may also be found.[151]

The *Epoch Times* used to be a free pamphlet. It was spruced up to look more like a newspaper in 2009. Actual journalists were hired in 2016, though their job was to produce stories that only looked like journalism while leaning hard right and promoting Donald Trump. This group was fired eight days before the 2016 election—waste not, want not. In 2020, NBC News had a close look at the publication and its expanding online presence because it spent more than almost any other group on pro-

Trump Facebook ads in 2019, in addition to spreading loony QAnon theories about deep-state leaders who kidnap and abuse children.[152] It has many millions of web views.

A paper version of the *Epoch Times* appeared in my mailbox one day in the fall of 2020, about a week after I viewed a documentary online that was produced by an associated television company. This seemed downright creepy to me—until my next-door neighbour mentioned he'd got one in his mailbox too. Apparently, *Epoch Times* thinks avid readers may be found in my neighbourhood. A friend who follows these things was sure that Canadian former publishing baron Conrad Black (pardoned by Trump in 2019 for a crime that earned him a significant jail term in the US[153]) has something to do with the *Epoch Times*' board. I couldn't find a listing for an *Epoch Times* board, but when I googled Conrad Black+*Epoch Times* I found, to my surprise, that Black had been writing a weekly opinion column for them since January 2020. Many of those articles focused on the re-election of Trump.[154]

But back to Perry Diaz in the *Gold Star Daily*. He quoted from Nyquist's *Epoch Times* story, which referred to a "secret" speech given by Chinese general and defence minister Chi Haotian to high-level Communist Party cadres in 2003. In this speech, according to Nyquist, Chi explained that biological weapons would ensure a Chinese national renaissance. The US, according to the general, stood in the way of China's two main goals: solving the problem of too many people stuck in a degrading environment, and conquering "new lands, in which a 'second China' could be built by 'colonization.'" China needs living space, the general allegedly said, and without more space it would undergo social strife, even civil war, resulting in the deaths of hundreds of millions. The US would not allow it to invade neighbouring territories, which in any case were also crowded. Yet the US, Canada, and Australia have lots of open space, so the question became: how to colonize them? The answer did not lie in nuclear weapons, according to the general. "Only by using non-destructive weapons that can kill many people will we be able to reserve America for ourselves," he allegedly said. He asserted that the reason Deng Xiaoping (former commander-in-chief of the People's Liberation Army and paramount—though not formal—leader of China) had prioritized biological weapons over ships and planes was because they are "lethal weapons that can eliminate mass populations of the enemy country.... We still emphasize economic development as our center, but in reality, economic development has war as its center."

Yes, I looked up General Chi, and yes, he did serve as China's defence minister in 2003. He was also involved in the Tiananmen Square massacre, like Deng, and his career seems to have had more to do with Party politics than with his generalship. But did he lay out this Hitleresque *Lebensraum* policy in a secret speech? Or was this made-up nonsense? Deng had certainly put a lot of money into China's 863 Program, aimed at pushing the country forward through government investment in information technology, telecommunications, space, and biotechnology.[155] That program would be accused of encouraging Chinese scientists and others working abroad to steal trade secrets and intellectual property from foreign companies.[156]

In every other way, the story seemed crazy. Even if one took seriously the claim that China is desperate for space, it still made no sense. Why would China's leadership focus on open space in Canada, the US, and Australia when Russia's vast emptiness lies right next door? The allegation that Chinese interest in biotechnology constituted preparation for a war of colonization was just the kind of thing a cult leader in exile would put into the mouths of his enemies to spur outrage. I almost stopped reading right there.

But then Perry linked China's goals with the National Microbiology Laboratory in Winnipeg. The Canadian in me said: What's this?

His opening question, about whether SARS-CoV-2 could have originated in a lab, then morphed into a flat statement that SARS-CoV-2 is a weaponized virus that escaped from the Wuhan Institute of Virology's BSL-4. He tied the National Microbiology Laboratory in Winnipeg to the WIV through that very strange 2019 shipment of dangerous pathogens to China that the CBC's Karen Pauls had revealed. Diaz wrote: "in July 2019, a group of Chinese virologists led by a Chinese Bio-Warfare agent Dr. Xiangguo Qiu were forcibly removed from Canada's NML...Dr. Xiangguo Qiu is married to another Chinese scientist Dr. Keding Cheng. The couple is responsible for infiltrating Canada's NML with many Chinese agents posing as students from a range of Chinese scientific facilities directly tied to China's Biological Warfare Program. Dr. Xiangguo made at least five trips to the Wuhan National Biosafety Laboratory, which is part of WIV...."[157]

The rational part of me said: Garbage! Diaz's leap from a small pile of known facts to a mountain of unverifiable allegations marks the divide between fair reporting and its opposite. What would a columnist in Sacramento know about people working in a lab in Winnipeg anyway? Dump it!

Yet I couldn't completely wash it from my mind. I had

read three articles published by knowledgeable authors who suggested that offensive biological warfare research goes on in China, one insisting that China is the world leader in bioweapons development.

And what about that paper by those Chinese military virologists who'd collected 344 horseshoe bats in Zhoushan City, Zhejiang, China, between 2015 and 2017? They'd pulled from those bats viral sequences distant from SARS but the next closest to SARS-CoV-2 after RaTG13. Though the group had been unable to grow viruses in cell culture, they had been able to infect suckling rats by inoculating intestinal tissue from one bat, ZC45, into three-day-old baby rats' brains. The rats developed SARS-like symptoms, and electron microscopy showed virus-like particles in their brains (although the microscope could not see the spikes: the authors figured preparation of the samples in a centrifuge may have knocked them off). The paper was published by the same journal that published the second paper Shi submitted on January 20, saying these viruses were the closest to SARS-CoV-2. Of the fifteen authors on the paper, ten were with the People's Liberation Army's Nanjing Command, others were with the Third Military Medical University in Chongqing, and still others were with civilian universities, including New York State-based Stony Brook University.[158] Clearly, the Chinese military was also deeply involved in the hunt for dangerous SARS-like viruses in bats, just like Shi Zhengli at the Wuhan Institute of Virology. And they had connections to civilian researchers in the US. In fact, in June 2020, *Science* magazine would report that, after an investigation by the US National Institutes of Health, dozens of scientists who won NIH grants were found to have also taken payments from institutions in China which they failed to disclose. Fifty-four were either fired from their institutions or resigned.[159]

Apparently, cooperation between academic and military researchers, both in China and across international borders, had been going on for some time. In September 2020, *Nature* would publish a story about how, under Xi, military and civilian researchers in Chinese universities had been told to work together, particularly in biotechnology. Some Chinese graduate students affiliated with Chinese civilian universities carried these military ties with them to Western universities and institutions.[160] Some hid those ties when they applied for US visas. The FBI had arrested a handful of people and charged them. Mary Gallagher, a political scientist specializing in China-US relations at the University of Michigan, Ann Arbor, explained to *Nature* that many top hospitals in China are affiliated with the military. "'And so by default, if you're a doctor at one of those hospitals, you're going to have an affiliation with the Chinese military.' That affiliation doesn't automatically mean that if you're collaborating with a US researcher you're engaging in espionage, she says."[161]

The same month the *Nature* piece came out, the US announced that President Trump had revoked, by proclamation, the visas of about 1,000 Chinese graduate students (there are over 300,000 in the US) suspected of transferring important technologies to China.[162]

What about in Canada? At about the same time, Richard Fadden, former head of CSIS and, for a short period before he retired, national security advisor to Prime Minister Trudeau, testified before the Canadian House of Commons' Special Committee on Canada-China Relations inquiring into Canada's relationship with China. Canada has about 140,000 students from China. Fadden told the CBC after the session that China views Canada as an easier target than the US.[163]

In other words, it was neither impossible nor improbable that Xiangguo Qiu, her husband, Keding Cheng, and their students from China might have had connections to both the WIV and the Chinese military. Public Health Agency of Canada officials had handed over information about them to the RCMP in May 2019, three months before their security clearances were revoked and they were escorted out of the NML. They were still under suspension with no explanation ten months later. Why would a mere administrative or policy issue—as their suspension had been characterized by a Health Canada spokesperson—take so many months to resolve, either in charges or reinstatements with apologies? And, come to think of it, since when do police deal with administrative matters or policy issues?

I thought I should check if gain-of-function experiments with pandemic potential were being conducted at the National Microbiology Laboratory or anywhere else in Canada. I wasn't sure which concerned me more: possible Chinese military infiltration of Canada's only BSL-4 lab or gain-of-function work in insecure labs that might lead to a Canadian-made pandemic.

Serious lab accidents that could lead to a pandemic aren't just possible, they've happened. Alison Young's findings in stories published in *USA Today* in 2014 and 2015, which roused a second round of public concern, leading to another NIH pause of gain-of-function funding in the US, were recounted in May 2020 by Rowan Jacobsen in *Mother Jones*. In addition to the CDC researchers exposed to live anthrax, the vials containing smallpox discovered in an NIH lab in a cold storage unit where they weren't supposed to be, and the un-killed Ebola sent by one CDC lab to another whose researchers thought it was safe, Jacobsen mentioned ruptured protection suits and rats "making nests out of biohazard bags and used lab supplies outside a UCLA lab.... On multiple occasions, mice carrying either SARS or H1N1 flu escaped from University of North Carolina at Chapel Hill...."[164] Pathogens had been stored at Fort Detrick without records. [Fort Detrick was closed in July 2019 by the CDC for insufficient "systems in place to decontaminate wastewater" and other issues. It did not reopen until November.[165]] Most troubling to Young in 2015 was that the US had no nationwide listing of labs doing dangerous work and no public reports of accidents, other than those that occurred in the BSL-4's handling of listed pathogens like plague and tularemia.[166]

Jacobsen reminded *Mother Jones* readers that the global history of lab accidents is more disturbing still. He cited the flu epidemic that swept around the world in 1977, noting that the virus that caused it turned out to be genetically identical to a flu virus that had caused an epidemic in the 1950s. Flu viruses in the wild mutate constantly: to remain unchanged, this virus must have been held in a lab fridge for 20 years, and then accidentally released. The Russians were accused, even though the epidemic erupted in China. No official has ever publicly admitted responsibility. And how about the anthrax spores that escaped from a bioweapons lab in the Soviet Union due to a faulty air filter? That accident killed at least 66 people in Sverdlovsk. It was only after the fall of the Soviet Union that Russian President Boris Yeltsin invited independent scientists in to investigate.

And just as SARS-CoV-2 began to spread in Wuhan in November 2019, brucella bacteria were released in waste fumes from the Lanzhou Biopharmaceutical plant in northeast China. The plant had used expired sanitizer while making a brucella vaccine. About 200 researchers in a nearby veterinary institute tested positive, and at least 4,500 people living nearby were eventually infected.[167]

These stories raised so many questions: Did Canada's BSL-4 lab have the same history of accidents as its counterparts in the US? Looking through the press coverage, it was clear the National Microbiology Laboratory has had some serious issues, though the Public Health Agency of Canada, its parent, tries hard to keep them quiet. In 1999, right after it opened, there was a serious escape of wastewater[168] that led to the creation of a liaison committee to keep the citizens of Winnipeg comfortable about having such a lab in its midst. That committee does not seem to be active. I tried in vain by telephone and email to reach it to find out how things are going now. No response. In 2017, Katie Pedersen, of the CBC investigative unit, reported on the results of an access to information request about incidents at the NML between January 2015 and October 2016. She found that malfunctioning equipment had led to staff exposures to HIV and Ebola. Ebola exposure happened due to a pressure-suit leak, though the suit had been repaired and passed a safety test only two days earlier. Exposure to avian flu occurred as a result of supposedly uninfected wild ducks being brought to the lab for experiments. The employees did not wear protective suits, assuming the wild animals were virus-free. But they were carrying avian flu. Then there were the employees who failed to wear puncture-resistant gloves and were pricked.[169]

With 45 incidents over 22 months, the rate of accidents at the NML appeared to be about two per month, much lower than in the US. But of course, there are many more high-security labs in the US, and the greater the number of labs, the greater the risk.

In 2019, Dylan Robertson of the *Winnipeg Free Press* reported that somebody at the NML had dropped a petri dish with active tuberculosis bacteria in it, cracking the lid. The dish remained, undiscovered, beneath a work bench for a year, exposing everyone who worked in that space, though no one later tested positive. (How come no one noticed it was missing? Wasn't the floor beneath the work bench cleaned in all that time?[170])

Most lab accidents are the result of human error. As the *Winnipeg Free Press* discovered, there has been a rising crescendo of complaints by staff to the laboratory's unions concerning oppressive management practices and workloads leading to serious problems of morale. A woman named Sky Soule ran screaming down the lab's hall one day in late 2016, pursued by managers. Soule, a bacteriologist of great skill, had had her duties changed—made more menial—coincidental with the arrival of a new manager in her department. She had begun to come to work smelling of alcohol. After the screaming incident, she was escorted out of the building and put on leave. At home, she fell down the stairs, went into a coma, and died, leaving two children behind. Another person came to work drunk one time too many and had to be put on leave. And yet someone who smoked marijuana in the building received nothing more than a reprimand. (Frank Plummer, the former head of the NML and a great scientist, confessed after he left that he had struggled with alcoholism for years, although he insisted it never interfered with his work.[171]) After Soule's death, signs were placed around the building making it clear that NML leadership were concerned about employees' mental health. And so they should be. It was a scientist working on an anthrax vaccine at Fort Detrick who the FBI believes sent anthrax to the offices of various US politicians in 2001. Five died, 17 were sickened. When the investigation was opened against Bruce I. Ivins, who had won the Department of Defense's highest civilian award in 2003, he killed himself.[172] Some managers at the NML were referred to by a former employee as "mean girls."[173] Sources told one reporter, and under a promise of anonymity another source told me, that there is a state of warfare between scientists in Winnipeg and their bureaucrat bosses running the Public Health Agency of Canada in Ottawa. Non-scientist bureaucrats (like the ones who neutered PHACS GPHIN early-warning system in May of 2019, and who failed to maintain a strategic stockpile of PPE) make endless demands for paperwork that the scientists in Winnipeg struggle to meet. According to my source, the relationship between scientists and bureaucrats became so bad that the best and brightest scientists left.

Yet, other than the tragic death of Soule, these problems still seemed small when compared to events at Fort Detrick or to the vaccine plant in China that managed to infect thousands of citizens with brucellosis. No one is permitted to handle either form of brucella in Canada outside a licensed BSL-3 lab. To enter a BSL-3 lab at the NML requires a security clearance, or the presence of a person with a security clearance.[174]

On the other hand, maybe more serious problems had not been revealed. Canadian bureaucrats have the magic touch when it comes to sweeping unfortunate events under dusty rugs. The Human Pathogens and Toxins Act of 2009, which finally came into force in 2015, regulates all activity involving human pathogens and toxins in Canada. The act requires that nobody—whether in government, private sector, university, or hospital labs—may handle listed human pathogens or toxins outside a licensed facility. Licensees and researchers are required to report all accidents and there are sanctions for failure to do so. But it appears that people working under license within the Department of National Defence may refuse to answer questions from the minister of health (who can grant and retract licenses) if the minister of defence believes the information might be harmful to the national security of Canada or of a state allied or associated with Canada, which is a pretty big loophole.[175] The act divides pathogens and toxins into five categories of risk. Risk Group 5 pathogens may not be handled by anyone: only smallpox is in that group. (Smallpox samples are kept, in case of need, by two repository countries: the US and Russia). The act categorizes "Severe acute respiratory syndrome-related coronavirus," otherwise known as SARS, as Risk Group 3. Risk Group 3 pathogens may be handled in BSL-3 labs, which offer less in the way of containment than BSL-4 labs but still are much safer than BSL-2s, the kind of lab Shi Zhengli used to study coronaviruses from bats.

According to the act, Risk Group 3 means:

> ...a category of human pathogens that pose a high risk to the health of individuals and a low risk to public health and includes the human pathogens listed in Schedule 3. They are likely to cause serious

> disease in a human. Effective treatment and preventive measures are usually available and the risk of spread of disease caused by those pathogens is low.

The SARS pandemic may have been small, but it was nevertheless worldwide. It killed hundreds, sickened thousands, made economic havoc in China and Canada. SARS CoV-2 kills a smaller percentage of those infected than SARS or MERS, but from a public health point of view, the spread of disease has been catastrophic globally. The public health risk is anything but low.

The Public Health Agency of Canada's Centre for Biosecurity publishes an annual report concerning human exposures to pathogens and toxins in all of Canada's 935 licensed labs. Their reports are not timely. The 2018 report did not appear on the website until September 2019. That report pointed out that the number of accidents in Canada had doubled over the previous year. The report's authors had no explanation as to why. They listed the kinds of accidents but did not name the labs, their classifications, or even their general locations such as, say, a BSL-4 in the centre of Winnipeg, or a university-based BSL-3 lab in downtown Toronto. By far the most frequent form of accident involved inhalation of an agent, followed by being stuck with "sharps." Most such events took place in BSL- 2 labs located mainly in universities and hospitals. Most university lab accidents were caused by technicians or students who did not follow the rules. The NML and its sister lab run by the Canadian Food Inspection Agency in Winnipeg are both licensed, so their 2018 accidents must be included in the report, but it's impossible to segregate NML accidents from all the rest. Even the section dealing with the biological agents and toxins involved in accidents is opaque. The graph in this section listed several agents by name but also included a category called "Unknown," which shows six reported incidents and six exposures.[176] The report for 2019, which appeared in September 2020, was even less specific. It asserted that the number of incidents was lower than in 2018, citing 60 exposures involving 86 individuals. Most accidents took place in hospitals among technicians who inhaled something noxious, but 37 percent happened in academia. Private license holders had the lowest number of incidents, public health the highest. This document said nothing specific about the nature of the accidents, listing them all under the heading "procedural issues" and calling human interaction the root cause of most incidents.[177]

In other words, there is no public data available about safety incidents and exposures at the NML.

So, in mid-July 2020, I filed an access to information application for all accidents that had occurred at the NML since 2015. To keep it simple, I excluded the neighbouring complex run by the Canadian Food Inspection Agency

The opening paragraphs of the Access to Information Act give any Canadian citizen or permanent resident *the right* to any record in the hands of the government of Canada or its agencies unless these records fall under listed exemptions.[178] After going three months without a response, though a response is required within 30 days, I was at last told the delay was because the incident reports I wanted—all 8,000 pages of them—had to be shipped physically from Winnipeg to Ottawa to be scanned, reviewed, and redacted first.

That sounded like quite a lot of incidents, way more

than two per month. And did the NML only keep such reports on paper? Really? In 2020?

They would be sent to me, I was assured. Sometime. Eventually.

I'm still waiting.

While I was setting that request in motion, I found myself wondering why I could not recall reading about any Canadian public debates on gain-of-function experiments, or the synthesis of dangerous viruses. In the tongue-clotting language deployed by bioethicists who specialize in these arguments, these experiments are known as dual-use research of concern. You'd think I'd remember reading phrases like that if reports of public meetings on this subject had made it to the newspapers. I could recall none. Was it because I hadn't been paying attention? Or was it because no public debates took place? Surely there must have been some—if only because there have been so many public arguments raised for and against gain-of-function experiments in the US—arguments that raged again as suspicion grew that the SARS-CoV-2 pandemic might be the result of an accident in a lab doing gain-of-function experiments on SARS-like viruses.

I checked to see if these questions had been brought up during committee hearings in the House of Commons on the Human Pathogens and Toxins Act. Fouchier's very public experiment was published at the end of 2011, so you'd think the issue would have been raised by someone as the proposed act was examined in committees of the House and Senate. I could find nothing to show that it had.

Then I looked for Canadian academic papers. I found only one. In 2014, three Canadian researchers searched to see who had rules and regulations and policies on dual-use biotechnologies. They checked Canadian university websites. They found only three with policies and none that were specific to gain-of-function experiments or the synthesis of organisms. Some government departments (National Defence, Industry Canada) had web pages dealing with dual-use, but not specifically with biological issues. Health Canada had nothing. The only Canadian regulations regarding dual-use were confined to control of exports. There were no specific rules about what scientists can get up to in Canadian labs. They are free to make organisms from scratch (synthetic biology) or to make them more virulent[179] (gain-of-function) so long as the biosafety officer of their licensed institution is satisfied the work will be safely done in a lab with appropriate containment technology. According to these researchers, there were no regulations set out by the WHO or by the EU either.

The lack of rules didn't mean no one had talked about this at conferences or at public fora. I made calls, sent emails. I asked researcher friends. I wrote to Gary Kobinger, Xiangguo Qiu's former boss at the NML, after I found he had co-signed a letter to *Science* in 2012—whose first author was Ron Fouchier—along with several other principal investigators including Heinz Feldmann, then the leader of BSL-4 at the NML, and Chen Hualan, director of the second BSL-4 in China at Harbin. In the letter, the scientific virtue of making avian flu viruses more transmissible was placed in the context of how it helped scientists understand the virus and what makes it transmissible to humans in order to prevent pandemics. The letter asserted that "responsible research on influenza virus transmission using different animal models is conducted by multiple laboratories in the world using the highest international standards of biosafety and biosecurity practices that effectively prevent the release of transmissible viruses from the laboratory. These standards are regulated and monitored closely by the relevant authorities." Note that the group said the *standards* are monitored, not the labs. The group offered a 60-day voluntary pause in experiments that could make avian flu virus more transmissible.[180] Kobinger told me in an email that he had participated in a few teleconferences on the issue, but he wasn't specific about where, or when. He said he was more concerned about rescue (synthesizing extirpated viruses, or viruses for which only a sequence is known) than gain-of-function. He was referring to the experiment done by David Evans at University of Alberta in which a cousin of the smallpox virus was synthesized from its genome alone.

So I filed another access to information request. Never say die.

I asked the Public Health Agency of Canada for all records of any such debates or policy discussions within its own ranks and with other government departments, such as Global Affairs and Public Safety.

As I waited for that request to be processed, I thought maybe I was barking up the wrong institutional tree. In the US, the debates about whether to fund or forbid gain-of-function research had been organized by the National Institutes of Health and then by the White House. Wasn't it more likely that the Canadian Institutes of Health Research (which is modelled after the NIH and is the major funder of biomedical research in Canada) had organized conferences or debates? The CIHR is (theoretically) an arms-length agency of the government of Canada. It reports how it spends money on its operations and grants to Parliament through the minister of health.

After numerous emails and searches I could find no evidence of any conferences held on this subject by CIHR, but I learned from its spokesperson, David Coulombe, that there had been a discussion among members of a subcommittee of the CIHR's Standing Committee on Ethics. Committee members are experts in their fields, and volunteer their services. A subgroup had looked at certain issues classed as disruptive technologies sometime between 2015 and 2018. Their deliberations had been worked into a "learning module" available on PHAC's website that researchers could consult to evaluate the risks of experiments that might have a dual use.

Coulombe sent me the names of the people on the subcommittee who had "briefed the newly appointed Scientific Director of the CIHR Institute of Infection and Immunity on these themes in July 2018."

And what about public meetings? I asked.

The subject of "dual use" had been taken to the Standing Committee on Ethics, according to Coulombe, as "part of CIHR's interest in broad dialogue and engagement on issues relevant to the field of ethics." The CIHR ethics office had then "collaborated with PHAC by drafting the ethics-related content."

In other words, Coulombe did not directly answer the question.

I contacted some members of the subcommittee by email. The first, Paul Garfinkel, professor emeritus in the Department of Psychiatry at University of Toronto, had minimal recall of the discussions, had no notes in his files, but suggested that Vardit Ravitsky, professor of bioethics at l'Université de Montréal, might remember more. Ravitsky suggested Judy Illes, professor of neurology and Canada Research Chair in Neuroethics at the University of British Columbia. Illes suggested Lisa Schwartz, professor of Health Research Methods, Evidence, and Impact, Faculty of Health Sciences, McMaster University.

According to Schwartz, when the subcommittee heard that PHAC was working on a module "we teamed with them for a while. Eventually we were told to end the work of the subcommittee. There was no explanation." She also made clear that "anything can be dual use—fire is a dual use, knives...." The approach "preferred and encouraged by PHAC," she wrote, is to "recognize these risks and to encourage an atmosphere of ethical practice including transparency and accountability. CIHR practices this as an agency by publishing information about the research they fund."

In other words, the CIHR had basically left this matter to researchers who have a vested interest in the work, without reference to the rest of us stakeholders, such as the neighbours of BSL-3 labs in downtown Toronto, or those who live close to Edmonton's University of Alberta, or close to the NML in Winnipeg.

I was curious as to why the subcommittee was shut down without explanation. Schwartz suggested I get in touch with Dr. Marcello Tonelli, Senior Associate Dean of the Cummings School of Medicine and Associate Vice-President, University of Calgary. Tonelli is also the chair of the Standing Committee on Ethics and a member of both the CIHR's governing council and its executive committee.

I tried to get ahold of Tonelli. When I called his office at the university, I was referred to University of Calgary's

PR people. When I got through to them, I was told they couldn't help because my questions related to Tonelli's role at CIHR.

I wrote an email to Tonelli through the office of his dean, setting out Schwartz's explanation of events. I got back an unsigned letter, ostensibly from Tonelli, sent from the CIHR Office of Governance. The letter explained that the subcommittee wrote a report, which completed its work. The report "was not made public or discussed at conferences" as "SCE products are internal documents developed to provide CIHR with strategic advice on ethical, legal and socio-cultural dimensions of the agency's mandate."

I was surprised that David Coulombe, the CIHR spokesman, hadn't mentioned this report. I phoned Marta Arnaldo, the governance person at CIHR who had sent the letter to me. I asked her for a copy. She explained that it had taken too long to get the report done and it had not responded to the original question put to the subcommittee about synthetic biology but had gone off into broader fields. Then there had been a change in leadership at CIHR and the new person replacing the old person did not feel this report would be of use. She said she'd send it.

When no copy of the report appeared in my mail, I wrote David Coulombe to ask for a copy.

He said he would just run it first by the CIHR's ATIP (access to information) people "informally" and then would send it right out.

Warning bells rang. Why? I had already made an informal access request with PHAC (see next chapter) and the result had been less than satisfactory.

I think I'll file, I told him. Which I did.

About a month later, the report in question plus a lot of emails and draft minutes of two meetings of the Standing Committee on Ethics, heavily redacted, were produced. And what did I find? Lisa Schwartz had made a presentation to the Standing Committee on Ethics and had asked if the report might be made public or its findings used in academic publications.

A series of draft letters on this question had been prepared for the signature of Dr. Tonelli. The final version said that when a subcommittee has finished its assigned task, that's it: it's up to the Standing Committee on Ethics to decide what next steps there may or may not be. "I assure you we will carefully consider how best to use the report to achieve the CIHR's goals," he wrote.

The committee had recently reviewed its roles and responsibilities, his letter continued, "recognizing that CIHR is an agency of Canada's federal government." Under its terms of reference, the Standing Committee on Ethics cannot inquire into something on its own but can only respond to requests "formally submitted by the CIHR President, Science Council/Scientific Directors and the Governing Council." Any documents developed by a subgroup are owned by the SCE, not the subgroup members, and are for the SCE's final action/decision, not the subgroup's. When work is complete, "once approved by the SCE" it is given "to the requestor and should be treated as 'advice to government,' i.e., not to be widely distributed (or published) as per the Access to Information Act." Only the governing council may decide to publish work that might be of interest to a broader audience, and even then "may mandate the SCE to produce a version of the document which is written for a broader audience and which does not disclose information meant exclusively for the government." He finished off with: "No further work will therefore be required from the SCE DT [Disruptive Technologies] sub-committee as outlined above and no publication or broad distribution of this document is permissible."

"Advice to government" is one of the permitted exemptions under the Access to Information Act. Anything deemed to be advice to government "may" be withheld from Nosy Parker Canadians, such as *moi*, who are interested in what the government has been doing or has failed to do.

So: not only had the CIHR failed to hold any public debates or conferences on the issue of gain-of-function/synthetic biology/dual-use experiments done in Canada, it had refused to let subcommittee members publish the results of its limited internal discussion.

This was disappointing, annoying, irresponsible, undemocratic, call it what you will. But it was the report itself that sent my blood pressure over the top. The first redaction occurred at the beginning in a section labelled "Objective of the Guidance Document." Apparently, the objective of a report is the same thing as advice to government.

The report outlined the history of the government's interest in dual-use technologies. An interdepartmental steering committee had been set up in 2013, led by PHAC, that including Agri-Food Canada, CIHR, the Department of National Defence, Environment Canada, Foreign Affairs, Health Canada, Industry Canada, and the National Research Council. But: "By 2016," the report said, "the federal focus shifted with education becoming the main strategy to deal with DTS and DUP [disruptive technologies and dual-use potential]. As a result, PHAC developed an online training tutorial, which was published in the summer of 2017, and into which the SCE provided substantive input."

The phrase "gain-of-function" did not appear in this report. It only offered two illustrations of "disruptive technologies" with "dual-use" potential. One involved the prospect of intrusive surveillance due to improvements in facial-recognition software. The other involved the rescue of a virus closely related to smallpox. This was David Evans' experiment.

In November 2016, according to the report, Dr. David Evans of University of Alberta told the WHO Advisory Committee on Variola Virus Research that by using commercially available "information, technology and tools" he had synthesized horsepox virus. He had not used a grant from CIHR or any other government agency to do this. Rather, he had entered into an agreement with a vaccine manufacturer in the US. He spent $100,000 to have the one known horsepox genome sequence synthesized in large fragments by a commercial supplier in Germany.

> The research findings were published in PLOS One... on January 19, 2018.[181] The research is the first demonstration of the ability to synthesize a functional pox virus; a dual use potential exists with respect to this science being applied to create smallpox, a deadly human disease that has been eradicated and is prohibited in Canada. This presents ethical and security concerns related to biological weapons and health security in general.

I knew from reading about Evans' experiment (magazines in the US carried stories about it, Canadian newspapers ignored it), and speaking with him, that he was interested in the history of the first smallpox vaccine made by Edward Jenner. He wondered if Jenner might have made it from horsepox, a smallpox relative, and if he, Evans, could make a better vaccine with fewer side effects if he based a new one on horsepox. But there was no horsepox virus to be had. Only one strain had been sequenced. The sequence was publicly available. It took about six months for a German supplier to synthesize the entire genome (minus the hairpin telomeres at both ends) in ten large DNA fragments. Employing a complex series of maneuvers in the lab involving making a chimera, Evans was eventually able to grow horsepox virus in cell culture. He and his colleagues prepared a vaccine from the virus and Evans proved it protected mice from a lethal dose. He ended his paper with a short discussion of the virtues and dangers of synthetic technology. Aside from producing a better smallpox vaccine, he thought it might also be useable in cancer treatment to provoke "potent anti-tumour immune responses." On the other hand, he pointed out that if he could do this in a university lab (a BSL-2), then somebody with wicked motives could do it as well, presenting biosecurity issues. He and his colleague wanted a debate on the dangers and benefits. "Our hope is that this work will promote new and informed public health discussions relating to synthetic biology...."[182]

The CIHR subcommittee's report listed the means available to regulate such work (not many, especially if it is privately funded) and suggested that researchers should have to demonstrate to funders "the steps they are taking to meet their responsibilities for managing DUP and DT."

It also pointed out the need for enhanced governance mechanisms—involving oversight, education, and communication of risks, harms, and benefits—when dealing with dual-use experiments.

> ...stakeholders (researchers, research institutions, government of Canada, funders, oversight bodies, general public) should be made aware of the potential harms and benefits of an asset [such as a synthetic biology experiment] through effective communication channels. Stakeholder input is integral for ensuring adequate governance and communication channels are in place.... Research is a social activity. Consequently, the potential risks associated with an asset may affect all stakeholders and therefore should be adequately communicated to and/or by them.

That sounded promising.

But there was a section at the very end of the report titled "potential role of key stakeholders in Dual-Use Potential." Infuriatingly, this section was entirely redacted. The exemption claimed, Section 21(1) (a) of the Access to Information Act, says that an institution's leader "may" withhold information that might have been developed by or for a government institution or a minister of the Crown. This is the advice to government exemption.

I wanted to know what the subcommittee had recommended on that score. So I appealed this redaction to the information commissioner on the ground that the report had been prepared by a subcommittee of volunteers and shelved, which meant it could not constitute advice to a government institution or a minister of the Crown.

My appeal failed. When I asked by email for the jurisprudence supporting the information commissioner's decision, the young man handling my case did not reply.

I ALSO FILED an access request for all PHAC's policy deliberations concerning gain-of-function/synthetic biology experiments and any communications on these issues with other departments like National Defence or Global Affairs and Public Safety. Clearly, they'd had such discussions: the CIHR report said that in 2013 PHAC led an intergovernmental steering committee investigating these issues until shortly after the Trudeau government was elected. I sent my request in July 2020. It was directed to the Infectious Disease Prevention and Control Branch of PHAC. On September 18, I got a response saying records had been found. The cover letter said certain things had been redacted under exemption 16 (2) (c) of the Access to Information Act.

Section 16 (2) (c) refers to matters of security. The head of a government institution "may" withhold any record that might reasonably be expected to facilitate an offence. Specifically, a record may be withheld under sub-subsection (c) "on the vulnerability of particular buildings or other structures or systems, including computer or communication systems, or methods employed to protect such buildings or other structures of systems."[183]

Whoa, I thought. My questions relate to the security of buildings or computers?

Included in the documents found and sent was a proposed agenda for a meeting of the PHAC's Public Health Ethics Consultative Group (acronym PHEGG, not to be confused with CIHR's Standing Committee on Ethics, acronym SCE). This meeting was held by teleconference on February

16, 2016. The document was labelled DRAFT. Beside agenda item 4, the phrase *dual-use scenario* PHECG *Secretariat* appeared. The teleconference ID number was redacted.

Attached to this agenda was an ethics consultation report labelled Case Study: Inadvertent Discovery with a Potential for Nefarious Use presented to Dr. Pascal Michel, Chief Science Officer, Public Health Agency of Canada by Public Health Ethics Consultative Group [PHEGG]." It too was labeled DRAFT and dated April 6, 2016.

The document made clear that on February 8, 2016, PHAC had sent PHEGG a list of questions concerning the following case study: a researcher working on dengue fever finds by accident a way to infect mosquitoes, female mosquitoes, with dengue virus much more efficiently and effectively than would occur in nature. This could be used as a bioweapon, so possibly the work should not be published. But if it isn't published, the researcher will lose his job and be unable to advance in his career. What to do?

On February 16, the committee discussed the case and PHAC's questions in a conference call. The page labelled "summary of recommendations" was blank. Not redacted, just blank.

The body of the document made it clear that the committee had concluded that the researcher should publish the results, as only a person well-educated in methods and means would be able to replicate the work and produce those scary female mosquitoes. It would be a very unlikely bioweapon though, as mosquitoes are indiscriminate about whom they suck blood from.

I wrote back to the person handling my access to information application and said there must be some mistake: these documents are drafts, so there must be finals in a file somewhere, and I cannot believe this is all PHAC did to generate policy on dual-use/synthetic biology/gain-of-function experiments.

The officer could see there was something odd about all this and promised to go back to the department and ask for more. A little more time went by. Surprise! PHAC found more documents. This time the cover letter said certain redactions had been made pursuant to sections 15 and 19 of the act. Section 15 says the head of the department "may" refuse to disclose information that relates to foreign affairs, security, cryptography, etc. Section 19 deals with personal information. Section 19 is a "shall" refuse section.

Two documents were sent to me.

The first was a record, in a different format, of the discussion and recommendations from that teleconference among members of PHEGG on February 16, 2016. It showed a little more information, such as that conflicts had been declared by two members. These had been redacted, even though members' conflicts are shown on the PHAC website. The first item had to do with a Médecins Sans Frontières proposal to create an information-sharing system related to Ebola outbreaks. The PHEGG members pointed out that this problem had already been addressed in earlier advice to PHAC about whether to permit the export of Ebola samples gathered during an outbreak. That advice was not included in this memo.

The case study regarding dengue fever was apparently dealt with swiftly. It was pointed out by one committee member that Britain's Royal Society had discussed these issues in 2012, and that it would be difficult to suppress publication "in the current climate of oversight."

Climate of oversight, I muttered to myself. What oversight would that be? And since when does oversight have a climate?

The rest of the document related to other business having nothing to do with my request.

The second document sent to me was a record of a teleconference meeting of PHEGG on April 12, 2016. This document was also labelled DRAFT. Most of the issues discussed were not germane to my application. Again, the dengue fever case was referred to. This time committee members wondered whether, to do justice to PHACS ethics questions, the case study should be more specific, or more general? The committee decided more general was best.

And that was it. These were all the documents sent to me from CIHR and PHAC concerning all the policy discussions held publicly, or internally and with other government departments, about gain-of-function/synthetic biology/dual-use experiments. While the NIH and NIAID in the United States had responded to researchers (like Marc Lipsitch and the Cambridge Working Group) who were worried about the dangers of gain-of-function/rescue experiments and had created some rules (wink/nudge rules, but still) at the prodding of the White House, nothing similar had happened in Canada. Instead, Canadian regulators and funders suppressed an attempt to inform the public about the risks and possible benefits of these experiments, and limited discussion to a learning module for students and researchers.

Whose interests did that serve?

7. On Nailing Jelly to the Wall

I SOMETIMES COMPARE doing good journalism to doing good science. I am not trying to raise journalism above its station—it's just a trade—but it is as important to a well-functioning democracy as good plumbing to a five-star hotel, or immunology in the face of a pandemic. Good science requires deep field knowledge, intuition, teamwork, precision, and the ability to follow a trail no matter where it leads, even if that means overturning a mentor's most beloved theory. Above all, it requires imagination. Journalism also demands wide knowledge, intuition, teamwork. We too juggle hypotheses to determine which are worth pursuing and which should be left in the file drawer, which leads to occasional shouting matches with mentors. Most of all, we must imagine what might have happened in order to find out what did. For both journalists and scientists, luck and accident play roles as significant as method and grit. But unlike scientists, journalists are constantly publicly humbled. What we published yesterday will be overturned tomorrow when a competitor/colleague peers under a more productive rock or just happens to know somebody who knows something important. Creating a lasting, truthful narrative is as tricky as nailing jelly to the wall.

Yet we persist.

Which brings me once more to Xiangguo Qiu, her husband, Keding Cheng, and the Wuhan Institute of Virology.

A science journalist interested in Ebola might have heard of Xiangguo Qiu during the Ebola epidemic that hit Liberia in 2014. But ordinary folk would not have encountered her name (let alone her husband's) until May 2018, when she and Gary Kobinger, her former boss in the Special Pathogens Program at the National Microbiology Laboratory in Winnipeg, were awarded the Governor General's Innovation Prize. Kobinger had by then moved away from NML to Université Laval, though he maintained his appointments at the University of Manitoba and at the University of Pennsylvania. Kobinger is a well-educated, busy, and talented man who has advised the WHO and served, until he resigned in protest, on the federal government's SARS-CoV-2 Vaccine Advisory Task Force. (He was disturbed by its failure to be transparent about the possible conflicts of interest of its members.[184]) Xiangguo Qiu has an MD and an MSC in immunology from Chinese universities, but no PhD. Kobinger and Xiangguo Qiu got the prize for an antibody cocktail to treat Ebola devised in collaboration with Mapp Biopharmaceuticals Inc. of San Diego. It's called ZMapp.

The Innovation Prize is relatively new: its first awards were given out in 2016. Its selection committee (which includes several senior civil servants) only accepts nominations from partner foundations. Qiu and Kobinger were nominated by the Ernest C. Manning Awards Foundation, which had given them its own $100,000 prize the year before.

NML scientists had worked very hard on Ebola for years. Heinz Feldmann, now chief of virology at Rocky Mountain Laboratories, an NIAID-owned BSL-4 in Montana, developed an Ebola vaccine when he led the BSL-4 in Winnipeg. But no vaccines work perfectly, so treatments are still required. Ebola explodes into human cells so fast that it overwhelms the immune system. The ZMapp antibody cocktail was developed to slow Ebola's replication long enough for the immune system to kick in. While single monoclonal antibodies had been tried by several groups, none had worked. Qiu believed that several different antibodies together might do the trick.

She started on the project in 2005, two years after she was hired at the NML as a biologist. Despite draconian cuts to the Special Pathogens departmental budget, she and Kobinger managed to keep the work going with a grant from a Department of National Defence program "for high-risk research linked to bioterrorism."[185] After nine years of struggle, ZMapp was administered on an emergency basis—it still wasn't approved for use by the FDA—to two American missionary doctors in Liberia, who became ill with Ebola while treating people during the 2014 outbreak, and to 26 other first responders. Out of the 28 who received ZMapp, 25 survived.

There is a short video about the creation of ZMapp on the governor general's website. Kobinger does most of the explaining while Xiangguo Qiu speaks mainly toward the end. In one sequence she's moving around the BSL-4 lab in a blue pressure suit. In another she's sitting on a white couch beneath a window in what appears to be her living room. She is very small, round-faced, a bit pudgy, with shoulder-length dark hair parted in the middle, and dark-rimmed glasses on her nose. She wears a red-and-orange dotted silky blouse and a string of dark pearls. The man sitting beside her is probably her husband, Keding Cheng, but he's never named and does not speak. He is about the same size as her, also wears dark-rimmed glasses, but there's no pudge on him at all. The couple are flanked by their two grown sons (at least I think they are their sons). Xiangguo Qiu says how excited her younger son was to discover that his mama had developed "a cure for Ebola." Her accent when speaking English is strong, a sure sign she learned it as an adult. Neither parent looks like a biowarfare agent. But what would biowarfare agents look like? They look like comfortable, high-achieving new Canadians. One can draw no inferences about nature or character from these sequences.

KAREN PAUL'S INITIAL story on July 14, 2019, about how Xiangguo Qiu and Keding Cheng had been removed from the NML and suspended at the University of Manitoba, over what PHAC's spokesperson called "administrative matters," was followed the next day by an article in the *Winnipeg Free Press* by Dylan Robertson. It appeared that the administrative matters being reviewed by the RCMP might concern the "sharing of research" that had somehow led the Public Health Agency of Canada, the NML's parent organization, to call in the RCMP in May. Robertson also reported that the University of Manitoba, one of the top 15 research institutions in Canada, had been warned the previous year by CSIS about intellectual property theft by Chinese scientists working in Canadian institutions.[186]

On July 23, 2019, Pauls added to her initial report. She called Gary Kobinger for comment concerning why Qiu and Cheng might have been removed. Was this a case of economic espionage? Kobinger told Pauls that Xiangguo Qiu is a solid scientist and that if he were in her place, after this was all over, he'd pack up and go somewhere else. Scientists of her calibre are in demand everywhere, he said. It was probably just some sort of breach of government policy "created by bureaucrats who don't know how science works."[187] Rumour had it, Pauls explained, that their removal had something to do with the shipment of materials to China. Kobinger thought that was nonsense.

"The Chinese—they have so many scientists, it's unreal," Kobinger told Pauls. "What we can do in six months, they can do in a month. There is nothing, nothing, nothing that I can see from my side that they would benefit from us in terms of knowledge, in terms of re-agents. They have better access to pathogens, everything else, the vaccine, therapies, everything." He thought Qiu and Cheng's removal might have to do with the diplomatic situation between Canada and China, that they were being punished over the arrest in China of the two Michaels, Spavor and Kovrig. "Everyone benefits from working together, that's the nature of science. Again, I think there is clearly other issues that are completely unrelated to scientific research," he said.

A week later, Robertson published another story. PHAC had confirmed that Risk Group 4 samples, which may only be handled in a licensed BSL-4 lab by a person with a "secret" security clearance, according to the Human Pathogens and Toxins Act, had been sent to China. Xiangguo Qiu was a senior scientist in the NML's Special Pathogens Program, where she worked with such pathogens. A PHAC spokesperson told Robertson that the shipment of samples had nothing to do with the administrative investigation of the couple and that the public had never been at risk. However, other sources told Robertson that the samples shipped to China had been sent without a material transfer agreement, contrary to policy and good sense. Exporting strains of dangerous pathogens without such an agreement is like telling the receivers they can do whatever they want with them, including patenting proteins sent to them and reaping the rewards. Robertson added that Qiu and Cheng had not just been suspended by the U of M, as Pauls reported, but that the relationship had been severed, and all information about them had been scrubbed from the university's website.[188]

It was this story that was picked up months later by those, like Perry Diaz, who thought the shipment of deadly viruses from the NML to China might have contained SARS-CoV-2.

On August 2, 2019, Pauls published another story saying that the NML had sent Ebola and Henipah (the genus that includes viral species such as Nipah and Hendra), from Canada to China.[189] These are Risk Group 4 viruses, capable of killing a significant percentage of infected humans—as high as 80 percent—the kind of viruses handled in the BSL-4 lab by Xiangguo Qiu. They were shipped on an Air Canada commercial flight. Sources told Pauls that they may have been shipped to the Chinese Academy of Sciences in a way that "circumvented the lab's operating procedures, and without a document protecting Canada's intellectual property rights." The story implied that this shipment might have been the reason why Xiangguo Qiu, her husband, Keding Cheng, and their students from China had been removed from the NML. Health Canada and PHAC's spokesperson Eric Morrissette felt compelled to say that it is routine for the lab to share samples of pathogens and toxins with "partners in other countries to advance scientific work worldwide," and that the public was never at risk.

On October 3, 2019, just before the pandemic began in Wuhan, Karen Pauls reported the results of a CBC access to information request. She had obtained travel documents showing that Xiangguo Qiu had made at least five trips to China between 2017 and 2018, "including one to train scientists and technicians at China's newly certified Level 4 lab, which does research with the most deadly pathogens." Xiangguo Qiu, reported Pauls, had been invited by "the Wuhan National Biosafety Laboratory of the Chinese Academy of Sciences [whose deputy director is Shi Zhengli] twice a year for two years, for up to two weeks each time." The Wuhan National Biosafety Laboratory is the correct name for the new BSL-4 lab at the Wuhan Institute of Virology, which also has BSL-2 and BSL-3 labs. The name of the organization that paid for Qiu's travel to China, and the names of the people she visited, including those in Beijing, had been redacted. While one source said they'd heard that Qiu and Cheng would soon return to the NML, another staffer said: "It's not right that she's a Canadian government employee providing details of top-secret work and know-how to set up a high-containment lab for a foreign nation." The same person told Pauls that the RCMP had not been in contact with key members of the NML and that NML staff had not been allowed to contact them (the RCMP).[190]

Nothing more was heard from official or unofficial sources about this shipment, the trips, or about the administrative investigation of Xiangguo and Keding Cheng for months.

On May 5, 2020, as the pandemic's first wave passed its peak in Canada, Matthew Gilmour, assistant professor at the University of Manitoba's Department of Medical Microbiology and scientific director of the NML as well as at the Laboratory for Foodborne Zoonoses in Guelph, Ontario (and before that, Keding Cheng's boss), resigned from his posts. This became known thanks to a tweet from Theresa Tam, PHAC's chief public health officer, and the federal government's chief communicator on the pandemic. She said Gilmour was leaving the NML to join a British research organization, Quadram Institute.[191] This exit was surprising given that two of the NML's senior researchers were still suspended under a cloud of suspicion, and the NML's researchers were buckling under the huge volume of work due to the pandemic. Gilmour must have felt the need to explain himself, because five days later he told a Winnipeg radio host that the decision to take this new job was all about his young family. It would be a big adventure. He had done a lot of training in the UK. He had friends there. An offer had come along a few months earlier. That's why he was leaving the NML.[192]

It's usually politicians, not scientists, who cite family reasons for leaving their posts. Later, I tried to reach Gilmour by phone at Quadram. I left a message. No response. I sent an email. Then another. No response.

On June 14, 2020, almost a year after Keding Cheng and Xiangguo Qiu were removed from the NML, the CBC's Karen Pauls reported on the results of another access to information request. It turned out that more than a few samples of Ebola and Hendra had been sent to China: 15

strains, two vials each, according to Pauls. The email traffic about this transfer, which Pauls had also reviewed, made it clear that Matthew Gilmour had asked questions about the shipment, and particularly about the material transfer agreement. No material transfer agreement was produced among the documents Pauls received. Eric Morrissette, Health Canada and PHAC's spokesperson, insisted again that "the administrative investigation [into Xiangguo Qiu and Keding Cheng] is not related to the shipment of virus samples to China. In response to a request from the Wuhan Institute of Virology for viral samples of Ebola and Henipah viruses, the Public Health Agency of Canada (PHAC) sent samples for the purpose of scientific research in 2019."[193]

An NML shipment of Group 4 pathogens to China probably wouldn't have raised eyebrows before the new Cold War between Canada and China over Meng Wanzhou and the two Michaels, and had the recipient not been the Wuhan Institute of Virology, now at the centre of concern about whether SARS-CoV-2 leaked from a lab. But by June 2020, Canadians no longer considered China a friendly power just popping in to borrow a cup of Ebola with a dash of Hendra. Amir Attaran, a law professor and epidemiologist at the University of Ottawa, and a frequent critic of government policies, wasn't happy. "It's suspicious. It is alarming. It is potentially life-threatening," he told Pauls.

As the documents obtained by Pauls also showed, the organization of this shipment had been less than perfect. There had been confusion over proper packaging of the vials, and the samples had been shipped to China on an Air Canada commercial flight. Attaran was concerned that Xiangguo Qiu and Keding Cheng had been removed from the NML almost a year earlier, yet there was still no explanation as to why. He didn't like that multiple varieties of these dangerous pathogens had been sent, increasing the genetic diversity for Chinese researchers to work with. In addition, he didn't like the recipients. According to Attaran, the vials had been sent to a laboratory in China "that does dangerous gain-of-function experiments. And that has links to the Chinese military." Attaran explained to Pauls that gain-of-function experiments aren't prohibited in Canada, but that they're not done because they're considered too dangerous.

As I would later learn, something close to a gain-of-function experiment was being conducted at that time at the NML. Researchers wanted to know whether SARS-CoV-2 could spill over from humans and infect deer mice, which are widespread in North America. Researchers at the NML had captured deer mice and made them inhale human SARS-CoV-2. They found that the virus replicated very nicely in the mice and was excreted in their urine and feces. Not content with this result, they tested to see if the infected mice could give the virus to healthy mice by direct contact. The answer was yes.[194] It would have become a gain-of-function experiment if they'd passaged the virus in the mice until it could infect via indirect contact.

Attaran had one more concern. He mentioned a paper on Ebola posted online on December 12, 2018, of which Xiangguo Qiu was last author—the project's most senior person. He pointed out that the paper's first author, Hualei Wang, was affiliated with the Academy of Military Medical Sciences in Beijing.[195]

CONTRARY TO KOBINGER'S contention that China has more of everything a scientist might need, the virus list that

Pauls reported on was long enough to suggest the NML had helped populate China's first BSL-4 lab with strains for study.

I filed two access to information requests to find out more about what Xiangguo Qiu had sent to the Wuhan Institute of Virology, if in fact that's where the samples went. I was worried that Xiangguo Qiu had been working with more than one Chinese military scientist. I was worried that Perry Diaz in Sacramento wasn't completely off base after all.

I was advised in this process by a nice young man at PHAC. He explained that, due to the pandemic, it would take a long time to answer my requests and that it would be faster if I divided the first application I'd sent in two. He thought I should ask to see the material already turned over to the CBC, in effect to piggyback on the CBC's work. I didn't like the idea. For one thing, I didn't know exactly what CBC had asked for, and exactitude in requests has a big impact on results. But I reasoned that, if I did what he suggested, I might find information in the documents that Pauls hadn't reported on because it didn't seem important to her at the time. I took his advice, which I came to regret. The second request asked for all correspondence between PHAC, Global Affairs, and the Ministry of Public Safety about the export of those pathogens.

I also wanted to be certain no coronavirus samples—no SARS-related viruses—had been exported to China or imported to the NML from China. So I filed a third request asking PHAC to find all documents concerning all shipments of coronaviruses sent from China to the NML, or from the NML to China, between 2015 and 2020.

TIME WENT BY. Canada's daily infection numbers came down as summer bloomed. For a brief stitch in time, it was possible to laze at a restaurant table set out along one of Toronto's main roads and pretend life was normal. Or might be again.

THE NICE YOUNG man got back to me about my requests. PHAC had identified 2,500 emails and 30 documents concerning what other government departments knew about Qiu's shipments of pathogenic strains to China. He told me that the emails would take a very long time for the departments to screen and redact. He suggested I could speed things up by asking for just the 30 documents. I would get them relatively quickly. I agreed to that, which I also came to regret. PHAC refused to send the 30 documents, insisting that what I'd already asked for regarding the CBC request was all I was going to get. My complaint to the information commissioner is still pending.

THE THIRD REQUEST, about coronaviruses sent between Canada and China, came back first. PHAC sent a letter via Canada Post's electronic document system to say that nothing relating to any coronavirus shipments had been found. I was pleased. At least I'd closed the door on the spurious allegation that the NML had been involved in some way with a precursor to SARS-CoV-2.

I'D BEEN RELUCTANT to try to speak to Qiu directly until I had all the documents I'd requested in hand. But time is the journalist's dominatrix and too much time had passed without results. That's the trouble with access requests, especially those made to a government agency under political fire and with demonstrated contempt for public debate about matters of public safety. The books I'd read on the 1918 pandemic had made clear that as soon as that virus burned itself out, even though it killed over 50 million worldwide, it fell out of mind entirely. Only a few scientists pursued the question of its origin. Journalists and fiction writers turned their backs on it for most of the next century. I was certain that as soon as sufficient vaccines were shot into sufficient arms to crush SARS-CoV-2, no one, including me, would want to read one more word about its origin, about the government of China's duplicity, about the WHO's complicity, about PHAC's terrible failure to see the obvious until it was too late, about the attempt to suppress appropriate questions by virologists with certain interests, or the case of Qiu and Cheng. Intuition told me these things were all connected, and it was my job to get as much of it on the public record as I could. Somehow.

You might never get any documents, I said to myself one morning. Pick up the phone and call Xiangguo Qiu. Ask her why she sent those pathogens. Ask her where they went. Ask her about her Chinese military connections.

I phoned the NML: I figured that, since she'd only been suspended, not fired, Qiu might still check her voice mail, that she might still have voice mail to check. And sure enough, the NML operator connected me to her office. A woman's voice identified herself as Qiu and asked callers to leave a message. I left a message. No response.

I tried to find a home phone number for her. Nothing.

I looked up her husband, Keding Cheng. Bingo! I found a street address and a phone number. I left a message on that voice mail. No response.

I pulled out some of Qiu's papers and found an email

address for her at NML. I sent her an email. No response.

And then I remembered. I have a lifelong friend, Dr. Larry Gelmon, a medical doctor who is an expert in tropical medicine and who has consulted for major institutions like UNICEF, the World Bank, the Bill & Melinda Gates Foundation, as well as for many countries. His CV fills 18 pages of small type. He's also an assistant professor with the Department of Medical Microbiology at the University of Manitoba, the same department where Xiangguo Qiu and Keding Cheng had enjoyed adjunct appointments. Larry works out of Nairobi, where he leads a 20-plus year relationship/partnership between the University of Manitoba and Kenya through the University of Nairobi, investigating many aspects of HIV/AIDS, especially why some Kenyan sex workers are resistant to it. The project was created by Frank Plummer, the first scientific director of the NML, who also taught at the University of Manitoba. Plummer had hoped to find clues that might lead to an AIDS vaccine. I would have called Plummer, but he had just died.[196]

Larry didn't know Xiangguo Qiu or Keding Cheng. He'd only heard the gossip about their suspension from the NML and termination by the U of M. Scuttlebutt was that it had to do with smuggled reagents to get viruses to grow in cells. I was pretty sure this was a garbled version of a different story. In 2009 a former NML researcher named Konan Michel Yao was caught at the North Dakota border on his way to the US carrying "genetic material from the Ebola virus" in 20 vials wrapped in aluminum foil inside a glove inside a bag. He said he was working on vaccines for Ebola and HIV and that he had taken this material from the NML lab, where he no longer worked, because he was starting a new job at the NIH in Bethesda and didn't want to have to start from scratch. He was charged with "failure to present merchandise for inspection," pled guilty, was fined $500, and got 17 days in jail with credit for time already served.[197] Nobody in Canada responded to a query from *Nature* about the case and no one at the NIH had anything to say about it either.[198] I told Larry that the story was about viruses, not reagents, and they'd been officially shipped to China, not smuggled. Was there anybody at his department in Winnipeg that I could speak to who could tell me why Qiu and Cheng were terminated? Did he know anyone at the NML?

He allowed me to use his name in an email to Keith Fowke, the head of the Medical Microbiology Department, and to another man at the U of M who's also on staff at NML. I emailed both, and, while I was at it, sent similar queries to others in the department. The only person who responded was Fowke. He wrote to say that Xiangguo Qiu and Keding Cheng's severance from the U of M had been automatic as soon as they were suspended at NML. (A check with the U of M's policy on when adjuncts must be let go was not at all clear on that point). As to why NML suspended them, Fowke said he only knew what was in the media.

I started hunting for anything on the record about the relationship between NML and the Wuhan Institute of Virology, about Qiu helping the latter to train staff and to set up its BSL-4.

I soon found an article published in the US Centers for Disease Control's *EID* [*Emerging Infectious Diseases*] *Journal* in 2019 on the way staff were being trained to work in the WIV's brand new BSL-4. Han Xia wrote that the SARS disaster and the Belt and Road initiative had led to China's first BSL-4 and that there would be five to seven more by 2025. "According to China's 'One Belt, One Road' initiative, the chance that exotic pathogens could be brought into the country has dramatically increased," wrote Han. The WIV's new BSL-4 would be a place to diagnose, research, and develop antiviral drugs and vaccines "while additionally preserving highly pathogenic BSL-4 agents for future scientific research." WIV staff had been sent abroad for training in other BSL-4 labs, including the Jean Mérieux-INSERM Laboratory in Lyon; the Galveston National Laboratory BSL-4 at the University of Texas Medical Branch; and the BSL-4 in Geelong, Australia. This article made no mention of WIV staff being sent to the NML for training. Nor did Han Xia say that Canadians were training people in the new BSL-4 simulator the piece described. The simulator had been finished in 2017, the same year "the National Health Commission of China approved research activities involving Ebola, Nipah, and Crimean-Congo hemorrhagic fever viruses."[199]

Contrary to Xia's statement that permission had been granted for research activities by 2017, according to a January 2018 cable sent to the US State Department from the American embassy in China, the new BSL-4's 32,000 square feet of space, finished in 2015, still had no strains to work with because the required permissions from all the Chinese regulatory agencies involved had not yet been granted. According to that cable there were two problems with the WIV's new BSL-4. They needed help to get enough trained staff to work in the lab safely. But they also needed permission to import pathogens. While one Chinese agency had given permission for the WIV to work on Ebola, Nipah, etc., another agency had not given permission to import Ebola. They didn't even have permission, despite Shi Zhengli's work on SARS-like coronaviruses, to study the original SARS.

A second cable to the State Department of April 2018 described a visit to the still empty BSL-4 made in March by Rick Switzer, environment, science, technology and health counsellor at the embassy, and Jamie Fouss of the US Consulate in Wuhan. The second cable quoted from the WIV's English brochure, which defined its national security role as "an effective measure to improve China's availability in safeguarding national bio-safety if [a] possible biowarfare or terrorist attack happens." It also mentioned that the WIV officials who gave them the tour, whose names were redacted, told the Americans they needed more help from the US institutions and funders who had helped in the past. Emphasis was placed on how China wanted to lead the planned Global Virome Project aimed at locating and sequencing all the viruses that might possibly infect humans extant on earth. Some skeptics told these US officials later that, while they did not doubt China could pull all that data together, they did doubt whether China would share it.[200]

The US officials weren't the only ones to write about that March 2018 meeting. When I searched the WIV's English website for information about any relationship it had with Canada's NML and Xiangguo Qiu, I found the WIV's own report of that US visit. It said Shi Zhengli was at the meeting as deputy director of the new BSL-4. It said the WIV's director general had acknowledged the long and helpful institutional partnership between the WIV and several US groups, including EcoHealth Alliance, the Galveston National Laboratory, and the US National Science Foundation. (This item has now been scrubbed from the WIV website.) But no mention was made of any role played by Canada. In fact, there were few references to Canada on any page I viewed on the WIV website.[201] But I did find a reference to a Canadian advisor to the BSL-4, Stefan Wagener, who had formerly been the chief administrative officer of the NML.[202]

I also found a description of the WIV's "8th International Symposium on Emerging Viral Diseases." More than 300 people had attended from 12 countries including China, the United States, France, Singapore, Australia, Canada, Japan, the United Kingdom, Germany, Estonia, Kenya and Pakistan. But there were no Canadians listed as plenary lecturers. The speakers were mostly people who'd worked on coronaviruses with Zhengli Shi, including Linfa Wang of Duke-NUS Medical School in Singapore, Ralph Baric of the University of North Carolina, Peter Daszak of Eco-Health Alliance, and Shi Peiyong from Galveston National Laboratory. If Xiangguo Qiu attended that meeting, nothing on the website said so. Ebola, her area of expertise, wasn't the main topic. Coronaviruses were the main topic.

One day I clicked on the link to the WIV's peer-reviewed journal, *Virologica Sinica*, whose editor is—you guessed it—Shi Zhengli. I found Xiangguo Qiu listed as a member of its editorial board. Her place of work was described as the National Microbiology Laboratory. In January 2021, when I checked the site again, this had been changed to the Public Health Agency of Canada.

Had Xiangguo Qiu helped train staff at the WIV's BSL-4 or not? And if not, what was she doing on those five trips to China? With whom did she meet? Who paid for her trips? If the WIV's BSL-4 had not yet received official permission to import Ebola and Hendra by April 2018, even though the lab had been officially open since January 2018, surely that meant the pathogens sent by Xiangguo Qiu from the NML in March 2019 were essential in order for such work to be done in China.

I GOT AN email telling me that documents responsive to my first access request—the copy of the CBC's access request—were available to download from Canada Post. When I printed the documents, the pile was more than an inch thick. I lost my temper on page one, which was a copy of the shipper's Declaration for Dangerous Goods. Xiangguo Qiu's name and address at the NML were given, but the name and address of the consignee, the institution in China where the samples were sent, was blacked out. This made no sense: there was no place other than the WIV's BSL-4 where they could be legally sent.[203] WIV was at that point China's only known functioning BSL-4 (though there is another BSL-4 for veterinary science under construction in Harbin). The exemption permitting this redaction under the Access to Information Act was Section 19 (1), which says the director of an institution "shall refuse to disclose any record requested under this part that contains personal information."[204]

Since when is the name of a foreign institution receiving an export of dangerous Group 4 pathogens from Canada personal information?

I flipped through transfer documents, requests for transfer documents, emails between the remarkable number of senior staff at the NML who were involved in this shipment of pathogens. I read questions, answers, more questions, explanations and counter explanations, apologies for things forgotten and then remembered, even a request from China to video the sample-packing process and to send that along too, which was refused. The exchanges went on for months before the samples were at last dispatched from the NML on March 29, 2019. It was not clear if the samples were sent that day to Toronto or if they sat in the Winnipeg airport or some other holding zone waiting for a plane. There were emails regarding the special team that had to be dispatched from Guelph to stand ready at Pearson Airport in case something bad happened to the samples before being loaded on the Air Canada flight to Beijing on March 31. There were a series of emails asking: Can we stand down yet? When?

By the time I worked my way through to the bottom of the pile, it was apparent that Xiangguo Qiu had made her first request to transfer these samples to China many months before, likely just after the WIV finally got permission to import them. The earliest email in the file was dated in mid-September 2018. Later, a trove of redacted documents handed over to the House of Commons Special Committee on Canada-China Relations by PHAC would show that she made her first request on May 30, 2018.[205] But the first email in the file sent to the CBC was from Xiangguo Qiu to a person at the NML named Tracy Drew, copied to David Safronetz, chief of special pathogens and Xiangguo Qiu's boss, plus Anders Leung, who seemed to oversee such transfers:

> Subject: viruses to be exported to China
>
> Hi Tracy,
>
> Pls see attached for the request letter from [name of persons and institution blacked out] and let me know if there are any concerns/questions.
>
> Dave, we had discussed this some time ago and you are agreed to send the viruses once all the documents are in place.
>
> Thanks!
>
> Qiu

Matthew Gilmour wrote an email to David Safronetz about the proposed transfer on the morning of September 14, 2018. Gilmour said:

> I have some concerns here. No certifications are provided, they simply cite they have them.
>
> What is the nature of the work, and why are our materials required? (ie. these are surely available from other, more local labs).
>
> MTA's [material transfer agreements] would be required, not generic "guarantees" on their storage and usage.
>
> Are there materials that they have that we would benefit from receiving? Other VHF? High path flu? Good to know that you trust this group. How did we get connected to them?

David Safronetz replied 42 minutes later:

> We would of course not send anything out without appropriate paper work [*sic*], MTAS, certifications, letters from the BSOs etc. They are requesting material from us due to collaboration with dr. Qiu [*sic*]. Historically, it's also been easier to obtain material from us as opposed to US labs. I don't think other, closer labs have the ability to ship these materials. We can certainly ask about their stocks and see if we can get some.
>
> Dave

In other words, if China didn't get these strains from the NML, they might not be able to get them from anywhere else, either. Oddly, in her first request of May 2018, Xiangguo Qiu had asked David Safronetz what documents she needed to effect this transfer, specifically suggesting a material transfer agreement in her email of May 30. Safronetz had replied:

> Personally I don't believe in MTAs for these materials but we would need to look into Agency regulations.[206]

The documents sent to me did not include certificates, a material transfer agreement, or emails showing that anyone at NML had asked the recipients in China if they had something to send to the NML in exchange. (Redacted documents sent by PHAC to the Special Committee on Canada-China relations would show that David Safronetz also had no interest in signing a material transfer agreement sent by the Wuhan Institute of Virology, which asked Qiu to sign it. She replied that she had no authority to do so.[207]) There was a transfer document signed by the proposed recipients (names redacted) on October 18, 2018, and by those shipping the samples from the NML on October 26, just after the 8th International Symposium in Wuhan came to an end. David Safronetz and the NML biosafety officer (the signature was illegible) signed it. Xiangguo Qiu was named as contact person.

However, the access documents made it clear that Qiu's request to transfer the pathogens still hadn't gone anywhere by January 2019, when Qiu asked for a list of the documents she needed to complete to send the viruses out. In late January, Jay Krishnan wrote to Allan Lau, copying Catherine Robertson and Qiu, asking that Qiu be given the list. He adds:

> Cathy, do you have the phrase "the crown doesn't regulate the crown" in writing from the Foreign Affairs please? Qiu seems to remember something like was [*sic*] included in previous exports from the CL4. Thanks.

Minutes later, Qiu wrote:

> Thanks! Jay.
>
> Cathy, I thought you said something like that "the crown doesn't regulate the crown" and we should include a sentence like that somewhere (I forgot where) when the shipment to be sent out during our discussion in the hallway (outside Jay's office) sometime ago.
>
> Thanks a lot everyone!
>
> Qiu

To this Robertson replied:

> We don't have anything in writing but when we try to get an export permit, we are told we don't need one.
>
> Cathy.

That went some way to explaining why there were no documents in this pile showing interdepartmental correspondence or an export permit for the shipment. If shipments sent to China from the NML did not require export permission from Global Affairs, that would also explain why it has "historically" been easier to ship listed pathogens to China from Canada than from the US. Xiangguo Qiu seemed to have sent samples out before: why else would she have remembered that line about the Crown not regulating the Crown, a doctrine that the lawyer friends I consulted had never heard of?

Finally, at the bottom of the pile, I came upon the full list of the samples sent to China in 2019.

They were: Ebola Makona (three varieties from West Africa); Mayinga (the first known version of Ebola dating back to 1976, when it was discovered in the Democratic Republic of the Congo, then called Zaire); Kikwit (an Ebola strain from another outbreak in 1995 in the Congo); Ivory Coast (a strain of Ebola that infected chimpanzees in 1994 and then one person); Bundibugyo (an Ebola strain from an epidemic in Western Uganda in 2007); Sudan Boniface (first described in 1977); Sudan Bulu; MA-Ebov (mouse-adapted Ebola); GP-Ebov (the glycoprotein of Ebola critical to its entry into cells); GP-Sudan (the glycoprotein for the Sudan Ebola variety); Hendra (a lethal virus harboured by fruit bats and first identified in Hendra, Australia in 1994, sequenced by Linfa Wang); Nipah Malaysia (first outbreak in 1998 also sequenced by Wang); and Nipah Bangladesh (discovered in a 2012 outbreak). The WIV officials had complained they didn't have permission to import Ebola and Hendra when speaking with US Embassy and consular officials in March 2018.

I could see why the PHAC spokesperson had said repeatedly that this shipment had nothing to do with Xiangguo Qiu and Keding Cheng's suspension from the NML. If there had been a policy fault or administrative problem with the way this shipment was handled, most of the NML's senior management would have had to be investigated as well. So why had Qiu and Cheng been suspended?

I still wasn't sure where the shipment was sent. I knew

it went first to Beijing, which doesn't have a BSL-4, at least no *known* BSL-4. There was a comment in one email that: "we have an office there." The WIV is part of the Chinese Academy of Sciences, which has offices all over the country as well as in Beijing. I decided to appeal the redaction of the recipient's name and address.

You can't appeal, I was told by the nice young man who had advised me to do things this way.

Why not?

Because you didn't make the original request.

I WANTED MORE information about Safronetz's statement to Matthew Gilmour that the group receiving the pathogens had been collaborating with Xiangguo Qiu. I assumed that meant Shi Zhengli's group, as Shi was the deputy director of the WIV's BSL-4. To confirm that, I searched for any peer-reviewed articles they had done together. I found none. But I did find the December 2018 paper Amir Attaran had referred to, the one he said Xiangguo Qiu co-authored with someone in the Chinese military, a paper she would have been working on when she was trying to export those Ebola samples to China. I was surprised to find that the lead author, Hualei Wang, was not the only person in the Chinese military who had worked on that paper. In fact, there were 11 co-authors affiliated with the Chinese Institute of Military Veterinary, Academy of Military Medical Sciences, of Changchun. There were seven NML authors listed as well, including Keding Cheng, Xiangguo Qiu's partner. The second author, Gary Wong, had been at the NML and at the University of Manitoba for several years. But by late 2018, when this paper was submitted, he was a member of Gary Kobinger's lab at Laval and was also at the

Institut Pasteur of Shanghai. More surprising still: George F. Gao, the current head of China's Center for Disease Control and Prevention, was listed as a co-author.

George F. Gao had spent a brief period at the University of Calgary in the lab of Dr. Robert B. Bell, a neuroscientist, in 1994. From there he went on to Harvard and Oxford, but returned to China after SARS. By 2017, Gao had become the head of China's CDC and had also acquired many other important positions in China's scientific institutions. On this paper, Gao was only identified with the Key Laboratory of Pathogenic Microbiology and Immunology, Institute of Microbiology, Chinese Academy of Sciences, Beijing. At the bottom of the paper he was also described as "a principal investigator of the National Science Foundation of China's Innovative Research Group," another way of saying he had significant grant money at his disposal. Like US politicians who indirectly control successful fundraising PACs and can throw money to political allies, principal investigators who have large grants can also help favoured colleagues. Grant money equals power, and the greater the sums, the greater the power. Gao's name on the paper demonstrated that Xiangguo Qiu and Keding Cheng were directly connected to the beating heart of China's civilian/military/political science system.

This paper was published online in the *Journal of Virology* three months before Xiangguo Qiu sent those samples to China. It was submitted on September 5, 2018, just days before Xiangguo Qiu's email reminding David Safronetz about his promise to send the samples, just six months after the staff at the WIV complained to American embassy and consular officials about not yet having permission to import pathogens to their BSL-4. This paper described using horses to produce immunoglobulin fragments to protect nonhuman primates from Ebola. This method of holding off Ebola while provoking an immune response turned out to be faster and cheaper than using a monoclonal antibody cocktail like ZMapp: antibody cocktails are expensive to produce, and ZMapp hadn't passed muster with the FDA anyway. The article showed that the nonhuman primates (species not identified) survived even if they were not given the fragments until five days after exposure to Ebola.[208] The authors concluded that this method was so successful that human trials should follow, a suggestion no doubt taken up by Chinese military doctors concerned with protecting their troops from Ebola, a scourge in African countries where China has major Belt and Road operations.

When I read the acknowledgements at the end of the paper, I was shocked to see that these experiments had all been done at the NML in Winnipeg. Funding had come from various Chinese granting agencies as well as from "the Public Health Agency of Canada, partially supported by grants from the National Institutes of Health...and a CIHR grant... to X. Qiu." But the CIHR grant to Qiu had been for a different Ebola study to be done with a researcher in Montreal, a person not listed as a co-author on this paper. George Gao was credited with helping to conceive the study.[209]

I searched for more papers Qiu had done with Gao. I found another published one month earlier called "Equine Immunoglobulin F(ab')2 fragments protect mice from Rift Valley fever virus infection." Rift Valley fever occurs in East Africa and the Arabian Peninsula, not in China. Djibouti, where China has a large PLA base, is only 20 kilometres across the Red Sea from Yemen, the southernmost country in the Arabian Peninsula. There is no known treatment for

Rift Valley fever, which kills a small minority of victims. The first author on the paper was Yongken Zhao of the Institute of Military Veterinary Medicine, Academy of Military Sciences, Changchun. Five of the authors, including the last, Xianzhu Xia, were with the same military institute. Hualei Wang, the first author on the other Ebola paper described above, was also a co-author. This paper was submitted to the journal on April 9, 2018, shortly after American officials wrote up their memo about their visit to the WIV.[210] So the relationship between Xiangguo Qiu and Chinese military scientists dated back at least to 2017, when she began making her trips to China.

I found another paper, published in May 2018, titled "Cellular-Beacon-Mediated Counting for the Ultrasensitive Detection of Ebola Virus on an Integrated Micromagnetic Platform." The first author was Shao-Li Hong of Wuhan University. A co-author was Jiangun Chen, who is affiliated with Shi Zhengli's Special Pathogens and Biosafety Department at the WIV. George F. Gao was listed as a co-author too. His role at China's Center for Disease Control and Prevention was not mentioned, but two other institutions were: the same organizations with which Gary Wong, another co-author, was said to be affiliated. Gary Wong's connection to the NML and Laval were not mentioned.[211]

It's reasonable for Canadian academic researchers to work with military scientists from allied countries on subjects of mutual concern, though academics often find that to be a bridge too far. But it's a whole other thing for scientists employed by the government of Canada, in a facility that requires a secret security clearance for employees who have access to the BSL-4, to work with members of the military of a country that is no friend and certainly not an ally of Canada. The experiments, by and large, seemed to have been performed inside the NML.

It was as if George F. Gao, his military colleagues, and the WIV had all reached into Canada's National Microbiology Laboratory, thanks to Xiangguo Qiu, and used it as their own.

Was it the first paper that caused someone at PHAC to call in the RCMP? The second? The third? None of the above? How long had the relationship been going on between Xiangguo Qiu, Keding Cheng, George F. Gao, Chinese military researchers, and the Wuhan Institute of Virology?

SAFRONETZ'S EMAIL TO Matthew Gilmour suggested the NML had previously sent samples to China because it was easier to send pathogens from Canada than from the US. I thought I should check that statement with James Le Duc, director of the Galveston National Laboratory of the University of Texas Medical Branch, which has close connections with the WIV. But one cannot write or phone James Le Duc and expect him to answer questions. The public relations officer who got back to me asked for an email explaining why I wanted an interview. I sent it. No response.

I also emailed a query to Dave Safronetz asking for clarification. I expected no response, and that's exactly what I got.

8. Who Are Xiangguo Qiu and Keding Cheng?

I WAS DETERMINED to find out more about Xiangguo Qiu and her husband. Emailing people at the NML had proven fruitless. But Gary Kobinger wasn't at the NML anymore, he was at Laval, so I called his lab. No answer. He is also CEO of a company called Guard RX, so I emailed the communications person. No response. Eventually, I got a better email address for Kobinger from David Evans at University of Alberta. I approached him carefully—okay, craftily—given his earlier statement to Karen Pauls that Xiangguo Qiu's suspension was likely the fault of bureaucratic nonsense, and his insistence that China needs nothing from Canada. I asked if he had participated in any public conversations in Canada concerning gain-of-function experiments. When I made it clear that these sorts of experiments concerned me, he made it clear he was much more concerned about rescue, the kind of experiment

David Evans had done with horsepox, and Tunney with the 1918 flu virus. After a bit more of such back and forth, I asked if we could speak on the phone. He did not respond. Which all goes to show, being crafty gets you nowhere.

I looked up the company that had planned to make the ZMapp antibody cocktail that gave rise to the Governor General's prize. I figured people there must have gotten to know Xiangguo Qiu quite well.

Larry Zeitlin, the president of Mapp Biopharmaceutical Inc., explained that the company was started in 2003 to develop antiviral medications for all the baddies: HIV, Zaire Ebola, Sudan Ebola, Marburg, Congo-Crimean Hemorrhagic fever, Nipah, Hendra, Hantavirus, respiratory syncytial virus, eastern equine encephalitis. He said he'd been introduced to Kobinger in 2012 by a scientist at the NIH. If he remembered correctly, it was Gene Olinger (then an advisor/contractor with NIH/NIAID specializing in high-containment pathogens). Zeitlin's group and Kobinger and Xiangguo Qiu had published papers on antibody cocktails at the same time. Olinger thought they should know each other and possibly work together. The cocktail they worked out was comprised of a humanized antibody that Mapp had developed, and two others developed by Kobinger and Xiangguo Qiu at the NML. Mapp made a deal with a Canadian defence contractor called Dreyfus Inc. (the company seems to have disappeared), which had licensed the antibodies from PHAC. Mapp made the antibodies in tobacco plants, something that can be done in a matter of months versus years in cells and can be scaled up quickly. They tested the cocktail[212] in rhesus macaques at NML.[213]

Zeitlin described Kobinger as a bit of a cowboy who, during the West African Ebola epidemic of 2014 to 2016, took an earlier version of the cocktail to Liberia and left it there in case of emergency. ZMapp was later shown to be only 91 percent effective, not quite good enough for the FDA, he explained, so would not be manufactured. As to what he knew about Xiangguo Qiu and her husband, Keding Cheng, he said that the few times he'd met her in person had been at scientific conferences. She'd seemed to him to be "a very sweet, curious, and smart scientist." He didn't think he'd ever met her husband. When the news came out about them being removed from the NML, Zeitlin had emailed her his sympathies. He got no response. "Gary couldn't either," he said. (Kobinger told the CBC she'd thanked him for his concern.)

That's all he knew about either of them.

I HAD QUESTIONS only Xiangguo Qiu and Keding Cheng could answer. What if they hadn't received my emails, or heard my phone messages? Fairness, never mind curiosity, required me to keep trying. Because of the pandemic, I couldn't hop on a plane to Winnipeg and knock on their door. So, at the end of the summer, I asked a friend who lives in Winnipeg to go to their house with a letter from me asking if I could please speak with them. It was a friendly letter. If I'd received such a letter, I would have answered it. He knocked on their door. No response. He put my letter in their mailbox, took a picture of that letter in their mailbox, and a few more of the house. Then he sent the pictures to me.

What can you tell about people from their house?

It was a two-storey of white brick and stucco set in a narrow crescent of similar houses, part of a development that likely dated to the early 1990s, when faux-mullioned bay windows were still a thing. A swatch of ivy ran up beside the front window, reaching for the second floor. Prairie roses lined the walkway to the front steps. It was the kind of garden meant to keep neighbours happy by not attracting a second glance. The lawn was green and there was a hose lying beside the walk, signifying someone was taking care of the place. There were two traditional black Chinese lions, one female, one male, set out on either side of a cement step. A clay flowerpot sat in the grass. It was empty save for a lot of cigarette butts and a discarded can of Coke. My friend thought this signified that someone who lived there was very anxious. He also said the house seemed unnaturally dark, as if no one had been there for a while. Yet there were two cars, no dust on them, parked in the driveway in front of the double garage. One was a Lexus, suggesting someone enjoyed a nice income. When I pored over the pictures later, I was pretty sure someone had been standing behind the second-floor window curtain, watching as my friend took his pictures. If I was right about that, my friend was right too. Whoever had been peering from behind the curtain was too anxious to demand to know why a stranger was standing on their sidewalk taking pictures of the property. Maybe frightened would be a better word. As it turned out, this was not Xiangguo Qiu and Keding Cheng's only house in Winnipeg: there was another bigger, newer one, as well as a recreational property in Gimli on Lake Winnipeg. But this house was associated with Keding Cheng's phone number. Who knows who'd been hiding behind that curtain?

I got no reply to my nice letter, so I was forced to rely on what others had written, or what the two of them had published about themselves. I found no reports at all with information about Keding Cheng, and nothing that explained how or why either of them had come to Canada and ended up at the NML. Most stories about their suspension mentioned that Xiangguo Qiu arrived in Canada in 1996 to do her PhD at the University of Manitoba. But I found no evidence that she had acquired a PhD. There is no thesis bearing her name on deposit in the U of M library, but there is a thesis in immunology in that library by Keding Cheng submitted in October 2000 as part of the requirements for an MSc from the University of Manitoba.

I found them both on LinkedIn.

Xiangguo Qiu's bio said she's from Tianjin (a port city of about 15 million on the northern coast of China, established in 340 BC). She got her MD in 1985 at Hebei Medical University, a long-established school where both modern and traditional Chinese medicine are taught. Her age wasn't given, but graduating from medical school in 1985 suggested she was born in the early to mid-1960s, making her a 50-something. Two years after becoming an MD she began to study for an MSc in immunology at Tianjin Medical University, a degree she received in 1990. Her LinkedIn page said nothing about what she did for the next 13 years. It did not say she entered a PhD program at the University of Manitoba, but it did say she was hired as a biologist at the NML in June 2003, right at the end of the SARS pandemic and just as that Chinese general, according to the *Epoch Times*, was giving a speech to cadres about using biotechnology in aid of a war of colonization. According to her LinkedIn profile, she remained a biologist at the NML until February 2015, when she was at last promoted to research scientist, no doubt due to her work on the antibody cocktail for treatment of Ebola. She had three endorsements, all by Gene Olinger, whose own LinkedIn page describes him as an American with a long interest in biosafety, containment, and in viruses like Ebola, etc. He commended her skills in molecular biology, life sciences, and infectious diseases.

I sent her a LinkedIn message. No response.

Her husband, Keding Cheng, had two LinkedIn listings. One said that he works as a "professor" at PHAC—just the acronym, not the full name of the agency. That was the only information on the page. The other listing said he was a biologist at the University of Manitoba.

If I'm right about her age, Xiangguo Qiu and Keding Cheng (who looks just a little older) grew up in China in a period of extreme volatility almost unimaginable to people born in Canada. Peace, order, good government, the great expansion of the middle class, and the boomer revolution are what Canadians experienced in the 1960s and '70s. In China, the terrible disaster of the five-year plan known as the Great Leap Forward resulted in a famine so profound that as many as 50 million people died of starvation. This was followed in 1966 by ten years of vicious, lawless anarchy known as Mao's Cultural Revolution, in which the educated and powerful were pulled from their posts and publicly humiliated, their children set upon on the theory that if the parents were dangers to the Revolution, their children must be too. In the first few years, thousands were tortured and murdered in Beijing by roving bands of Red Guards. Millions were sent to the countryside for "re-education" by peasants.

Xi Jinping was shaped by this period, according to a wonderful profile of him in the *New Yorker* by Evan Osnos, which I rely on here. His family had been very prominent, thus bigger game for Mao and the Red Guards than the average hapless professor. Xi's father had been Mao's chief propagandist, and, for a time, the vice premier, but in 1962 he was kicked out of his post over a novel Mao did not approve of and sent to work in a factory, his wife to do hard labour on a farm. As the Cultural Revolution gathered steam, Xi Jinping's elite Beijing school was closed, and for a time he and his former princeling school friends wandered the streets getting into fights. In 1968 his father was imprisoned, and Xi Jinping was sent to do labour in the countryside—the same countryside his father hailed from, and which had been his political base. While Xi was hard at work there, sleeping on a kang like all the rest, his half-sister committed suicide after enduring years of persecution. From these experiences, apparently, Xi developed an iron determination to get back to the centre of power. He applied and reapplied—seven times—until he was finally allowed to become a member of the Communist Party. He went to Tsinghua University in 1976 due to his political merit, the same year Mao died, the same year his father was finally released from prison. His smarts, and his father's friends, helped him climb the Party ladder. Two of his siblings went into business in Hong Kong. His sister is said to have made her way to Canada.[214]

Xi now occupies all the positions that were once spread among several individuals to avoid a repeat of Mao's catastrophic and highly personalized dictatorship. He has also gotten rid of the term limits that would have otherwise ended his leadership, and he has brought back the cult of personality. I found a YouTube video shot on China's National Day ceremonies in 2019. The camera focuses mainly on Xi, who stands tall on a vast balcony overlooking the vast Tiananmen Square among the Party's top leadership (all men). One of the oldest is so frail he is held upright by two younger men. Xi, dressed in a perfectly tailored Mao suit, stands supreme, his head tilted, as always, to one side, his expression hovering between worldly sadness and a forced smile. Row on row of perfectly dressed, perfectly synchronized troops, each soldier a perfectly beautiful specimen of humankind, march below, their eyes turned up to him. The video isn't quite *Triumph of the Will*, but it's trying.

When Xiangguo Qiu and Keding Cheng entered high school, revolutionary tumult was dissipating, but the main subject on offer in their classes may still have been Mao's little red book (*Quotations from Chairman Mao Tse-tung*). When Deng Xiaoping came back to power in 1977, after the overthrow of the Gang of Four (Mao's wife and friends), he reinstated the national competitive university entrance exams shut down by Mao during the Cultural Revolution. Students were assigned to specific universities and courses based on their performance in those exams. George F. Gao was sent to veterinary school, though he never wanted to be a vet. Linfa Wang had wanted to be an electrical engineer, but his math results weren't good enough—owing to a deficient high-school education—so he studied biology.[215] It is probable that exam results assigned Xiangguo Qiu to that medical school and that Keding Cheng had no choice in where he was sent. It is also probable that the State paid for them to study abroad as part of Deng's cure for the years of darkness, though by the middle 1990s, thanks to Deng's economic policies, some Chinese families were wealthy enough to pay their children's way.

RESEARCHGATE LISTS PUBLISHED works authored by scientists. When I last checked, it said Qiu had co-authored 171 papers and Keding Cheng co-authored 72. Either he is much less prolific, or he took a significant time out from doing public science. ORCID, a site where scientists post information about themselves with links to their papers and their associates, had listings for both Xiangguo Qiu and Keding Cheng. But all public information except for their names had been removed and the last inputs of private information recorded in November and December 2020. On my first quick scan of Google Scholar, I found academic papers that Keding Cheng co-authored concerning SARS, HIV, Creutzfeldt-Jakob, and one on ACE2, the receptor SARS-CoV-2 uses to enter cells. I was looking for more when my imagination at last kicked in and I figured out how to trace their movements from China to Canada. All papers in peer-reviewed journals cite each author's professional affiliations. By looking up their early papers, I could follow in a rough way where they had worked and when.

Google Scholar showed me that Xiangguo Qiu had published in Chinese scientific journals in 1990, 1991, and 1993. These were in Mandarin, which I did not try to translate, but they established that she was working in science in China after she received her MSc. Cheng also had a paper published in China in 1990, followed by several more in 1992, 1994, and 1995.

In 1996, both their names appear on a paper published in an American journal called *Blood*. Cheng was the first author, Xiangguo Qiu one among several, which suggested that Cheng was in the lab on some sort of postdoctoral fellowship while Xiangguo Qiu might have had a less formal position. But if that was the case, why did his LinkedIn profile claim no PhD? The last author on the paper was Moshe Talpaz, a well-known leukemia expert then at the

University of Texas MD Anderson Cancer Center. A co-author on the paper, Razelle Kurzrock, is now a leading clinician at the University of California, San Diego. I called her to see if she could tell me anything more about them. She was too busy to take a call on this subject. Ever. Moshe Talpaz was also unavailable for comment, and believe me, I tried. So I was left with surmise. I surmised that for their names to appear on one of the Talpaz group's papers, Xiangguo Qiu and Keding Cheng were likely in his lab at least for several months before the paper came out.[216]

There was another paper in the same journal, *Blood*, with Cheng as lead author and Talpaz the last, also published in 1996.[217]

Then I found their names in the acknowledgements section of a master's thesis submitted to the Faculty of Graduate Studies, Department of Physiology, at the University of Manitoba in May 1998. Tao Fan thanked "co-workers," including Xiangguo Qiu and Keding Cheng, for help with his study of Vitamin K1-dependent growth regulatory pathways during embryogenesis. In other words, by 1998, they were both students and/or technicians at the University of Manitoba. I found nothing more listing Cheng as a co-author until a paper submitted on July 23, 1999, and published January 7, 2000, in the *Journal of Biological Chemistry*.[218] At that point, Cheng was affiliated with the Institute of Cell Biology at the University of Manitoba. His co-authors included people at the Dana-Farber Cancer Institute at Harvard Medical School. He was also the first author on a paper published in the same journal with a similar group of co-authors submitted in early January 2000 and published in July 2000. For that paper, however, he was said to be with the Lovelace Respiratory

Research Institute in Albuquerque, New Mexico. It was a very accomplished paper for someone who had not yet submitted his master's thesis to the University of Manitoba.

On a paper submitted one month later, in February 2000, and published in the journal *Leukemia* on August 31, 2000, Talpaz was listed as first author, and Cheng and Qiu as co-authors.[219] No mention was made on that paper of the Lovelace Respiratory Research Institute. Qiu and Cheng were both said to be with the Institute for Cell Biology at University of Manitoba. And yet, on another paper, submitted in July 2001 and published in July 2002 in the journal *Cancer*,[220] on which Cheng was first author, Qiu was the third, and Moshe Talpaz last, all were affiliated with the Department of Bioimmunotherapy at University of Texas MD Anderson Cancer Center, which suggested the work had begun while they were still there, but not completed until 2001, long after they'd left. But if that was the case, why did their affiliations say University of Texas/MD Anderson?

One thing was clear: the two of them had worked together in Talpaz's lab and then moved to Winnipeg to study together by 1998. By 1999, they were studying and working together at the University of Manitoba's Institute of Cell Biology. Why did they move to Manitoba? Perhaps coincidentally, 1999 is the year the National Microbiology Laboratory, the first BSL-4 in the world to combine both animal and human pathogen research under the same roof, opened its doors. While the University of Manitoba is a fine research school, it is not up to the standards of University of Texas/MD Anderson. However, there was no BSL-4 in Texas at that time, and there wouldn't be until 2003.

In 2003, Keding Cheng co-authored a paper describing how to use mass spectrometry to characterize proteins from SARS. It was published in May, at the end of the SARS pandemic, in the *Journal of Molecular and Cellular Proteomics*.[221] By then, Cheng was affiliated with the Centre for Proteomics and Rheumatic Diseases at the University of Manitoba. Co-authors on that paper included the very distinguished Frank Plummer and Heinz Feldmann. Plummer was then both a professor at the University of Manitoba, in a different department from Cheng, as well as director general of the NML. Feldmann was chief of the NML's special pathogens unit. In other words, Cheng was working with the NML's leaders before the spring of 2003 on a matter of great importance to both Canada and China. One month after that paper was published, Qiu got a job at the NML as a biologist. Cheng stayed at the University of Manitoba.

In October 2003 Cheng worked on another paper that involved mass spectrometry.[222] In 2004 he worked on a paper on the ACE2 receptor in rats.[223] On a 2005 proteomics paper he was listed with the Proteomics and Rheumatic Diseases department at the University of Manitoba. In 2007 he was with the Manitoba Centre for Proteomics and Systems Biology.

In 2008 he co-authored another paper with Frank Plummer. It was called "Identification of differentially expressed proteins in the cervical mucosa of HIV-1-resistant sex workers."[224] Cheng was said to be affiliated with the University of Nairobi, Plummer with the University of Nairobi and the University of Manitoba. There was an icon for the National Microbiology Laboratory on the article, but it was not attached to any author's name.

I found no papers listing Cheng as an author of published science in English for the next four years. Then, in 2013, he was the first author on a paper co-authored with

Matthew Gilmour.[225] He was listed as working at the NML and the Department of Human Anatomy and Cell Science at the University of Manitoba Medical School. After that, his papers declared him to be at the NML working mainly on bacteria until 2018. That year he again co-authored papers with Qiu. By then, Qiu had become the head of vaccine development and therapeutics within the Special Pathogens Program at NML. When I went to the Chinese equivalent of ResearchGate, there were also several citations for him between 2015 and 2016.

But where was he between 2009 and 2013? In Nairobi? I wrote to my friend Dr. Larry Gelmon. I said: I've got this paper that shows Keding Cheng working with Plummer on the sex workers project. It says he was at the University of Nairobi.

Larry wrote back to say he would know of him if he'd worked on the sex workers project. He asked for the title of the paper.

I sent it.

He wrote back to say he'd asked a colleague for help and forwarded to me the colleague's response. The colleague said the affiliation notation on the paper must be a mistake. "Keding must be one of the Chinese from NML associated with UoM," he wrote. "Thought he was expelled from NML due to the samples debacle."

I wrote back to correct the word "expelled" to "suspended." But that still left me with the question: I had found no papers he'd co-authored between 2009 and 2013. Where was he during those four years, and what was he doing?

I DID THE same searches for Qiu. In October 2002, she was with the Institute of Cell Biology and the Department of

Pediatrics and Child Health at the University of Manitoba. She stayed connected to that department until at least 2004, though she had started work at the NML by June 2003. By 2005, she was working on the world's worst infectious disease—Ebola. Yet there was nothing in her earlier published work in English to suggest an interest, never mind expertise, in Risk Group 4 pathogens. In North America her first papers were on blood cancers. There was even a neurology paper involving a mouse model. But by 2005 she was a co-author on an Ebola paper concerning the possible use of antibodies derived from a particular species of Ebola-exposed macaques as a human therapy.[226] The paper was submitted in June 2005, which meant she must have been assigned to the BSL-4 some time before that. The lead author and the last author worked for Cangene Corporation in Mississauga, Ontario. But two other co-authors were with the US Army Medical Research Institute of Infectious Diseases at Fort Detrick. You can't get more connected to dual-use research than the Army Medical Research Institute of Infectious Diseases at Fort Detrick.

By the time Qiu started work on Ebola, the mishandling of the SARS epidemic by China's leadership was well-known worldwide. Officials had been fired for misbehaviour. China had called upon leading scholars in the US, like W. Ian Lipkin, and Chinese scholars with significant success abroad, such as George Gao, to help reorganize its public health system, especially its disease surveillance methods. That year, a group of scientists from Guangdong, the centre of the SARS pandemic, arrived in Winnipeg to tour the Canadian Science Centre for Human and Animal Health, which includes both the NML, now managed by PHAC, and the animal diseases lab managed by the Canadian Food

Inspection Agency. China was negotiating its agreement with France, which led to the construction of the BSL-4 at the Wuhan Institute of Virology.

Natalie Salat, a reporter for *Legion*, a magazine about Canada's military history, came to Winnipeg to write about the facility after the visitors from China had come and gone. Their attendance was presented to her as a mark of the NML's global importance. Salat was told that the NML had partnerships with the Department of National Defence, the University of Manitoba, the US Centers for Disease Control and Prevention, the US Bioterrorism Response Network, etc. Stefan Wagener, whom Salat identified as the centre's scientific director for biosafety and environment, told her that everyone who worked at the lab for more than ten days had to have a rigorous security clearance. He jokingly said that if he told her anything else about the detailed security measures in place, he'd "have to kill her."[227]

Reading the piece, I found myself wondering: How could Qiu have qualified for even a low-level clearance by 2003? She would need secret clearance to work in the BSL-4 by 2005. I'd read somewhere that to get a secret clearance one must have been a citizen for ten years, have no criminal record, and no other obvious exploitable problems of a criminal nature, like an addiction to drugs or gambling, etc. Surely Qiu had not been in the country long enough to qualify when she was first hired at the NML, though she and her husband might have been permanent residents by then. When I checked, I learned those are the requirements for a top-secret security clearance. It is less onerous to get a secret clearance.

I remembered Stefan Wagener's name from the Wuhan Institute of Virology's website. He'd been listed there as a member of the WIV's Scientific Advisory Committee of the Center for Emerging Infectious Diseases. I looked him up. The bio on his company's website said he had worked at the NML from 2001 until 2013, then been appointed to the Grain Commission in Winnipeg, and that he now ran an international bio-risk consultancy. Surely he would know Qiu? We did a Zoom call rather than a phone call because he likes to see the people he's speaking with. My video died within the first three minutes, but I got a glimpse of him: an older man sitting at a desk with a headset over his ears, waving at me.

He said he started at the Canadian Science Centre for Human and Animal Health in 2001 as chief administrative officer. His responsibilities included human resources, but mainly the safety and security of the building and its complex systems. In 1999, before he was hired, there had been a leak: about 2,000 litres of insufficiently sterilized wastewater was released into Winnipeg sewers. Nothing was said about this to the public until the CBC published a story about it a month later. Ebola and Lassa fever had not yet arrived at the lab, so there was no danger, just like there was no danger the next year when there was another wastewater leak. That one never got out of the building.[228]

Are you still an advisor to the WIV BSL-4 in Wuhan? I asked.

He said he was not. He had visited Wuhan when the BSL-4 lab was in the design stage, but that was some years ago.

Did you meet Shi Zhengli? I asked.

He was pretty sure he had. He remembered her face when he saw the press stories about whether SARS-CoV-2 might have leaked from her lab. But he hadn't worked with the WIV for quite some time: his advisor role had lapsed a couple of years after his visit.

Do you remember Xiangguo Qiu from the NML? I asked.

"I don't remember her," he said. He explained that there were about 400 people working there in his day, and that he had been more concerned with the physical plant's operations. He certainly didn't know everybody.

Keding Cheng?

"Did I know Cheng? Might have had conversations," he said. But he had no clear recollection of "these people."

Yet Wagener's position at the NML was eliminated in 2013, which is when Keding Cheng apparently started work there. So why did he remember conversations with him, but nothing about the woman who'd been toiling for years in the NML's special pathogens lab?

"What can you tell me about the security clearance process at the NML?" I asked.

He explained that it was a graded system, requiring different levels of security clearance depending on which labs a person had access to. Canadian citizenship or landed status had nothing to do with those clearances, it was not a requirement to work at the NML. The Level 2 labs required lesser clearance than the Level 4s. People tried to arrange their work so that the bulk of it could be done in the Level 2s. Genetic material or proteins can be worked with in the BSL-2 because they don't pose a risk to anyone.

How long did it take to get a security clearance?

"It could take a long time to get access to data. It could take six to eight months before a full security request to work in a containment environment," he said. He explained that the point is to bring the best scientists from all over the world to the lab and to do the best work possible, to be constantly training so that if infectious-disease outbreaks occur everybody is ready to deal with them. He himself was a German citizen when he worked at the NML. Before that, he'd spent 12 years working at the University of Michigan.

So how were people cleared? I asked.

There would be background checks and criminal record checks organized by the RCMP, who would deal with the country of origin, he explained.

You mean the RCMP would ask Chinese police about the status of a Chinese citizen?

He could see where I was going with that question: If China wanted to slip someone into Canada's only BSL-4, Chinese officials contacted by the RCMP would of course send a glowing but not necessarily complete report.

I told him about Qiu and Cheng collaborating on papers with members of the Chinese military. I told him that the experiments had been done in the BSL-4. How could that have happened?

"I don't have an answer to [that]." he said. But he wasn't fussed by it. He asked if I'd have a problem if people in the NML worked with the American military.

The two situations are not comparable, I said. "Canada and the US are allies and we protect the North American continent together. China is not a friend of Canada." And I am not alone in that view. On February 10, 2021, the director of CSIS, David Vigneault, gave a public talk in which he described China as a "strategic threat."[229]

But Wagener saw things differently. Pathogenic agents can be looked at with two eyes, he said, from the point of view of public health, but also from a safety and security perspective, biowarfare and bioterrorism. "Almost any country with a Level 4 will have an interest in these agents in a military perspective. You prepare your armed forces in the event someone might have a weapon against troops.

Just by publishing it, it tells you this knowledge is in the public domain."

What he meant is that, if something is published, it no longer constitutes a threat. Everybody can read it. I wasn't so sure about that. And it raised another question: Had all the experiments done in the NML lab by Qiu and Cheng, along with Chinese military personnel, been published? And did the two of them retain their access to data even after their security clearances were withdrawn?

FROM WHAT WAGENER said of the NML's clearance process, it seemed to me more than possible that the security investigation of Keding Cheng and Xiangguo Qiu had been less than rigorous. As Wagener had also explained, people working in Level 4 labs organize their work so that they don't have to be inside them any longer than necessary. The bulk of experimental protocols are conducted safely in labs with lesser containment. People coming from China to work with Qiu might not have needed security clearances—so long as they were working with someone in the lab who did have one.

I went back to Google Scholar. I wanted to find out exactly when Qiu and Cheng began working with co-authors from the Chinese military, the Wuhan Institute of Virology, and with George Gao.

I dug out Qiu's paper on treating Ebola in nonhuman primates with "ZMapp." It had been published in *Nature* online in August 2014, in final form in October. She was the lead author, Kobinger the last, Larry Zeitlin was in the middle. Other co-authors, such as Gene Olinger, were affiliated with the NIH, still others with the United States Army Medical Research Institute of Infectious Diseases at

Frederick, Maryland. But there was one author, Heiyan Wei, who was affiliated with both the NML and the Institute of Infectious Diseases, Henan Center for Disease Control, China. This seemed odd given that money for this paper had come from the US Defense Threat Reduction Agency, the US National Institutes of Health, PHAC, and the Canadian Safety and Security Program.[230] Could a person working for a Chinese state-run institution qualify for a Canadian security clearance? And if not, how could Heiyan Wei have been affiliated with the NML?

I found another paper co-authored by George Gao which appeared to be the first Xiangguo Qiu published with him. It was called "Molecular Characterization of the Monoclonal Antibodies Composing ZMab: A Protective Cocktail Against Ebola Virus." It was submitted to *Scientific Reports* in September 2014, just as the *Nature* piece appeared online. It was published in November, just after the *Nature* piece appeared in its final form. Kobinger was one of the co-authors. Gao and two others were with the CAS Key Laboratory of Pathogenic Microbiology and Immunology, Institute of Microbiology, Chinese Academy of Sciences, Beijing. Qiu was the corresponding author.[231] The acknowledgements said the NML authors were supported by the Canadian Safety and Security Program, while George F. Gao was supported by two grants from the government of China, the China National Grand S & T Special Project, and the Special Foundation of President for Ebola Virus Research from Chinese Academy of Sciences. He was credited, along with the other colleagues from China, with performing the "surface plasmon resonance assay and data analysis." Gary Kobinger was credited with writing the paper.

That same year, contrary to Amir Attaran's assertion

that gain-of-function experiments are permissible in Canada but not done, Qiu and Kobinger did a gain-of-function/passaging experiment at the NML involving mice and Marburg. The paper was called "Establishment and characterization of a lethal mouse model for the Angola strain of Marburg virus." Marburg normally does not infect mice, but after 24 passages in immuno-deficient mice it adapted itself sufficiently to give the mice severe disease. The money for that experiment also came from the Canadian Safety and Security Program, but no one affiliated with a Chinese institution worked on it.[232] But when Qiu and Gary Wong did a similar experiment in 2015 that entailed making guinea pigs susceptible to Sudan virus by adapting the virus to them through multiple passages, Heiyan Wei of the Henan Center for Disease Control and Prevention, China, was again a co-author.[233]

A paper published in 2016 compared the pathogenicity of Kikwit and Makona strains of Ebola in rhesus macaques. Gary Wong worked on that paper as first author. George Gao worked on it too.[234]

I found three articles focused on the creation of devices for rapid detection of Ebola. George Gao was a co-author on all three. The first appeared in the journal *Biosensors and Bioelectronics*, and the other two in *Analytical Chemistry*. The *Biosensors* paper was published at the end of 2015.[235] The first *Analytical Chemistry* paper appeared in December 2016, the next in May 2018. A co-author on both *Analytical Chemistry* papers was Jianjun Chen of the Laboratory of Special Pathogens and Biosafety at the Wuhan Institute of Virology, where Shi Zhengli was deputy director.[236] Gary Wong worked on those papers too. He was no longer affiliated with the NML but with the same institutes in China as George Gao: one in Shenzhen, one in Beijing.[237] The only co-author from the NML on the May 2018 paper was Xiangguo Qiu. But work with Ebola must have been done at the NML. It could not be done in China at the WIV's BSL-4 because the latter had not yet received permission to import Ebola to China.

And what about the Chinese military?

Members of the military began working with Xiangguo Qiu about a year after George Gao. I found a paper titled "An Adenovirus Vaccine Expressing Ebola Virus variant Makona Glycoprotein Is Efficacious in Guinea Pigs and Nonhuman Primates." It was published online on October 4, 2016, in the *Journal of Infectious Diseases*.[238] The first author, Shipo Wu, was affiliated with the Beijing Institute of Biotechnology, an organization known to work closely with the Chinese military. Several other co-authors were with the same institute. The most important among them was Chen Wei. Her name is given on the paper as Wei Chen, but that's the Western way of naming in which the family name comes last; in China it's the other way around. Chen Wei, China's lead expert on Ebola, had made the vaccine used in the study. She was then and still is a major general in the People's Liberation Army, though that affiliation was not mentioned in this paper. Xiangguo Qiu, the last author, was said to be with both the Special Pathogens Program at the NML and with the University of Manitoba. Gary Wong was listed with both the NML and the CAS Key Laboratory of Pathogenic Microbiology and Immunology, Institute of Microbiology, Chinese Academy of Sciences Special Pathogens Program, which was George Gao's home institution. Two other co-authors, Xuefeng Yu and Tao Zhu, were with Tianjin-based CanSino Biologics Inc.[239]

Xuefeng Yu came to Canada from China in 1996 and worked mainly at Sanofi, the pharmaceutical company, until 2009. He returned to China and set up CanSino to develop vaccines. It is now listed on the Hong Kong stock exchange. The cell line used to develop Major General Chen Wei's Ebola vaccine is a variant on an original cell line crafted long ago by Dr. Frank Graham of McMaster University. The National Research Council (NRC) licensed its use to CanSino. This same cell line was used again by Major General Chen Wei to make China's first vaccine to fight SARS-Cov-2.

Canada was supposed to have gotten access to that SARS-CoV-2 vaccine for Phase 1 trials, along with the right to manufacture it at an NRC plant being adapted for vaccine production in Montreal. But in May 2020, shortly after this deal was announced, it fell apart, a fact not shared with Canadians until August. Chinese authorities would not permit the vaccine to be sent to Canada at all (adding vaccine diplomacy to hostage diplomacy in China's bag of pressure tactics).[240]

Major General Chen Wei was awarded a special prize for her SARS-CoV-2 efforts (not for the vaccine, but for a plasma treatment) in a heavily publicized ceremony hosted by Xi Jinping himself. If you google her name, you'll find an image of her resplendent in her military uniform with a very large decoration, like a mayor's chain of office, hung over her shoulders. She is standing beside President Xi.[241] There are several laudatory stories about her in the Chinese press, the kind that aim to show China's tremendous military precision in combating the virus. She told state broadcaster CCTV that "the pandemic is a military affair, and the affected areas are the war zones."

I also found a lovely picture of her being tossed in the air by her African colleagues during the West African Ebola epidemic of 2014–2016, as if to celebrate her efforts in helping to stem the outbreak. Xiangguo Qiu tested Chen's Ebola vaccine at the NML on guinea pigs and nonhuman primates: they were infected with Ebola to get a better fix on the vaccine's efficacy. To do such experiments in China, Major General Chen Wei would have needed access to a BSL-4 lab, but the new BSL-4 at the Wuhan Institute of Virology was not yet finished so Ebola experiments couldn't be done there.

The acknowledgements on the vaccine paper said that Xiangguo Qiu, Lihua Huo, and their colleague Wei Chen—Chen Wei—had conceived and designed the study. Funding came from the Public Health Agency of Canada, CanSino Biologics, and the Institute of Biotechnology in Beijing. In other words, by late 2015, when this experiment began, Xiangguo Qiu was already working with George Gao, the Wuhan Institute of Virology, and the Chinese military medical establishment at the highest level. Either the Public Health Agency of Canada approved these relationships or it just never noticed that China's leading military expert on viruses with pandemic potential was using Canada's only BSL-4 lab for China's purposes. And where was CSIS in all of this?

EVEN AFTER KEDING Cheng and Xiangguo Qiu were suspended from the NML in July 2019, they continued to co-author papers that must have involved long conversations, many emails, and data exchanges with their colleagues at the NML and in China. I wondered if they'd also been given direct access to the lab during their suspension, taken in by colleagues who still had secret clearances. In order to write the papers, they must have had access to data gathered before they were suspended. But how could that be if, as Wagener had pointed out, security clearances also cover access to data?

For example: I found one paper done by Qiu and her NML colleagues, along with a group at the NIH in Rockville, Maryland. It was published on the bioRxiv preprint server in July 2020, finally making it to *Nature Biotechnology* in February 2021. It detailed a method of using big data to seek out antivirals against SARS-CoV-2.[242] SARS-CoV-2 wasn't sequenced until January of 2020, six months after her security clearance was revoked. Yet Xiangguo Qiu worked with her NML colleagues on this paper. The first author, Ruili Huang at the NIH, refused to be interviewed about it.

I found another called "Atypical Ebola Virus Disease in a Nonhuman Primate following Monoclonal Antibody Treatment Is Associated with Glycoprotein Mutations within the Fusion Loop."[243] The paper had been submitted to *mBio* in May 2020, ten months after Xiangguo Qiu and Keding Cheng's clearances were revoked. It was accepted in November and published in January 2021. Qiu was the last author, meaning she supervised everything from experiment to publication. Nine co-authors were said to be with the NML's Special Pathogens Program, including the first author, Logan Banadyga, while Keding Cheng, also a co-author, was said to be with the NML's Science and Technology Core. Co-author Feihu Yan was said to be affiliated with both the NML's Special Pathogens Program and with the Key Laboratory of Jilin province for Zoonoses Prevention and Control, Changchun Veterinary Research Institute, Chinese Academy of Sciences, Changchun. Feihu

Yan would later be described by the *Globe and Mail* as a member of the Chinese military.[244]

But the most surprising paper done while both were under suspension was yet another with PLA Major General Chen Wei. It was called "Potent neutralizing monoclonal antibodies against Ebola virus isolated from vaccinated donors." It was published in *mAbs* online in March 2020. Scientists from the Special Pathogens Program at NML teamed up with scientists from the Laboratory of Vaccine and Antibody Engineering, of the military-connected Beijing Institute of Biotechnology. The first author, Pengei Fan, is with that institute. Co-authors from the NML's Special Pathogens Program included Logan Banadyga, Shihua He, and Xiangguo Qiu. Chen Wei and Qiu were tagged as senior co-authors, though mail was to be sent to Wei Chen (her name written in the Western style) whose name came last on the paper.[245] Her affiliation with the PLA was not mentioned.

That paper described taking blood samples from people inoculated in China in Phase 1 trials of the major general's Ebola vaccine and using their antibodies to synthesize a monoclonal therapy for Ebola-infected mice. This therapy was tested in the BSL-4 in Winnipeg. The article was submitted to the journal on January 18, 2020, as SARS-CoV-2 was having its way with the people of Wuhan. Twelve days after the article was submitted, Major General Chen Wei was sent to Wuhan, working first in a tent to generate her SARS-CoV-2 plasma treatment, then moving into the BSL-4 at the Wuhan Institute of Virology to create a SARS-CoV-2 vaccine using a cell line created in Canada and licensed to CanSino Biologics. When this paper was submitted, Xiangguo Qiu and Keding Cheng had been on suspension for six months. Did PHAC and the RCMP fail to notice that they continued to work with the leading military medical/biowarfare figure in China even as the RCMP investigated them? Or did PHAC approve this continuing collaboration even though their security clearances, and therefore access to data and the labs, had been withdrawn?

FROM TIME TO time, I called the NML's switchboard and asked for Qiu. Until January 20, 2021, I was always connected to her voice mail. But on that day I got another surprise. The NML operator said no one named Xiangguo Qiu, and no one named Keding Cheng, was in the phone system. That meant they had either quit or been let go. I called the home phone number associated with Cheng's name. It had been disconnected. I wrote an email to the RCMP media officer in Winnipeg to ask if the RCMP investigation into that "administrative" problem was ongoing. The officer wrote back to say yes. I wrote again to say Qiu and Cheng were no longer with the NML and that Cheng's phone had been disconnected. I thought they might have left the country. I asked the officer to please check with her colleagues again.

The media officer did not respond.

On January 21, I wrote to Eric Morrissette, the spokesperson for Health Canada, PHAC, and the NML. I asked him to confirm that Xiangguo Qiu and Keding Cheng had been let go. I asked to interview Mike Drebot, a senior researcher who'd been at the NML since 1999. He had answered his phone when I called, but said he couldn't talk to me without permission. I asked to speak with Logan Banadyga about the paper he'd published that very month under Qiu's supervision. The government had said that government scientists could speak to the media about their published work. I told Morrissette I wanted to know when the experiments were done, and I asked for an email address for the last author on the paper—Xiangguo Qiu. What I really wanted to know was if someone had let one or both of them into the lab and/or transferred data to them for analysis when they no longer had security clearances.

Morrissette promised to get right back to me, which I was sure meant within 24 hours, because finding out if his own department had let go of two people under RCMP investigation would surely require no time at all. The other requests were a bit trickier. Asking to speak to Logan Banadyga was roguish, but I had come to believe that no one at PHAC or Health Canada read papers published by their own staff. How else to explain experiments dating back to 2016 done at the NML and published in leading journals with a PLA major general's name on them?

The days dragged on without a response to my requests. I sent reminders. Morrissette wrote back to say I would get my answers soon, very soon.

Two weeks later, Saturday morning, February 6, 2021, I turned on the 9 a.m. CBC radio news. Karen Pauls had a story. It had been confirmed, she said, that Xiangguo Qiu and Keding Cheng had been let go from the National Microbiology Laboratory. Thanks to "sources," Pauls had learned that on Thursday, February 4, workers in the Special Pathogens Program had been called to a meeting and told that neither would be returning to the Laboratory. No explanation had been given. The rest of the report was a repeat of her earlier stories about the shipment of samples to China, except for a new bit about how Qiu and Cheng had been living comfortably while suspended. That turned out to be a reference to their other house in Winnipeg, a much bigger house.

I got out of bed and ran to my computer. I made a 50-cent bet with myself that I would find an email from Eric Morrissette that came in after I last checked my email at 10:30 p.m. Friday night. If you're a government or a corporation in trouble, Friday is the best day to release embarrassing news. It disappears in the weekend papers and might only make the second round of stories on Saturday newscasts, if it's noticed at all.

I won my bet. Morrissette had sent me an email at 10:49 p.m. Friday, about two hours before Pauls posted her story.

Morrissette's email confirmed that Qiu and Cheng had been let go on January 20, the very day I'd called to inquire about whether they were still under suspension. (And yet, as documents filed by PHAC with the Special Committee on Canada-China Relations would show, no "letters" were sent to them about being fired until January 22.[246]) He didn't respond to my request to speak to Logan Banadyga, but said the experiments detailed in the paper published in January had been done first in 2017 and that *in vivo* and *in vitro* experiments were done in 2018 and 2019—giving no specific dates in 2019. As he did not respond to my request to interview Banadyga, I'm sure some of that work was done after the suspension and that Qiu did have access to data, and possibly to the lab, while under suspension and investigation. He declined to pass on the email address for the last author on that paper—Xiangguo Qiu. He said he could not give out personal email addresses. He declined to make Mike Drebot available for an interview. He offered no explanation.

9. All About Shi Zhengli

EARLY SUMMER 2020. The first wave of the pandemic was winding down, the second not yet in sight, though I knew it would come, everyone with any knowledge of how pandemics work knew it would come. My cousins in Israel had endured a difficult lockdown. In Israel they did not refer to the virus as SARS-CoV-2, or to the disease as COVID-19. They called both "the corona." Their first wave of the corona had hit weeks before ours and subsided long before it waned in Canada. Their infection rate was higher, but the death rate lower. In Ontario and Quebec, it took the Army to clean up the abattoirs—otherwise known as privately owned long-term care facilities—where thousands of elderly people had died, either from COVID-19 or from appalling neglect, left helpless and unwashed and in unchanged diapers in their beds as bugs crawled the floors, as food rotted on trays, crying out for water and for their families. But their families had been banned from the buildings lest they bring the virus in.

It was the personal support workers (PSWs) who brought the virus in. These governments had allowed them to work in more than one facility because the companies that employed them save money by hiring them by the hour, without benefits, and giving them insufficient hours to keep their families fed, so they take on jobs in multiple locations. In a repeat of the mistakes of SARS, caregivers took the virus from one long-term care residence to another and then home to their own families, a daisy chain of disease. Personal protective equipment was in short supply in some facilities or locked away by penny-scrimping managers. Some PSWs were forced to re-use gowns and masks and gloves to the point of providing no protection at all. Some resorted to covering their clothes with garbage bags. Some were even ordered not to use the masks they paid for themselves and brought to work: how would it look to residents?[247]

In the United States it was one continuous wave of suffering and death. On May 28, 2020, the US had 30 percent of the SARS-CoV-2 infections in the world. Trump did not like wearing masks, disdained those who wore them. He held rallies where thousands attended without protection, though he had told Bob Woodward what Xi had told him, that it was bad, that it was airborne.

There was so much blame to go around that the problem was to figure out who to accuse first.

I was beginning to see patterns. But were they two-point curves or were they significant? The year 1996 kept cropping up. That seemed to be when many aspiring scientists from China arrived in the UK, the US, Australia, the Netherlands, to do their doctorates in important institutions, to learn from those at the leading edge of biological science. It should have struck me as a good pattern, a valuable development, but in the light of what I'd been reading about what had been going on at the NML, it now seemed almost sinister. Many fine scientists, such as Gary Kobinger, believe that science works best without regard to borders or national interests, that the best minds working with the best minds will always be of value to the whole world, politics be damned. Peter Daszak of the EcoHealth Alliance believes that too, or seemed to, which in part explained his determination to support Shi Zhengli and her colleagues in China against any who said this virus might have come from a lab—her lab.

The other pattern forming in my head had to do with China's official behaviour, its single-minded determination to stymie any real investigation into the origin of the virus in China. On May 18, 2020, after a vote of all its member states, the WHO had announced it would put together a joint committee of independent experts to inquire into the origin of SARS-CoV-2, starting in Wuhan. Chinese officials had been resisting such an investigation for months. They kept suggesting that SARS-CoV-2 could have come from somewhere else, anywhere but China. They did not explain why they weren't allowing outsiders to study samples from the earliest known infections, why they weren't publishing more detailed epidemiological information about the first persons known to have endured SARS-CoV-2. Instead, according to an analysis later published by the Associated Press, they were pushing out a wall of disinformation. There is an old Yiddish proverb that my paternal grandmother, the mother of seven children, used to invoke, probably because every one of her children tried to pull the wool over her eyes at some point. It loosely translates as: the thief, his head burns, meaning the thief knows what he did and works to deflect blame to others. According to a nine-month investigation by the Associated Press in concert with the Atlantic Council's Digital Forensic Research Lab, China worked hard to get spurious claims about the origin of SARS-CoV-2 circulating in Western social media like Facebook and Twitter (both banned in China) through newly established diplomatic accounts. These accounts spread evidence-free messages such as that the US Army engineered COVID-19. These claims were picked up by Chinese state media and "seeded and spread through established Kremlin proxies in the West. China, Russia and Iran also reinforced each other's messaging, cross-referencing reports and sources, deepening their echo chamber of authenticity."[248]

Which begged this question: when should journalists and scientists working in democratic countries take note of what is endured by scientists in authoritarian or totalitarian regimes and ask questions about the reliability of what they publish? Scientists are supposed to be objective, to inquire into things according to their skills and talents, and to report what they find. Don't we need to acknowledge that when a dictatorial regime asserts control over its scientists' output, that output may no longer be reliable?

How do we know that vital information discovered by scientists working in such regimes—information that might change what we think we know about the origin of a pandemic, for example—will be published at all?

I KEPT SEARCHING for information about Shi Zhengli. She is central to China's research into the SARS-like coronaviruses circulating in bats, rats, and other critters, and to the story of the pandemic's origin. She'd described the closest known viral relative to SARS-CoV-2—RaTG13—in *Nature*. She was last author on that paper, which meant she took responsibility for its contents. But she had also been last author on the paper submitted on the same day to another peer-reviewed journal, *EMI*. There she asserted that the closest sequences to SARS-CoV-2 were two described by a Chinese military group. *EMI's* editor, Shan Lu of University of Massachusetts Medical School, also published a commentary in *EMI* that same month (after a same-day turnaround for peer review), stating there is no credible evidence that SARS-CoV-2 was engineered in a lab.[249] Linda Saif, a microbiologist at Ohio State and a co-author of that *EMI* commentary, also helped Peter Daszak draft the *Lancet* statement declaring the idea of a lab leak to be a conspiracy theory.[250] She later joined *The Lancet's* task force, headed by Daszak, inquiring into the origin of SARS-CoV-2. This appeared to be co-ordinated political action by scientists in the US and China to shape the origin story of SARS-CoV-2. I was trying to figure out if Shi Zhengli was playing a political game, or if she was a straightforward scientist who'd just made a weird mistake. Was she engaged in willful deception, bending truth to serve power? I was leaning toward deception.

The story of Xiangguo Qiu and Keding Cheng was certainly about deception, a magician's trick involving the infiltration and use of Canada's National Microbiology Laboratory by members of China's medical/military leadership, performed in plain sight ever since 2014. Shi Zhengli and the unusable BSL-4 lab at the WIV were like ghosts in the NML's machine. Though Shi Zhengli's name had not appeared on Qiu and Cheng's published papers, one of her WIV colleagues had worked on two of them. Qiu had travelled to the WIV's BSL-4, of which Shi is deputy director, to help train staff, yet no acknowledgement was posted on the

WIV website about this help from an important foreign friend. Maybe Xiangguo Qiu was not considered foreign: a friend, yes, but not foreign. She was certainly important. Xiangguo Qiu performed experiments with Chinese co-authors at the NML that could not be done at the WIV. They involved the use of pathogen strains developed at the NML that Qiu then shipped to the WIV with no material transfer agreement. That freed the WIV to do what it wanted with the strains, even patenting anything the NML had failed to patent. Xiangguo Qiu was also a member of the editorial board of *Virologica Sinica*, where Shi is editor. It had published at least one of Qiu's papers.

Shi's role investigating SARS-like coronaviruses was more central to the story of the origin of the pandemic than that deceptive entanglement with the NML. For 15 years, Shi had searched out and manipulated SARS-like coronaviruses harboured by bats, testing to see which could infect humanized mice and human cells. She had noted in a published paper that one such ginned-up virus could unleash a pandemic if released from a lab.[251] Yet she presented this work as science in the service of humankind, suggesting that the generation of pathogens with pandemic potential is necessary to create a pan-coronavirus vaccine to save us from same. Shi and her colleagues downplayed the risk of gain-of-function experiments and did not refer to their dual-use potential. So this question buzzed in my head like a blowfly: did Shi believe what she published about imminent natural spillover doom from a virus harboured by a bat living too close to humans, something she said would most likely happen in China? Or was that convenient cover for militarily important dual-use experiments?

Maybe it wasn't either/or; maybe it was both.

In March 2019, eight months before the pandemic began, Shi published in a review (inserted brackets mine) that:

> China is the third [*sic*] largest territory and is also the most populous nation in the world. A vast homeland plus diverse climates bring about great biodiversity including that of bats and bat-borne viruses—most of the ICTV [International Committee on Taxonomy of Viruses] coronavirus species were named by Chinese scientists studying local bats or other mammals. The majority of the CoVs can be found in China. Moreover, most of the bat hosts of these CoVs live near humans, potentially transmitted [*sic*] viruses to humans and livestock. Chinese food culture maintains that live slaughtered animals are more nutritious, and this belief may enhance viral transmission. It is generally believed that bat-borne CoVs will re-emerge to cause the next disease outbreak. In this regard, China is a likely hotspot. The challenge is to predict when and where so that we can try our best to prevent such outbreaks.[252]

In fact, the US is the third largest territory on earth. China is the fourth. That fact is easy to check and publishing such an error suggests no one did. Sloppiness rather than deception could explain her failure to notice RaTG13 was the closest known sequence to SARS-CoV-2 until just before she submitted the paper to *Nature*. But if that's what happened, why was no correction made on the *EMI* paper? There would have been time. A correction was posted to that paper on March 5, 2020, but only to a table. RaTG13 was still not mentioned. Science publishing, like journalism, has rules: if you make a mistake in what you publish, you correct it as soon as you can so that others will not rely on something untrue.

And so, deception seemed more likely. When she wrote the paper excerpted above, Shi and her colleagues and their funder, EcoHealth Alliance in New York, had already acquired significant data that undercut it. Between 2013 and 2017, Shi worked on gain-of-function experiments with Baric's group as well as doing gain-of-function and rescue experiments in her own lab. But in the same period, Shi and several WIV colleagues, along with Peter Daszak and his EcoHealth colleagues,[253] went to the field to see if they could find evidence of spillovers among people living within five kilometres of bat caves. They took histories and blood samples from 1,497 people in three southern provinces (Yunnan, Guangxi, and Guangdong), most of whom were poor farm labourers and migrant workers. A grand total of nine (0.6 percent) showed evidence of a coronavirus infection at some point in their lives. "We did not find evidence supporting a direct relationship between bat contact and bat coronavirus seropositivity in the human population," they noted. But if spillovers are so likely, and people living near bats are most at risk, why so little evidence of infection? One explanation they offered for this low rate of spillover was that these infections have high mortality, so perhaps those infected had died, a statement for which they offered no evidence at all. This paper was published online on November 9, 2019, just before the first introduction of SARS-CoV-2 into the human population, not in Guangdong, Guangxi, or Yunnan, but many hundreds of miles away in the megalopolis of Wuhan, a city with at least two labs known to be working with bat-derived coronaviruses sampled in the field.

The same argument—that there will likely be a

pandemic caused by a coronavirus jumping from bats into a human population—underlay the now defunct $207-million USAID PREDICT program. Envisioned in 2006 after an avian flu scare and launched during the H1N1 flu pandemic of 2009, PREDICT set up surveillance relationships in 30 countries selected according to a 2008 paper that mapped viral hotspots. Yet over its ten years, the PREDICT program failed to predict any viral outbreaks at all, let alone coronavirus spillovers. When I asked Dennis Carroll, the former USAID official who got PREDICT up and running, why, for example, the organization had missed the Zika outbreak entirely though it had started slowly in the Pacific, he said it was due to financial constraints: "Zika we did not see coming at all. We shut down the Americas between 2012–2013." He offered the same financial-constraints explanation as to why PREDICT failed to warn of the 2014–2016 Ebola epidemic in West Africa. But PREDICT had spent lots of money in China. PREDICT helped fund the swap of spikes/gain-of-function experiments Shi did in her own lab,[254] the 2015 paper with Baric,[255] as well as the study on bat spillover infections Shi did with Daszak. It is interesting that only American institutions—USAID and NIH/NIAID—put up the money to get that spillover paper done. Not a penny came from China's funding agencies. That suggested to me that China's science leaders were not concerned about bats infecting people impinging on their habitats in China. By contrast, the Chinese science funding agencies did help pay for Shi's swap of spikes/gain-of-function experiments. Such experiments have dual-use potential.

ONE DAY, I downloaded Shi's picture from the WIV's website and searched it for clues, character clues. All journalists and non-fiction authors make judgments about whether they should trust the people they write about. I don't do it by any method known to science. I meet people, ask about their lives, listen for what they don't say as much as what they do, watch how they treat their colleagues, their families, and strangers. Some sort of process goes on in some dark recess of my brain, and I come to a conclusion about who I can trust and who I can't. Sometimes I'm right, sometimes I'm wrong. But I could not do any of that with Shi. She wouldn't, or couldn't, speak to me. As with trying to discern Xiangguo Qiu's character by watching her talk on a video and looking at a picture of one of her houses (which gave no hint she is a real estate maven), I was stuck with staring at Shi's picture.

The photo was of a very young woman, middle thirties at the most. She looked the way so many young female scientists do—enthusiastic and forthright. Her short black slightly wavy hair and bangs were parted in the middle. There was a tiny gap between her front teeth, revealed by a half-smile. She wore light pink lipstick and a beige turtleneck but no jewelry. I wanted to trust the young woman in that picture, to forgive her errors.

But the brief biography on the WIV website made it clear that this image was seriously out of date. Shi is at least twenty years past her thirties. A more recent picture that accompanied several news stories about the WIV showed her in a blue pressure suit in the lab. You can see her head inside its clear plastic bubble. Her chin is tucked tight to her neck in a determined, let's-get-to-it manner. She doesn't look sweet, she looks formidable.

She graduated with a BSc in genetics from Wuhan University in 1987, so she was probably born in or around 1966, of the same generation as Xiangguo Qiu. Did she have a partner? Children? Siblings? Elderly parents? The biography on the website did not say. In 1990, the same year Xiangguo Qiu earned her MSC in immunology, Shi earned hers in virology from the Wuhan Institute of Virology. Like Xiangguo Qiu, Shi was in graduate school in China as revolution boiled over once more. In 1989, students and working people in Beijing and Shanghai wrote big-character protest posters and hung them everywhere, just as they had during the Red Guard period, but these called for a very different revolution involving free speech, a free press, democracy, and an end to corruption. These aspirations were soon crushed by PLA tanks in Tiananmen Square.

Did Shi take part? Was she interested in politics? Shi's public profile on ORCID shows that she got a job at the WIV as a research intern in 1990, as soon as she had her master's. In 1993 she rose to research assistant. In 1996 (the year of Xiangguo Qiu and Keding Cheng's first publication with Talpaz's lab at University of Texas/MD Anderson) she went to Montpellier 2 in France, a university in the upper rank of French schools, for her PhD. It is unlikely that a Chinese state institution would support and promote a student who got involved in unacceptable political action.

By her fruits you will know her, I found myself thinking. Forget her pictures, forget her CV, examine what she's published.

The list of her publications on the Wuhan Institute's site begins in 2005, five years after she got her PhD. At first, I thought either she had published nothing between 2000 and 2005 or she did not want what she had written about to be known. When I searched Google Scholar it was clear she had published in that period, but mainly in Chinese journals, including *Virologica Sinica*, and not on coronaviruses. The first paper on the WIV website was the 2005 *Science* article that identified bats as the probable reservoir of SARS. She was not the first author. She was not the last author. She was in the middle among many others, including Peter Daszak and a group from Australia. Last-author honours went to Linfa Wang, still working in Australia at that time. By placing this paper first on her list, in effect Shi was saying Linfa Wang was the mentor who made her career. Without Linfa Wang, that paper never would have happened, and Shi would not have become the Bat Woman.

AFTER SOME BACK and forth by email, Linfa Wang agreed to speak with me on the phone. He explained that he was in the first group of Chinese students to benefit from Deng's reinstatement of the national university entrance examinations. He had to compete for a place against his own high-school teachers, people whose education had been on hold since 1966. His family story is like a miniaturized version of China's remarkable climb from deepest poverty and civil war to major power. His family comes from Shanghai. His father went to work at age seven after only one year of education. His grandfather was "a chef for rich people," as Wang put it. His mother, who came from a village near Shanghai, got no education at all. His family somehow survived the famine that followed the Great Leap Forward, and unlike his parents, Wang and his siblings got to go to school, but mostly what he was taught in high school came from Mao's quotations, the little red book. That's why, though he aspired to be an electrical engineer, he didn't do well enough in math on the university entrance exams and was sent to study biology at East Normal University in

Shanghai instead. He wasn't interested in biology, but he worked hard. When he graduated in 1982 as the highest achieving biology student in the country, the state decided to send him abroad for graduate school. But he won a full scholarship to do his doctorate at University of California, Davis, thereby, as he put it, saving the state a lot of money.

China had embarked by then on a strategy of sending its most talented students to the West to study. How else to catch up after so many lost years? The problem was how to get these students to come home and build science in China after having experienced the freedom to pursue their own ideas in places like the University of California, Harvard, Princeton, Oxford, the University of Texas, and the University of Manitoba. The government of China fixed on status as the lure, which proved to be no lure at all, and later, money. As soon as Wang got his PhD, he was made a full professor at East Normal University in Shanghai and provided with a lab and students. But a "friend" sent his resume to schools in Australia, and by 1989 he and his wife were both working there, glad to be away from the turmoil in China. He worked first at Monash University, then he got a job at CSIRO (Commonwealth Scientific and Industrial Research Organization), equivalent to Canada's National Research Council, in the Animal Health Laboratory, which contains Australia's BSL-4, in Geelong. He became an Australian citizen just as fast as he could because travelling on a Chinese passport was difficult: he'd have to wait months for visas that sometimes would not be forthcoming. He mastered gene sequencing techniques which were then still extremely finicky and time consuming. In the 1990s he sequenced the newly emerged Hendra and Nipah viruses and helped determine that they had originated in fruit bats, jumping to humans through intermediary animals—horses and pigs.

In 2003, Wang was invited by his boss, John McKenzie, to be part of an international group of scientists tasked by the WHO to try and determine SARS's origin. When he arrived in Beijing, China was sufficiently chagrined by the well-earned international opprobrium over the way it had handled the pandemic that the group could go anywhere they wanted, even to military hospitals, he explained. But when he asked leading figures in Beijing if they had considered bats as the possible reservoir for SARS, he got the brush-off. By contrast, in Wuhan at the WIV, he found interest from Zhihong Hu, then the WIV's director general, who had completed her PhD at Wageningen in the Netherlands in 1998, two years before Shi finished at Montpellier 2.

The research that led to the 2005 paper (crawling through caves full of bats, trapping them and developing a method to screen them for evidence of SARS-like viruses) was funded by Chinese, European Commission, Australian, and American grant money. Peter Daszak, then the executive director of the Consortium for Conservation Medicine in the US, got involved because he'd worked on projects in Australia and was well-known to Wang and his colleagues. In other words, it was Wang who brought Western scientists and those at the WIV together. And it was Wang who got that first, vitally important paper into *Science* after *Nature* turned it down.[256]

He left Australia after he was offered a major lab of his own in Singapore, but he continues to teach students in Shanghai. He and his two siblings, who also earned postgraduate degrees, have travelled an almost unimaginable distance from their parents' experiences. As he told me this story, I could hear so many conflicting emotions in his voice—resentment and frustration at having his life's course determined by an autocracy with no interest in what he might have wanted, pride in his personal achievements despite that, along with the desire to help lift others in the place he used to call home. When Wang spoke of his father, he choked up. When he spoke of his mother, and the ostracism she endured due to a parasite that prevented her from getting pregnant until a doctor in Shanghai helped her, he cried.

AFTER THAT 2005 *Science* paper, Shi's career shifted from studying the genetics of infectious disease in shrimp and crab to hunting all over China for bats that might be harbouring something like SARS. Year after year, Shi and/or her students ventured into beautiful caves and horrible caves, looking for bats, trapping bats, noting their species distributions, noting the seasonal variation in rates of infection, sampling their feces and urine, taking their blood, trying to isolate viruses or at least to capture sufficient traces of viral genetic information to re-create full genome sequences. A complete genome sequence is more than the information from which a virion will be assembled by the cell it infects. It is also a history of past recombination and mutation and reveals evolutionary relationships to viruses of the same species whose sequences are already known. As China's determination to be top dog in biological science led to ever larger grants on offer, and as PREDICT and NIH/NIAID, through EcoHealth Alliance, provided more money, new papers rolled out of Shi's lab and into leading Western journals, "high impact" journals.

Getting published in leading journals is a big thing for

any scientist. In the West, it supports a scientist's climb toward tenure and brings business opportunities. In China there are also monetary rewards. Cash gifts are given by Chinese science institutions to their researchers when they publish an article in a major journal. Journals are ranked, and the size of the cash award reflects that ranking. A publication in top journals like *Nature* or *Science* earns top dollar, an average award of $44,000 US but sometimes as high as $165,000 US.[257] This policy has been associated with a huge increase in the number of mistakes made by Chinese authors, mistakes that required corrections after their work was published. The number of corrections grew from two in 1996 to 1,234 in 2016. Did scientists suddenly get sloppy? Or did they craft submissions to be attractive to better journals, the results spun to look more interesting? Publication in big journals is a way of earning a much better income. As one source told me, leading scientists at the WIV live comfortably. Some have villas in gated communities as well as nice apartments in Wuhan.

In addition to cash awards, there's the acclaim. Chinese media have published many laudatory stories about Shi—the Bat Woman.

I kept watch for articles Shi published after those January/February papers on SARS-CoV-2. At first, there weren't any. Then there were a few evaluating therapies that might stop the virus—plasma, remdesivir, chloroquine. She also contended publicly with the rumours that her lab was the Pandora's box—the point of origin for the pandemic. She responded via a social media post in early February which, according to the *Wall Street Journal*, showed up in Wuhan's "main Communist Party newspaper." She said she could "guarantee with my life" that the virus had not "originated" in her lab. She further advised "those who believe and spread malicious media rumours to close their stinky mouths."[258]

Normally, leading scientists don't give advice concerning stinky mouths, at least not in public. The phrase smacked of Party speak from the Red Guard days. Some wondered if the Party wrote it for her. Others wondered if she'd run away. According to the *South China Morning Post*, a story began to circulate that Shi and her family had smuggled out hundreds of "confidential documents" and fled to Paris, where she sought asylum at the US Embassy.[259] This Shi denied on WeChat. "No matter how difficult things are, there will not be a 'defector' situation as the rumours had said," Shi wrote. The *South China Morning Post* said *Global Times*, a paper close to the Party, "confirmed that post had been written by Shi," again signalling some thought the Party wrote responses for her.

I found a study she embarked on with colleagues in the Wuhan Ministry of Agriculture and Huazhong Agricultural University. She began work on it in January, even as she submitted the paper to *Nature* on SARS-CoV-2 that declared its closest relative to be RaTG13. By then, Shi had spent years chasing bats to see if they harboured SARS-like viruses, and yet the first animals Bat Woman and her colleagues investigated as possible reservoirs of SARS-CoV-2 were not Wuhan bats, but Wuhan *cats*. This work was done while Wuhan was in the strictest lockdown, suggesting Shi considered it to be extremely important. The resulting paper appeared on April 3 on the bioRxiv site. Shi was credited with helping conceive and design it. It was called "SARS-CoV-2 neutralizing serum antibodies in cats: a serological investigation."[260] Eventually it appeared in *EMI*, the same journal that published the article that failed to mention RaTG13.

I was surprised she was so interested in cats as a possible reservoir of SARS-CoV-2. The *Science* paper of 2005 had found antibodies to SARS-like infections in 71 percent of horseshoe bats captured in Yichang, Hubei, only 286 kilometres from Wuhan. There are bats in Wuhan. It is built on karst (limestone), ideal geology for bat caves. If she thought bats harboured SARS-like viruses, why not go sample the local bats first?

The co-authors declared they did not think the seafood market was the source of the outbreak, which was clear from the early epidemiological reports. They argued that previous studies had shown that ACE2 receptors in ferrets and cats make them susceptible to SARS, so they reasoned both might also be susceptible to SARS-CoV-2. They tested blood samples taken from 102 Wuhan cats after the outbreak began. They also tested 39 blood samples taken from cats between March and May of 2019, long before the outbreak was thought to have begun. None of the samples taken pre-outbreak were positive. But 15 percent of cats sampled after the outbreak were positive for the receptor binding domain of SARS-CoV-2, and 11 of those cats had neutralizing antibodies to SARS-CoV-2, suggesting that they'd been infected and fought it off. The cats with the highest levels of antibodies were pets of patients who had COVID-19, suggesting to the authors that humans had infected their pets, although the data could also be interpreted the other way around, that pets infected their owners. Some stray cats in shelters were also positive. How to explain that? A group in Harbin, where a new BSL-4 would soon open, had shown that SARS-CoV-2 can be transmitted between cats by respiratory droplets. Shi and her coauthors proposed that the stray cats got it "due to the contact with SARS-CoV-2 polluted environment, or COVID-19 patients who fed the cats." In other words: humans must have given it to cats.

It seemed to me that indirectly, this paper advanced the lab-leak possibility. If the wet market was not the source of the outbreak, if bats in Wuhan were not considered to be possible sources of the virus, if humans were deemed to be the source of the cats' infections, not the other way around, and if at least two labs in Wuhan worked with coronaviruses sampled from bats in the field, it made the lab-leak thesis more likely, not less. Where else could the virus have come from?

IN LATE FEBRUARY, while the cat paper was still in the works, Shi was interviewed on the phone by writer Jane Qiu for a story in the June 2020 edition of *Scientific American*. It appeared online on March 11, a week before the proximal origin letter appeared in *Nature*. Since 1845, *Scientific American*, based in New York, has presented interesting articles to the public written by scientists describing their own work. Lately articles have also been written by journalists. The chance to have *Scientific American* do a friendly profile of Shi (most *Scientific American* profiles are friendly) must have seemed like a gift to those in China trying to shape the narrative on the origin of SARS-CoV-2. A nice article in a journal like *Scientific American* could help bury the argument that the pandemic might have originated in Wuhan. I could imagine the PR people on a Zoom call saying "How could it hurt?"

The author, Jane Qiu, is a freelance science journalist with a PhD in cancer genetics. She is based in Beijing. According to her listing on Muck Rack, she has written for *Nature*, where she spent two years as an editor, and also for the general readers of magazines like *National Geographic* and *Forbes*. Her article on Shi was true to *National Geographic*'s preferred story form. It was organized around dramatic moments and utterly free of investigative techniques, avoiding mention of the controversies surrounding Shi's gain-of-function experiments or that the US government was the source of some of Shi's funding. Instead, Shi was presented as a pandemic hero, as Bat Woman to the rescue. Her direct collaborators (save one) were the only other scientists quoted about her work. They included Wang, Baric, and Daszak, with Daszak at the greatest length.

The piece opened with Shi getting a call from her boss on December 30, 2019, while she was at a conference in Shanghai. She was told to drop everything and get right back to Wuhan because the Wuhan Center for Disease Control and Prevention had found a new coronavirus, something suspiciously like SARS, in the lungs of two patients with pneumonia. Shi was told to "deal with it now."

Shi told Qiu that as she rode home to Wuhan on a fast train, she thought there might have been a mistake. "I had never expected this kind of thing to happen in Wuhan, in central China," she said. Her studies, according to Qiu, had shown "that the southern subtropical provinces of Guangdong, Guangxi and Yunnan have the greatest risk of coronaviruses jumping to humans from animals—particularly bats, a known reservoir." Qiu apparently did not challenge that statement, though if she'd read the 2005 *Science* paper, she should have. In that paper, the greatest number of bats found with traces of SARS-like viruses were in Hubei province, not in Yunnan, Guangdong, or Guangxi. Shi told Qiu that she wondered if coronaviruses were the issue: "Could they have come from our lab?"

That statement would come back to haunt Shi. If her first thought was that the pneumonia might have been caused by a coronavirus studied in her lab, others who raised the same question could hardly be labeled conspiracy theorists. And yet Daszak and other colleagues had done exactly that. Her statement also made one wonder: what exactly had Shi been working on that she thought might infect people if it escaped containment?

The rest of the story described how Shi became Bat Woman, how she teamed with Daszak and Wang and others in 2004 to find the origin of SARS and spent five years sampling bats in one cave which showed that bats can harbour many viruses and not be made ill. Though Qiu spoke with Ralph Baric, she did not mention that Shi had isolated some of those coronaviruses and that Baric's lab had made chimeras from them which infected human cells and humanized mice and did not respond to any known therapeutics. The phrase *pathogens of pandemic potential* never appeared in this story.

Qiu moved on to discuss another cave Shi worked in, a mineshaft Shi said she visited in 2012 in Yunnan's Mojiang County, "where six miners suffered from pneumonialike diseases and two died." After sampling bats and their feces in the cave for a year, Qiu wrote, Shi's team discovered a diverse group of coronaviruses in six bat species. "In many cases, multiple viral strains had infected a single animal, turning it into a flying factory for new viruses."

> "The mine shaft stunk like hell," says Shi, who, like her colleagues, went in wearing a protective mask and clothing. "Bat guano, covered in fungus, littered the cave." Although the fungus turned out to be the

> pathogen that had sickened the miners, she says it would have been only a matter of time before they caught the coronaviruses if the mine had not been promptly shut.[261]

The mine may have been shut to miners, but it would soon become a home away from home for several scientific groups—including George Gao's—interested in the viruses the bats might be harbouring. Shi's lab would sample there many times. It would eventually be revealed that this mine was the source of the RaTG13 sequence, but in *Scientific American* no such connection was made, which would also come back to haunt Shi. Her statement that the miners died (three died, not two) from a fungal infection would be shown to be untrue. Shi would be forced to send an "addendum" (which should have been labeled a correction) to be attached to the February *Nature* article about SARS-CoV-2 which named RaTG13 as its closest relative, to set the record straight. That addendum would give rise to more questions still.[262]

Qiu reverted to the Bat Woman-to-the-Rescue drama. She described how Shi and her team isolated the virus from the patient samples sent to her lab, then sequenced the genome and sent samples out for independent confirmation, which they had by January 7. As she waited, Shi told Qiu, she frantically searched through her lab's sequence data to see if anything they'd been studying matched what would soon be called SARS-CoV-2. She also went through her lab notes to make sure nothing could have been mishandled, especially during waste disposal. (This should have poked Qiu to ask: if you were so sure nothing matched, why would you also check for mishandled waste disposal? But this was a friendly story.) Not until the new viral sequence was confirmed by the outside agency, Shi told Qiu, was she finally able to relax, because:

> ...none of the sequences matched those of the viruses her team had sampled from bat caves. "That really took a load off my mind," she says. "I had not slept a wink for days."[263]

After the piece was published online, Qiu called Shi again to do an update. By then, the *Daily Mail* had published the first of several pieces pointing at the WIV as a possible source of the pandemic.[264] Members of the British government were said to be no longer sure that the virus had not leaked from a lab.[265] The *Daily Mail* also mentioned that the NIH/NIAID had granted $3.7 million over five years to EcoHealth Alliance to study viruses in bats, some of which was to go to the WIV.[266]

To her original story, Qiu added:

> Back in Wuhan, where the lockdown was finally lifted on April 8, China's bat woman is not in a celebratory mood. She is distressed because stories from the Internet and major media have repeated a tenuous suggestion that SARS-CoV-2 accidentally leaked from her lab—despite the fact that its genetic sequences do not match any her lab had previously studied. Other scientists are quick to dismiss the allegation. "Shi leads a world-class lab of the highest standards," Daszak says.[267]

By then, Daszak's praise was less than helpful: his competing interest was obvious. By then too, some of the facts in this friendly story in *Scientific American* had spurred journalists, scientists, and other interested parties to an orgy of fact-checking.

On April 14, Josh Rogin of the *Washington Post* published an opinion piece about the 2018 memo US officials wrote to the State Department after they were told that the BSL-4 lab at the WIV needed more trained technicians and wanted US help.[268] On April 17, President Trump was asked at a press conference if he knew that the US had been funding the lab in Wuhan from which the virus might have leaked. Did he intend to continue? Trump said no. "We will end that grant very quickly." On April 24, the NIH informed EcoHealth Alliance that its recently re-approved $3.7-million grant to study emerging diseases (some of which was to go to Shi, who intended to do experiments involving chimeras that might have pandemic potential), had been cancelled.[269] Daszak raised a storm in the press and on Twitter. A group of 77 Nobel prize winners was moved to sign and send a public letter to Francis Collins, director of the NIH, decrying this awful precedent—this vile politicization of the independent grant process and therefore of science—saying the explanation for the grant withdrawal was "preposterous."[270] Thirty-one science organizations sent a similar letter asking for a review.

The grant was restored, technically, but Daszak claimed that NIAID sent EcoHealth Alliance a letter saying it would get no money until, among other things, EcoHealth Alliance got early samples of SARS-CoV-2 from the WIV and permission for American scientists to go to the WIV and examine Shi's lab notes. More outrage followed. But journalists were not as willing to lend a sympathetic ear as they had been over the *Lancet* statement. Some, like me, began to ask exactly how much taxpayer money Daszak's charity had pulled in and passed on to Shi. What had she done with it? Had the US funded the creation of SARS-CoV-2?

AT SOME POINT not revealed, Jon Cohen of *Science* submitted a list of written questions to Shi Zhengli. He received answers by email on July 15. On July 31, *Science* published them in the form of a story about the story and a Q and A devised from his email and her responses. Unfortunately, *Science* had not noticed Shi's publication of two articles with conflicting statements about the closest known sequence to SARS-CoV-2 submitted on the same day. There were no questions about that. *Science* was more concerned about the origin of RaTG13 and its relationship to a short gene sequence called 4991.

Cohen's explanatory story was titled "Wuhan coronavirus hunter Shi Zhengli speaks out."[271] He tried to be even-handed about the international tensions over the pandemic. He pointed out that President Trump kept calling SARS-CoV-2 the "China virus," but that China said it probably came from the US; that Trump had cut off that grant to EcoHealth Alliance, but that China was exercising tight oversight over any articles about the virus's origin. Cohen mentioned as well that the WHO had finally been allowed to send two people to China to negotiate the terms under which an independent committee would investigate the virus's origin. (The terms negotiated would turn out to be extraordinarily China-friendly.)[272]

Cohen explained some of the controversy surrounding

RaTG13. A gene of about 400 base pairs in length is found in all beta coronaviruses like SARS and SARS-CoV-2. The full name for this conserved segment is the RNA-dependent RNA polymerase gene, usually referred to as RdRp. Finding this RdRp sequence in a bat or in its droppings demonstrates infection with a beta coronavirus. An RdRp labelled 4991 had been discovered in a fecal pellet in the same bat-ridden mine/cave Shi had described to Qiu in *Scientific American*. Shi's group had referred to this RdRp sequence in a 2016 paper (labelled there as RaBTCoV4991 to signify that it was from a horseshoe bat of the *Rhinolophus affinis* species). According to that paper, the sequence had been submitted to GenBank, an open-source site used by biologists worldwide to compare sequences. By inputting any genome sequence into a software program called BLAST, it can be compared to other similar genomes. A controversy had arisen because some researchers had discovered that the RdRp labelled 4991 is identical to the RdRp of the virus closest to SARS-CoV-2—RaTG13—which surely meant that Shi had sequenced RaTG13 years earlier, that perhaps it *was* the ancestor of SARS-CoV-2. This history had not been mentioned in Shi's *Nature* publication.

Cohen's story explained that Shi's lab had not tried immediately to get a full genome sequence from the sample containing the RdRp 4991 because it was distant from the RdRp of the original SARS. The whole genome labelled RaTG13 had only been sequenced recently. RaTG13 had never been isolated and grown as a live virus, so no one could have been infected by it, giving rise eventually to SARS-CoV-2. (Isolation of a virus means the virus is extracted from a sample, put into culture where it grows and replicates. That is hard to do, worthy of a *Nature* publication.)

Wait a minute, I had found myself muttering. Shi didn't have to isolate RaTG13 to experiment with it: she could have rescued it from its genome sequence alone, then passaged it in human cells and humanized mice sufficiently for it to become SARS-CoV-2. As David Evans had shown with horsepox, and as the group in Switzerland had shown with SARS-CoV-2, if you know the whole sequence, you can order it in small fragments from a commercial supplier, and after a series of moves too complex to list here, a live virus will emerge. Shi certainly knew how to do that. A 2017 paper from her lab described how her team assembled 11 full genomes from samples gathered in another cave in Yunnan. They'd determined the sequences of their spike proteins. They'd made several chimeras by joining different spike sequences to the "backbone" of a SARS-related virus Shi had isolated earlier called WIV-1. Two of those chimeras had replicated very nicely in cells (Vero cells and HeLa cells) expressing the ACE2 receptor.[273] And why would she be interested only in beta coronaviruses with RdRps close to the original SARS? If she was worried about coronaviruses circulating in bats that might infect humans, she should have been looking for any coronavirus that could exploit the ACE2 receptor, never mind its propinquity to SARS. And come to think of it, why wasn't Daszak's name on that 2016 paper that described the RdRp called 4991? The acknowledgements said it had been funded in part from a previous five-year, $3.7-million grant from NIAID to EcoHealth Alliance.

So, I read the Q and A that followed with a skeptic's eye. Shi avoided answering some questions entirely. She used the same stock answer for others. She even pushed China's line that perhaps the virus had originated somewhere other than China. She said she thought the virus came from a bat, but probably jumped first to an intermediary animal, then to humans, and not necessarily in Wuhan or Hubei. She said Hubei had been surveyed for years but no one had found "that bats in Wuhan or even the wider Hubei Province carry any coronaviruses that are closely related to SARS-CoV-2. I don't think the spillover from bats to humans occurred in Wuhan or in Hubei Province."[274]

Cohen asked if it was possible that infected people who live near mines became the index cases and travelled to Wuhan.

"I guess you are referring to the bat cave in Tongguan town in Mojiang county of Yunnan Province," she wrote. "To date, none of nearby residents is infected with coronaviruses. Thus the claim that the so-called 'patient zero' was living near the mining area and then went to Wuhan is false."

She dodged a question about whether anyone had sampled animals in the Huanan Seafood Market looking for SARS-CoV-2. She asserted that no SARS-CoV-2 had been discovered in farm animals in Hubei but offered no information about the scale of that search. She ducked a question on when the first case was discovered. She asserted that the history of pandemics is such that their origin is rarely where the first cases are observed, citing HIV as an example. As to the distance in evolutionary time between RaTG13 and SARS-CoV-2, she relied on E.C. Holmes' estimation of twenty to fifty years. She said she did not know whether veterinarians had been contacted about viral illnesses in animals. When asked about the cat study, she repeated that she thought humans infected the cats but did not explain why she believed that, and said she was doing no more work on cats without explaining that either. She was asked about Chinese studies on the origin of SARS-CoV-2, and answered that studies were going on but did not say where, how, or by whom.

How many bat samples in all did your lab collect and test? Do some remain untested for coronaviruses? Cohen had asked.

All had been tested, she answered, and 2,007 had tested positive for coronavirus. "We did not find any viruses whose gene sequence is more similar to SARS-CoV-2 than RaTG13."

And yet further along she contradicted that statement. Cohen had asked why she only had RdRp sequences for some samples and full sequences for others. At that point she replied: "Due to financial and manpower constraints, it is impossible for us to do the whole genome sequencing of all samples. We hope to conduct further full-length coronavirus genome sequencing in some other samples within the next two years...."

But that raised the question, how then did she know RaTG13 was the closest in her sample collection to SARS-CoV-2?

Cohen asked, could SARS-CoV-2 have leaked from your lab? President Trump, he told her, had said that he had "high confidence" that the virus came from the lab. Then he threw her a softball: he asked what impact Trump's statement had on the lab and on her personally.

She said no one in the lab knew this virus until it arrived in patient samples, and scientists around the world "overwhelmingly concluded that SARS-CoV-2 originated naturally rather than from any institution.... He owes us an apology."

Cohen pointed out that some scientists believe it could have "existed" in the lab and "accidentally infected a lab worker." He wanted to know how she could rule out such a possibility.

That produced the longest answer of all. Her lab, she wrote, had only isolated three viruses, all SARS-related coronaviruses. None are as close as RaTG13, which differs from SARS-CoV-2 by 1,100 nucleotides. She noted "five renowned virologists" [Andersen et al.] said its spike "diverges in the receptor binding domain." Further, as Holmes had pointed out, it would have taken "a very long time to accumulate sufficient numbers of mutations through natural evolution.... Therefore, RaTG13 evolving into SARS-CoV-2 is only theoretically possible."

All of which was true but did not really answer the question. Isolation was beside the point: she had the means to make a virus from its sequence alone, and by passaging it in human cells or humanized lab animals she could speed up evolution without leaving a trace. The same was true for other labs in China, specifically the military group in the Nanjing command who had published the next closest sequences and infected mice with one of them. Shi turned instead to the strict management of the BSL-3 and BSL-4 labs at the WIV. "To date, no pathogen leaks or personnel infection accidents have occurred."

Are bat coronaviruses grown in the institute? Cohen had asked.

Again, she said only three had been isolated, once more ducking the real question.

Do you do the sequencing or does someone else?

She said the sequencing "was done mostly in Wuhan." Again, a non-answer to a straightforward question. Wuhan is a big city with biotechnology companies that do sequencing.

Are any animal experiments done in the lab with SARS-related viruses? he'd asked.

She said yes. In 2018 and 2019 they had performed experiments with humanized mice and civets; the mice had been altered to express human ACE2. These experiments had been done in the "biosafety laboratory." She said the experiments were done with bat viruses "close to SARS-CoV." They showed that civets could be directly infected. They showed "low pathogenicity" in mice, none in civets. But the papers describing those experiments had not yet been prepared for publication.

Could someone with the institute have become infected while collecting or sampling bats? Cohen had asked.

"Such a possibility did not exist," she wrote. Sera had been tested of all staff and students, and the result was "zero infection."

Unfortunately, Cohen hadn't asked about former staff or students, which was what the rumours were about.

When had she finally sequenced the whole genome known as RaTG13?

She wrote that it was done in 2018. When new generation sequencing technology became available (very rapid and cheaper) and

> capability in our lab was improved, we did further sequencing of the virus using our remaining samples, and obtained the full-length genome sequence of RaTG13.... As the sample was used many times for the purpose of viral nucleic acid extraction, there was no more sample after we finished genome sequencing, and we did not do virus isolation and other studies on it. Among all the bat samples we

> collected, the RaTG13 virus was detected in only one single sample....

I read and reread that response. I was struck by the internal inconsistency. First she spoke of "further sequencing of the virus using our remaining samples," then she referred to "the sample" and how sequencing used it up. So was there one sample, or more than one?

Cohen asked, "Some people who suspect a lab accident...have suggested that BtCoV/4991, a bat virus you described in 2016, is SARS-CoV-2.... Why did you rename the virus?"

She explained that "Ra4991 is the ID for a bat sample while RaTG13 is the ID for the coronavirus detected in the sample. We changed the name as we wanted it to reflect the time and location for the sample collection. 13 means it was collected in 2013 and TG is the abbreviation of Tongguan town, the location where the sample was collected."

Again, this missed the point. There is supposed to be consistency in naming. The number 4991 referred to the RdRp of a sequence, not just to the bat it came from.

He asked if she had done or collaborated on any gain-of-function experiments with coronaviruses that had not been published. He was trying to get her to tell him if she had unpublished sequences that she was doing gain-of-function experiments with. He was trying to get at the synthesis question.

She wrote no. But RaTG13 *had* been published, so this answer did not exclude that it had been used to synthesize a virus for gain-of-function experiments.

Since coronavirus experiments are usually done in BSL-2 labs and the BSL-4 wasn't operational until "recently," he asked, why had she done coronavirus experiments under BSL-4 conditions?

She wrote: "The coronavirus research in our laboratory is conducted in BSL-2 or BSL-3 laboratories." However, they had also used "low-pathogenic coronaviruses as model viruses" to train researchers on how to safely work in a BSL-4. After the pandemic began, the government had decreed that any experiments using animals and SARS-CoV-2 had to be done in a BSL-3. But the WIV's BSL-3 lab didn't have the right equipment to accommodate work with nonhuman primates (such as rhesus macaques) so that work was being done in the BSL-4.

In other words, the experiments published in 2017 or done in 2018—gain-of-function experiments using coronaviruses in human cells or humanized mice—might have been done in a BSL-2, the kind of lab that uses a hood to draw air away from a researcher, the kind of lab found in most university biology labs, the most minimal form of containment.

Toward the end, Cohen asked if there were any questions he had failed to ask. She wrote that she wanted the NIH grant reinstated, mentioning her close collaboration with Peter Daszak, and that the NIH's retraction of the Eco-Health Alliance grant was irrational. She also wanted to make an appeal for international cooperation among scientists "on research into the origins of emerging viruses. I hope scientists around the world can stand together and work together...."

Cohen got comment on Shi's responses from the usual suspects—Kristian Andersen, Edward Holmes, Richard Ebright, Peter Daszak. Holmes thought her answers made sense, that "the penny dropped" when he realized she'd failed to sequence RaTG13 for all those years because she was looking for SARS-like viruses and the RdRp portion was quite different from the same gene in SARS. Andersen thought her answers were logical, if "carefully vetted."

Daszak described Shi as "social, open, and something of a goodwill ambassador for China at meetings, where she converses in both French and English.... What I really like about Zhengli," he said, "is that she is frank and honest and that just makes it easier to solve problems."

Richard Ebright, on the other hand, thought her answers were "formulaic, almost robotic, reiterations of statements previously made by Chinese authorities and state media."

I thought the responses were anything but robotic. I thought they were both deceptive and sloppy, as if drafted quickly by a committee with conflicting PR goals. Her claim that her lab—working for 15 years to predict spill-overs of dangerous coronaviruses from bats—had waited six years to fully sequence a sample taken from a mine where six miners became sick and three died seemed extremely unlikely. And why, if she got that full sequence in 2018, didn't she publish it until 2020?

TEN DAYS AFTER the *Science* Q and A appeared, NBC aired an exclusive interview with the director general of the WIV, Wang Yanyi, and her colleague Yuan Zhiming, identified as the director of the WIV's Biosafety Laboratory. NBC got five hours to tour the labs and ask questions. Wang, a small, graceful woman, sat immobile in a chair facing NBC's inter-viewer, Janice Mackey Frayer. Wang wore no mask; her hands were clasped in her lap and never moved as she fielded questions about whether the virus could have escaped from the WIV's BSL-4. But the BSL-4 lab was a red herring. In fact, until after the pandemic began, as Shi told *Science*, coronaviruses had not been studied in the BSL-4 but in lesser containment, in the BSL-2 or BSL-3 labs. Wang was intent on undermining the State Department memo described by the *Washington Post*'s Josh Rogin, the one that referred to a WIV official who said the WIV needed more trained technicians to run the BSL-4 safely. Wang insisted US officials had only come to the lab once, on a day after that memo was drafted, a statement meant to imply that the memo was bogus.

The State Department memo Rogin saw was written after a conversation with Shi at an academic conference in 2017. US officials who attended heard Shi's talk on how her lab had made infectious SARS-like coronavirus chimeras and put them in civets and humanized mice to see what would happen. Shi had explained to the US officials that there were difficulties getting sufficient trained staff to operate the BSL-4 safely, and that the required permission had not yet been granted for the WIV to import pathogens like Ebola. Her experiments had sounded risky to the officials, which was why the first memo was sent to the State Department. A second was written after US officials got a tour of the WIV in March 2018, in part because they were concerned about the possibility of leaks.[275]

Wang's colleague Yuan told NBC that none of the WIV staff had tested positive for SARS-CoV-2 or had antibodies, trying once more to quell the internet buzz about that former WIV employee, thought to be the elusive patient zero.

NBC did not fully identify Wang and Yuan. They are not only scientists working at the WIV, but also Communist Party officials. In addition to running the WIV since 2018,

Wang was also deputy director of the Wuhan Municipal Party Committee. Yuan was the president of the Chinese Communist Party Committee within the Wuhan Branch of the Chinese Academy of Sciences to which the WIV belongs.[276]

THE *SCIENTIFIC AMERICAN* story about Shi Zhengli, the emailed answers to the questions from *Science*, and the NBC interview with Party people but no Shi were clearly part of a Chinese government disaster-PR offensive, an attempt to rewrite the pandemic narrative from blame China to praise China. They had their work cut out for them: the *Scientific American* story with Shi worrying right off the top that the pandemic might have resulted from a leak from her lab was referenced in most articles about the virus's origin that followed. Chinese authorities also had to worry about what was to come—a damning final report into the origin of the pandemic from the minority staff of the US House Foreign Affairs committee. While unable to say exactly where the virus came from, the minority staff report compared the Chinese government's behaviour during SARS to its behaviour at the outset of SARS-CoV-2. It would find the same pattern of cover-up and denial. It would show that the WHO had been complicit. It would show that this behaviour had allowed the virus to spread around the world.[277]

By then, I would also have the answer to my questions about Shi's published papers: were they merely sloppy? Or deceitful?

They were both.

10. So Many Suspicions, So Little Evidence

LATE AUGUST 2020. My kids and I had rented a cottage on the eastern shore of Lake Huron, an old rambler with water stains on the ceilings accumulated through generations of pounding rainstorms. Spiders were hard at work in every nook and cranny. Huron had attained a height not seen since 1986, swallowing the sand beach that once stretched 150 yards out from the cottage's frontage. A wall of concrete gabions filled with rock and sand had been set out along the edge of the garden in a last-ditch attempt to hold Huron back. As the waves crashed against them, the spray splattered on the cottage windows and cooled the deck.

There was no cell signal available, only a landline, and even that didn't work. There was no radio or TV either. I thought this was perfect. I would be able to pretend for a time that life was normal. I would stop telling myself to be grateful that my husband and my mother had both died before the pandemic hit, that they were not trapped alone in ICU beds, prone and intubated. I would escape the roll of doom on the hourly newscasts, the grim accounts of the pandemic's death-dealing swath, of long-haulers whose COVID-19-induced disabilities never seemed to fade away. I wouldn't even have to listen to political pundits braying about the impending US federal election upon which the future of the Western world hung.

I told myself to relax, to stop trying to prise open the secret of the pandemic's origin. I told myself it was a hopeless task.

But I couldn't stop thinking about where the virus came from—lab or nature—no matter how hard I tried to lock my curiosity in a box. One fact stood out. It had been nine months since SARS-CoV-2 burst into the human population of Wuhan. In that time many new strains had appeared with a nucleotide mutation here, a deletion there, but there had still not been a significant change that augmented the virus's capacity to infect or to kill. Yet many millions around the world had been infected.[278] If the virus was new to humans, it should have spread slowly at first, evolving rapidly as it struggled to get better at evading the human immune system. But that's not what happened. Its long-term genome stability demonstrated what epidemiologists had recognized almost from the beginning: that SARS-CoV-2 was already extremely well-adapted to humans when the first patients flooded into Wuhan hospitals.

While it may have had its origin in a bat long before the pandemic began, somehow, somewhere it got to be very, very good at invading human cells. Where? In Shi Zhengli's virology lab? In a Chinese military bioweapon lab? In a human population living close to bats but so far off the beaten track that no doctor noticed people getting sick and dying of pneumonia?

I thought about Xinjiang as a place where that might have happened. Millions of Uyghurs had been locked up there for years in what epidemiologists call congregate settings, ideal places for a spillover virus to adapt and spread. In camps where government forces stood accused of torture, and rape, and the forced sterilization of women, perhaps doctors would not notice strange pneumonias, let alone investigate them. And if it started there, it could have been transported around China. The *Globe and Mail* and the BBC would later report that as part of China's Uyghur assimilation policy, 20 percent of the Uyghur working-age population has been sent under armed guard as forced labour to factories all over the country, including Wuhan.[279]

The more China cast blame on others, the more suspicious I grew that someone high up in the government of China knew exactly where SARS-CoV-2 came from. Suspicion is a mind microbe that grows where there is no evidence, especially when evidence is known to have been deliberately withheld. I couldn't stop asking: If China had nothing to hide, why didn't it show samples from early patients to independent scientists? Why had the earliest samples been ordered destroyed? Why the clampdown on what its scientists were permitted to publish? Why did their publications have to be coordinated?

While my grandchildren jumped into the waves from a floating plastic rose, shrieking with joy, I sat on the cool cottage deck thinking over what I had read in the spring and early summer. I decided it amounted to circumstantial evidence against the nature-made-it thesis, but that there was no direct evidence for the counter-argument, that SARS-CoV-2 was the result of human manipulation.

A lot of people were suspicious that RaTG13 was just a

sequence that didn't exist in nature. Shi had explained to *Science* that she never isolated it, and that the sample she got it from—bat feces—had been used up. Some found that way too convenient. A man named Dean Bengston argued in a preprint that Shi's report of RaTG13 was so iffy—no mention made until the *Science* interview that the RdRp labeled 4991 and published in 2016 is a piece of RaTG13; no publication of RaTG13 until 2020 though it had been fully sequenced in 2018—that it should be removed from scientific consideration.[280] That would leave SARS-CoV-2 in a clade—a group of organisms with similar genomes because they share a close common ancestor—all by itself, with no close relatives at all. That fueled more suspicion that it had been ginned up in a lab.

But suspicion did not demonstrate that SARS-CoV-2 had been made, manipulated, or leaked. There was only evidence that Shi's publications on SARS-CoV-2 could not be trusted.

The few papers I'd read insisting their authors had evidence of human manipulation mainly relied on appeals to logic. They argued that lab-made was the more parsimonious explanation, and therefore better than the natural origin argument. Parsimony refers to a dictum known as Occam's Razor[281] which is often relied upon in scientific debates—when you have two conflicting theories, the simplest explanation is probably correct.

One paper arguing that a lab leak was the more parsimonious explanation of SARS-CoV-2's origin had appeared in early July via a link provided by the Norwegian magazine *Minerva*. The magazine did a story on the paper's lead author, biochemist and vaccine developer Birger Sørensen. The paper was the second of two written by Sørensen, Dr.

Angus Dalgleish, and Andres Susrud. They had worked for years to make a vaccine for HIV (also an RNA virus) and applied what they'd learned about HIV to make a vaccine for SARS-CoV-2.[282] Their first paper describing that vaccine had been published in a peer-reviewed journal. The second paper came from material that the journal's editor had insisted be cut—their argument about the lab-made origin of SARS-CoV-2.[283]

In their second, not-yet-published paper they explained that upon close examination of its biochemistry, they had discovered things about SARS-CoV-2 that could best be explained by human manipulation. They looked at what the virus's genes call up—the amino acids that combine to make the virus's proteins, such as the spike protein with its unusual receptor binding domain. Their argument turned on a group of amino acid inserts not seen in other related coronaviruses. These inserts modulate the biochemistry of the spike protein. They create increased positive charge on the top of the spike and negative charge on another portion of it. The accumulated charges allow the spike to adhere to two human receptors, not just ACE2. These inserts make the virus behave like a phagocyte (an immune cell which devours or engulfs invaders and usually produces inflammation). They argued the charges also allow the spike to bind "to the negatively charged phospholipid heads on the cell membrane." In other words, SARS-CoV-2 has three possible means of attachment and entry into human cells, not just one. In their view, three such useful tricks were very unlikely to be the product of natural evolution.

However, shortly after the Sørensen group's first paper was submitted for publication, an article reporting similar inserts in a bat-borne coronavirus had been sent to *Current*

Biology and published in early June. Its co-authors included E.C. Holmes, one of the authors of the proximal origin paper published in *Nature Medicine*, and co-author of the first published SARS-CoV-2 genome. Several of Holmes' co-authors on this new paper were at the University of Shandong and included some ecologists who went to Mengla County, Yunnan, in search of SARS-like viruses in bats. They captured 227 live bats and gathered samples from their livers, lungs, and feces between May and October of 2019, just before the first cases of SARS-CoV-2 appeared. "All but three bats were sampled alive and subsequently released," the paper said. (It did not say what happened to the three that were not released. Did they die on the spot? Were they taken back to a lab? Did they bite somebody?) They extracted and pooled genome fragments from these samples and set to work assembling whole genomes from the fragment reads. One assembled genome, labeled RmYN02, turned out to be 93 percent like SARS-CoV-2, though its receptor binding domain differed so much they thought it might not bind to the ACE2 receptor. They did not try to isolate the virus. In other words, they did not prove that the sequence they assembled was real, nor did they determine the receptor it might use. But they pointed out it had "multiple amino acid inserts" in the same place as SARS-CoV-2—where the dance begins between virus and host cell. "This provides strong evidence," they said, "that such insertion events can occur naturally."[284]

If Sørensen and his colleagues were aware of this *Current Biology* paper, they did not refer to it in their second paper.

After describing the biochemistry, Sørensen and his colleagues switched to deduction. They asserted that the accumulated charges on the SARS-CoV-2 spike were just the sort of novelty sought in gain-of-function experiments. They noted that Shi's lab, which had been doing such experiments for years, might have learned about SARS-CoV-2's third cell-entry method when studying a new coronavirus called SADS (swine acute diarrhoea syndrome). SADS had infected and killed thousands of piglets on four farms in China between 2016 and 2018. The disease appeared first on a farm about 100 kilometres from the site of the index case of SARS. Naturally, Shi and her co-workers, along with Linfa Wang and Peter Daszak, investigated. They determined the disease was caused by an HKU2 type coronavirus. They found more viruses like it in anal swabs taken from *Rhinolophus* (horseshoe) bats they trapped in caves not far from that farm. In the resulting paper, which landed with a splash as a letter in *Nature*,[285] they explained that they followed almost all of Koch's postulates to demonstrate the reality of the virus: they isolated it from sick piglets; they grew it in culture; they inoculated it into healthy three-day-old piglets to see if the same symptoms resulted. They only postulate they did not report fulfilling was extraction of serum from the piglets they'd infected and isolation and growth of the SADS virus from it. Their paper pointed out that none of the three receptors used by similar coronaviruses was used by this new one. But they did not identify which receptor the SADS virus uses to enter the piglets' cells.

Sørensen and his colleagues pounced upon this failure to identify the receptor. It gave rise to another deduction which amounted to an accusation. They reminded readers that through the experiments conducted by her lab and published between 2008 and 2018, Shi had figured out how to make SARS more infectious in humans, how to swap spikes between one SARS-like virus and another, and how to cause the resulting chimeras to infect human cells and humanized mice. They argued Shi had probably also discovered a previously unknown means of cell entry while studying SADS. They deduced that "a field test" likely led to SARS-CoV-2.

In other words, their lab-made thesis was based on speculation about what Shi might have done but had not published. Then they insisted that those who argue for a natural origin of SARS-CoV-2 should be forced to produce evidence or accept human manipulation as the more parsimonious explanation.[286]

Which I thought was a big stretch.

Sørensen, Dalgleish, and Susrud's arguments were treated respectfully. Thanking Dr. J.F. Moxnes of the Norwegian Defense Research Establishment in their first paper's acknowledgements hinted at connections to power. And Sir Richard Dearlove, former head of MI6 in the UK, was quoted in the *Daily Mail* (who got it from the *Telegraph's* podcast *Planet Normal*) saying their idea that this virus had been made in a lab and escaped had to be taken seriously.[287]

Later, I emailed Sørensen and asked for a telephone interview. He asked for questions in advance. I sent them, including whether he was aware of the *Current Biology* paper. I asked for a convenient time to call. Too busy on the vaccine, he replied. My emailed questions weren't answered.

ALINA CHAN WAS not treated so respectfully, though her credentials are impeccable. A Canadian born in Singapore, she is a post-doc at the prestigious Broad Institute, which partners with MIT, Harvard, and Harvard's five teaching hospitals. The institute is known for leading work on genomics. After her lab shut down due to the pandemic in April, Chan cast about for a project she could do in her apartment on her laptop. She pounced upon trying to determine whether SARS-CoV-2 had evolved more slowly than SARS at the start of the pandemic. She asked her friend Shing Hei Zhan of the Department of Zoology and Biodiversity at University of British Columbia to help. Her supervisor, Benjamin Deverman, also lent a hand. They compared SARS's evolution to SARS-CoV-2's evolution in the first quarters of both pandemics.[288]

Working from published sequences of SARS during the early, mid, and late stages of the epidemic, and from randomly selected SARS-CoV-2 genomes published in the first four months of the pandemic, they demonstrated that SARS had changed a lot as it first invaded human bodies, but then settled down to a stable form. SARS-CoV-2, on the other hand, hardly changed at all in the first quarter of the pandemic. Two figures published in their paper demonstrated in a very colourful way that SARS had evolved rapidly to fit itself to a human environment, while SARS-CoV-2 did not need to. "Even by April 28, 2020, the SARS-CoV-2 genomes...spanning 4 months exhibited modest genetic diversity," they wrote.

Though she was wary of fueling conspiracy theories and did not want to be branded as a Trump supporter, Chan believed these facts needed to be known, which is why the paper was posted as a preprint to the Cold Spring Harbor site bioRxiv on May 2, 2020. Chan and her colleagues never said they believed SARS-CoV-2 could have been whomped up in a lab. Instead, they pointed to the need to find the animal carrying this virus that was so clearly pre-adapted to humans, to prevent another spillover.[289] They asked these questions, my questions:

> Did SARS-CoV-2 transmit across species into humans and circulate undetected for months prior to late 2019 while accumulating adaptive mutations? Or was SARS-CoV-2 already well adapted for humans while in bats or in intermediate species? More importantly, does this pool of human-adapted progenitor viruses still exist in animal populations? Even the possibility that a non-genetically engineered precursor could have adapted to humans while being studied in a laboratory should be considered, regardless of how likely or unlikely.

That timid statement turned into a media event. Within days, *Newsweek* reported that serious scientists had produced a paper saying a lab-leak origin should not be ruled out.[290] Critical response from interested parties was swift and harsh. Jonathan Eisen is an evolutionary biologist, fellow attendee with Peter Daszak at the US government's Forum on Microbial Threats, and a professor at UC Davis, where Jonna Mazet directed the PREDICT program. Eisen told *Newsweek* that he did not find the study convincing; he said on Twitter that he did not find the study "remotely convincing." Peter Daszak's Twitter comments were worse: "This is sloppy research...a poorly designed phylogenetic study with too many inferences and not enough data, riding on a wave of conspiracy to drive a higher impact."

Her Twitter exchanges with Daszak led to Chan being

featured as a feisty truth seeker in a *Boston Magazine* profile by Rowan Jacobsen, whose reporting I rely upon here. His story's subtitle hinted that scientists might be critiquing this paper not because of its quality, but as a means of censorship.[291]

Jacobsen quoted some scientists without attribution who said they agreed about the necessity of keeping an open mind about a lab origin but were keeping their mouths shut to avoid being hated by their colleagues. Why would their colleagues hate them for speaking their minds? According to Richard Ebright, they wanted to avoid funding cutbacks and tighter oversight and regulation of their gain-of-function experiments. Science journalist Antonio Regalado of MIT's *Technology Review* told Jacobsen that if it turned out that this virus escaped from a lab, "it would shatter the scientific edifice top to bottom."

SITTING ON THAT deck beside Huron's heaving waves, it seemed to me that the scientific edifice had already cracked top to bottom. It had been battered by geopolitical competition so intense it amounted to low-level warfare, with truth as the first casualty. Normally unconstrained scientific debate had been funneled into a ridiculously narrow frame by those with interests to protect. If SARS-CoV-2 *had* leaked from a Chinese lab, Chinese virologists who had the evidence wouldn't tell—they wouldn't be allowed to tell. On the other hand, and worse: even if they could demonstrate that SARS-CoV-2 came directly from nature, they would not be believed. Peer-reviewed publishing had been irrevocably sullied by the errors and omissions and contradictions in the early SARS-CoV-2 reports out of China. In addition to its unconscionable delay in telling the world about the exploding pandemic, with its orchestration of all publications on SARS-CoV-2 and its secret order that no samples or information be given to outsiders, the Chinese state had dragged leading Western scientific journals into its propaganda machine. That these journals leant themselves to China's interests broke trust with readers who expect peer-reviewed articles to be objective, accurate, and fulsome. Trust had been further diminished by scientists in the US who'd published articles and statements in the same high-impact journals in order to constrain debate without declaring their interests. They amplified China's it-came-from-a-bat-or-a-pangolin-in-another-country-and-don't-you-dare-suggest-otherwise origin narrative. They had tried to make it the only legitimate narrative.

By the late summer of 2020, I had concluded that truth about the virus's origin could only come from a group that was scientifically competent but independent, with no need to enter China and no need for grants from major US agencies concerned about blowback over what they'd funded. They would also need access to a responsible, independent publication platform.

Never going to happen, I said to myself.

RIGHT BEHIND THE cottage a steep forested hill, draped in ivy and periwinkle and deadfall, rises to the garage where our cars were parked. I could get a decent cell signal on the road out past the garage, but I had to climb a rickety 93-step wooden staircase to get up there. I dragged myself up those steps once a day to call my office in case something needed my attention. Other than that, my phone lay silent on the table beside the bed, except on one perfect, blue-sky-sweet-wind August morning. As I walked into my bedroom, it bleeped. There was just enough signal to show that a new email had come in from a publisher friend, Marc Côté, who knew what I was working on. The subject line said: something on the origin you might not have seen. There was an attachment. But the signal wasn't strong enough for me to open it.

Curiosity broke from its cage. I ran up the 93 steps, only stopping once to get my breath, and walked a hundred yards down the road before I found a good signal. Côté had forwarded something published July 15 by a news outlet I'd never heard of—*Independent Science News for Food and Agriculture*.

And that's how I was introduced to the work of Jonathan Latham and Allison Wilson, both PhDs in biological disciplines, one a plant virologist, both without academic appointments and so independent of the need for grants, with no need to enter China to do their work, and with their own publication.

11. The Latham/Wilson Thesis

THE TITLE OF their article was "A Proposed Origin for SARS-CoV-2 and the COVID-19 Pandemic." Standing on the road, I tried to read it quickly, but it was too hard to follow the embedded links and to check the endnotes. As soon as I got home from the lake, I printed it out.

Latham and Wilson had published it on July 15, 2020, two weeks before the interview with Shi appeared in *Science*. It followed an article they'd published in June that had raised a lot of questions, such as: are BSL-4 labs safe, is China's BSL-4 safe, and what does it mean that China is unrolling a network of them? Such as: is trying to predict pandemics by bringing bat viruses from the distant countryside to a lab in the middle of a major city useful? Is doing gain-of-function experiments on those viruses helpful? (Or is it "the definition of insanity," as Ebright had put it?) If the closest ancestor of SARS-CoV-2 came from a very distant mine in a very distant province, how did its descendent make its way to Wuhan? Wuhan does have two major labs whose researchers make trips to bat caves in Yunnan hunting for coronaviruses year after year. What about that? It also pointed to various competing interests, especially Peter Daszak's. It was a long article, well sourced, but there was nothing there I had not read before.[292]

But the second article, the one Marc Côté sent me, was a revelation. After I read it, I even wondered if someone in China knew it was coming and Shi answered *Science's* questions to divert attention from it. *Science* is authoritative, one of the two best general science publications in the world: academics in China get the most money for landing an article in *Nature*, but *Science* is number two. Latham and Wilson published both their articles in their own newsletter, *Independent Science News*, which is usually read by environmentalists who don't like Big Pharma or GMOs. In a reasonable world, a publication in *Science* would always trump an article in such a newsletter. But a world crushed by a once-in-a-century pandemic is not reasonable. A world dependent on Big Pharma to produce the vaccines to save us from a pandemic that may have leaked from a lab is not reasonable. By the time I spoke to Latham, their thesis had attracted many thousands of reads. The posted comments stretched to the crack of doom and came from all around the world.

The article included a well-put-together review of the literature. It wasn't flashy. It wasn't clever. It didn't accuse. It was careful about unfounded speculation. Most assertions were endnoted by reference to articles published in peer-reviewed journals in the standard style of peer-reviewed journals, sending an unwritten message that this paper could have appeared in one. They based their case on Shi's publications, on preprint publications by others, and on a master's thesis by a Chinese medical doctor, Xu Li, published in Mandarin in 2013. They also referenced a 2017 PhD thesis by Canping Huang, supervised by George F. Gao, which had led to two peer-reviewed papers, one with E.C. Holmes as a co-author, both with George Gao as last author. The master's thesis presented fascinating facts which were supported and augmented by the PhD thesis. Together, they explained why Shi Zhengli's group went on numerous expeditions to that Mojiang County, Yunnan, mine where they found the bat fecal sample harbouring RaTG13, the sequence closest to SARS-CoV-2. The key point: the master's thesis, supported by the PhD thesis, said certain samples had been taken from the six miners who came down with SARS-like symptoms in that mine back in 2012. Both theses said some of those samples were sent to Shi's lab (and to others) for analysis because coronavirus infection was suspected. Yet Shi's publications made no mention of those samples. Shi did not refer to them in the *Scientific American* story in which she claimed that two miners had died of fungal infections. She didn't mention the samples in her reply to *Science*, either. I am sure of that because after I read Latham and Wilson's article, I double-checked.

In their opening paragraph Latham and Wilson said:

> In all the discussions of the origin of the COVID-19 pandemic, enormous scientific attention has been paid to the molecular character of the SARS-CoV-2 virus, including its novel genome sequence in comparison with its near relatives. In stark contrast, virtually no attention has been paid to the physical provenance of those nearest genetic relatives, its

> presumptive ancestors, which are two viral sequences named BtCoV/4991 and RaTG13.[293]

The master's thesis shone a bright light on that provenance. It told a much more detailed version of the story of the six miners than Shi had given in *Scientific American* or in her reply to *Science*. The thesis revealed that when clearing bat feces from the Tongguan copper mine in Mojiang County in late April and early May of 2012, the miners contracted a strange and deadly pneumonia. Three of them died, not two as Shi had told *Scientific American*. The description of the miners' symptoms, as recorded in the master's thesis by Dr. Xu Li, who treated them, were very familiar: they were the same as the symptoms of SARS-CoV-2. Dry cough, high fever, headache, sore limbs, severe pneumonia which showed up on X-rays as ground glass-like opacities, difficulty breathing, blood clots, etc. The title of the master's thesis ("The Analysis of 6 Patients with Severe Pneumonia Caused by Unknown viruses") made it clear that the cause of the miners' deaths was not a fungal infection, as Shi claimed in *Scientific American*, but viruses. The thesis itself said they were probably infected with coronaviruses.[294]

Latham and Wilson had been alerted to the question of provenance and the master's thesis by a reader and anti-GMO activist with a PhD who had sent them one preprint after another, including one[295] posted April 2, 2020, and another[296] posted May 20 (and reposted May 24). The latter paper, written by two researchers at research institutions in Pune, India, pointed out that the sequences of BtCoV4991 and the RdRp segment of RaTG13 were identical, and according to the data on the public sites where these sequences had been uploaded, both came from the mine where a master's thesis said miners had become sick and died. Shi Zhengli's paper in *Nature* had failed to mention BtCoV4991. It had also failed to mention where RaTG13 came from, which, given the story told by the master's thesis, was deceptive with a capital *D*. Clearly this preprint paper, and others like it, had pushed *Science* to ask, and Shi to explain, more about the origin of RaTG13.

The other preprint made a case about how SARS-CoV-2 could have been engineered without leaving a trace. This involved a technique called site-directed mutagenesis. (When I googled to check that claim, the first thing that came up was the website for a company called Gene Universal in Delaware. It offers such a service at $99 per mutation, recommending use of its bundled gene synthesis service as well. The company can do mutations of up to 8,000 base pairs in length in 15 to 20 days. The entire SARS-CoV-2 genome is only 30,000 base pairs long).[297] When RaTG13's raw sequence reads were finally uploaded by Shi's lab to a public repository in May 2020, four months after the *Nature* publication, they revealed that RaTG13 had been sequenced in 2018, which Shi later confirmed to *Science*, leaving plenty of time for such experimentation.

But who were these folks, and how did authors in India know about these theses, both written in Mandarin?

From the point of view of leading journals like *Nature* and *Science*, the authors were nobodies. But these nobodies had put out some very important information.

The preprint showing how SARS-CoV-2 could have been engineered without leaving a trace was written by Rossana Segreto, a microbiologist post-doc at the University of Innsbruck who specializes in fungi, and Yuri Deigin, an MBA based in Toronto. Her ORCID site doesn't show many previous peer-reviewed papers. He runs a company called Youthereum along with two Russian-based PhDs. Some of his earlier companies were based in Moscow. Youthereum has no offices. It claims to work with scientists at various universities to advance the company's mission—defeating aging. Segreto and Deigin were part of an informal group of investigators known as DRASTIC[298] who had been digging into Shi Zhengli's publications and what her lab had been doing since February 2020. And they continued to do so. Some members of the DRASTIC group posted an article in 2021 asking why the code-word protected—meaning you can't get access without the right code—part of the WIV's huge database of bat samples and viral sequences had been taken offline in September 2019. They pointed out that even the contents description had been removed on December 30, 2019, the night Shi Zhengli sat on that high-speed train from Shanghai to Wuhan, worried about a lab leak.[299]

Monali Rahalkar and Rahul Bahulikar, the authors of the other preprint, said in their acknowledgements that they had been steered to these theses by an anonymous Twitter user called theseeker. They had relied on translations of the theses by Francisco de Asis de Ribera Martin. They too were members of DRASTIC, which appeared to be something like a hacker collective. Some members hid behind pseudonyms.

Latham and Wilson had been nervous about that, worried that maybe the theses weren't real, or that the translations were inaccurate. They'd hired a Mandarin speaker to do a proper translation of the master's and PhD theses. They'd also checked to make sure both had been posted on the official Chinese website that publishes all postgraduate theses as well as everything that appears in China's journals. It is called CNKI and is sponsored by Tsinghua University, but also by the Propaganda Department of the Chinese Communist Party. They located the theses on the site.[300] They decided the master's thesis was real and built their argument on it. And what an interesting argument.

> We do not propose a specifically genetically engineered or biowarfare origin for the virus but the theory does propose an essential causative role in the pandemic for scientific research carried out by the laboratory of Zheng-li Shi at the WIV; thus also explaining Wuhan as the location of the epicentre.

They said that their theory could account for the origin of the furin cleavage site peculiar to SARS-CoV-2, "which greatly enhances viral spread in the body," and could explain the extreme affinity of the virus's spike to human receptors, the weird lack of evolution of an apparently new virus that was spreading rapidly through the human population, and why it targets the lungs, though that is unusual for a coronavirus. They said that thanks to the proximal origin paper in *Nature*, anybody who suggested a lab origin for SARS-CoV-2 had been shouted down as a conspiracy theorist. So the first thing they did was take apart that *Nature* article.

> It is also noteworthy that the Andersen authors set a higher hurdle for the lab thesis than the zoonotic thesis. In their account, the lab thesis is required to explain *all* of the evolution of SARS-CoV-2 from its

> presumed bat viral ancestor, whereas under their telling of the zoonotic thesis the key step of the addition of the furin site is allowed to happen in humans and is thus effectively unexplained.

They said that the co-authors further unbalanced their argument by failing to mention the papers from Shi's lab describing passaging live bat viruses in monkey and human cells and performing recombinant experiments with bat coronaviruses. They quoted from Shi's most recent NIH/NIAID grant—the grant to EcoHealth Alliance that was cancelled, then restored but not paid— showing they planned "virus infection experiments across a range of cell cultures from different species and humanized mice" with recombinant bat coronaviruses. They pointed out that Andersen et al. barely acknowledged that lab escapes have happened, never mind that they have been frequent and that escapes of SARS had occurred in China. They linked to the State Department memo raising concerns about the BSL-4 at the WIV. "It is hard not to conclude that what their paper mostly shows is that Drs. Andersen, Rambaut, Lipkin, Holmes and Garry much prefer the natural zoonotic transfer thesis. Their rhetoric is forthright but the evidence does not support that confidence."

They pointed to evidence published after the proximal origin article appeared. Several papers had shown that not only was the proposed intermediate animal (those pangolins) *not* the natural reservoir for SARS-CoV-2, and that the earliest cases were *not* connected to the Wuhan market, but a new paper had shown that SARS-CoV-2 does not replicate in bat kidney or lung cells, meaning SARS-CoV-2 could not be a recent or direct spillover from bats. (This argument was bolstered further by a study out of Flinders University, published in *Scientific Reports* in June 2021. Complex computer models analyzed the behaviour of the receptor binding domain of the virus with 12 animal and human ACE2 receptors. The virus bound extremely well to humans, poorly to bats, less poorly to pangolins. The paper showed the virus could not have spilled over to humans from any of those animals.)[301]

Then they turned to the important facts about the former copper mine in Yunnan. Dr. Xu Li had reported in his master's thesis that the sick miners were treated in the No. 1 School of Clinical Medicine at Kunming Medical University, in the capital of Yunnan. The miners had worked at the mine from late April 2012 through early May. The sickest miner had been hard at it for about 14 days, the less sick for less time. The physicians were immediately concerned that this was an outbreak of an infectious disease and behaved accordingly: it was reported up the chain. The miners' symptoms included dry cough, sputum, high fevers, difficulty breathing, sore limbs, hiccups and headaches. Patients one through four also had low levels of blood oxygen and damage indicative of viral infections, and two patients had "a tendency for thrombosis" (blood clots). The severity of symptoms was related to age. The tests the hospital ran eliminated Epstein-Barr virus, SARS, influenza, dengue, hemorrhagic fever, Japanese encephalitis, and Hepatitis B, though patient two was positive for Epstein-Barr and hepatitis. Two patients were placed on ventilators, all were treated with steroids, and five got antivirals and blood thinners. Antibiotics and antifungals were also administered due to fear of co-infections. The doctors were stumped as to which virus was at work. They consulted with Zhong

Nanshan, the pulmonologist who had spoken truth to power about SARS, the same Zhong who told the press that he didn't think the fallout from SARS-CoV-2 would be like SARS. "Samples from the miners were later sent to the WIV in Wuhan and to Zhong Nanshan," the thesis said. Some of the samples tested positive for a coronavirus although the thesis did not make clear how many.

That Zhong Nanshan was consulted indicated the doctors treating the miners suspected a SARS-like viral infection, picked up in that horseshoe bat-infested mine, which had jumped directly from bats to humans. When Shi's group went to the mine to investigate (while the miners were still in hospital), the bats they trapped and sampled turned out to harbour many coronavirus infections. They found two new beta coronaviruses (a group that includes SARS, SARS-CoV-2, and MERS), just the kind Shi's group was looking for, including BtCoV4991, alias RaTG13.[302] They had been warning of beta coronaviruses' pandemic potential for some time, wrote Latham and Wilson, so to find "RaTG13 where the miners fell ill was a scenario in perfect alignment with their expectations."

Latham and Wilson did not point this out in their article, but the paper by Shi's lab describing the discovery of bats carrying many viruses, one of which was BtCoV4991, had only mentioned that the bats were sampled in an abandoned mine. They shared no information at all about the miners who got sick there or about samples taken from them for analysis by Shi's lab. The paper wasn't published until 2016. It appeared in the WIV's own journal, *Virologica Sinica*, the equivalent of a stealth publication because the journal is not well read in the West. By contrast, a 2017 paper by Shi's lab describing the many viruses harboured by horseshoe bats sampled during a five-year survey of another mine near Kunming, Yunnan, was published in PLOS *Pathogens*, a high-impact Western journal. Co-authors on the PLOS *Pathogens* paper included Linfa Wang and Peter Daszak. But Daszak's name was not on the 2016 paper reporting the RdRp sequence called BTCoV4991, alias RaTG13, though the acknowledgements referred to the support of the NIAID grant issued to EcoHealth Alliance in 2014. Those who get US research grants and subcontract part of the work to others, as EcoHealth Alliance had done with Shi, are responsible for how that money is spent. Daszak should have been consulted on the work that led to this paper. Yet his name didn't appear on it.

The abstract for that NIAID grant says:

> We will sequence receptor binding domains (spike proteins) to identify viruses with the highest potential for spillover which we will include in our experimental investigations.... In vitro and in vivo characterization of SARSr-Cov spillover risk...to identify the regions and viruses of public health concern. We will use S protein [the spike] sequence data, infectious clone technology, in vitro and in vivo infection experiments and analysis of receptor binding to test the hypothesis that % divergence thresholds in S protein sequences predict spillover potential.... We will combine these data with... human survey of risk behaviors and illness, and serology to identify SARSr-Cov spillover risk hotspots across southern China...to prevent the re-emergence of SARS or the emergence of a novel SARSr-Cov.[303]

Sequencing the RdRp of 4991 fell perfectly within the aims of that grant. Yet Daszak was not involved. It was as if Shi's lab ran parallel experimental streams, one involving outsiders like Daszak, and one for insiders only.

The WIV was not the only group that took samples from that mine but remained close-mouthed about the miners. The PhD thesis by Canping Huang, published in 2017, stated that at least three groups investigated the cause of the miners' illness. He mentioned that Zhong Nanshan asked that blood samples be sent to the WIV and that the WIV also received bat dissection products. He mentioned that the WIV found antibodies to something SARS-like in four of the samples from miners. The Chengdu Military Region Disease Control Center got blood samples.[304] Yet the two papers Huang published in Western journals before his dissertation was accepted did not mention where in Yunnan the bats he'd sampled came from, nor did they mention the miners and their illnesses. George Gao supervised Huang's thesis. He was also the last author on both papers published in Western journals.

While Segreto and Deigin had written a complex argument about how Shi's lab could have engineered SARS-CoV-2[305] without leaving a trace, and Rahalkar and Bahulikar had raised sharp questions pointing to the mine as the place where SARS-CoV-2 originated,[306] Latham and Wilson presented a different case for how SARS-CoV-2 emerged. They suggested that something like passaging had taken place during which the original bat-borne virus became acclimatized to humans. But they thought it happened in the miners' lungs, not in a WIV lab.

Latham and Wilson pointed out that the miners would have been exerting themselves hard, breathing hard, as they tried to clear away the bat guano. They thought the virus that afflicted them probably hitched a ride on the small particles of bat feces stirred into the air as they worked, finding its way deep into their lungs. A typical coronavirus infection mainly settles in the upper respiratory tract. Because of the exertion and the constant exposure over many days, the miners would have had high viral loads. Their lungs would have provided ample cells in which the virus could replicate, way more than in the upper respiratory tract.

> The human aerodigestive tract is approximately 20 cm in length and 5 cm in circumference, i.e., approximately 100 cm^2 in surface area. The surface area of a human lung ranges from 260,000–680,000 cm^2. The amount of potentially infected tissue in an average lung is therefore approximately 4,500-fold greater than that available to a normal coronavirus infection. The amount of virus present in the infected miners, sufficient to hospitalise all of them and kill half of them, was thus proportionately very large. Evolutionary change is in large part a function of the population size. The lungs of the miners, we suggest, supported a very high viral load leading to proportionately rapid viral evolution.

The miners were sick for a long time. One spent 117 days in hospital before being discharged as cured. Two were ill for more than four months. Those miners who survived had impaired immune systems for a long time: their infections may well have continued at a low level.

And why does that matter? The virus was not well-adapted to humans when it infected the miners, so it was under intense selective pressure to adapt, to change, just as it would have been if placed in human cells in a lab and passaged repeatedly. The master's thesis reported that the infections seemed to move from one part of the miners' lungs to another, clearing in one area, occluding the next with the same ground-glass opacities later seen in SARS-CoV-2 patients.

> In such a situation the particularities of lung tissues become potentially important because the existence of airways (bronchial tubes, etc.) allows partially-adapted viruses from independent viral populations to travel to distal parts of the lung (or even the other lung) and encounter other such partially-adapted viruses and populations. This movement around the lungs would likely have resulted in what amounted to a passaging effect without the need for a researcher to infect new tissues. Indeed, in the Master's thesis the observation is several times made that areas of the lungs of a specific patient would appear to heal even while other parts of the lungs would become infected.

Co-infections would have made things worse. As Shi's group remarked in their 2016 paper about the viruses found in the mine's bats: "we observed a high rate of co-infection with two coronavirus species and interspecies infection with the same coronavirus species within or across bat families."

Latham and Wilson thought the miners may have been co-infected with different virus strains as they worked.

> Combining these observations, we propose that the miners' lungs offered an unprecedented opportunity for accelerated evolution of a highly bat-adapted coronavirus into a highly human-adapted coronavirus and that decades of ordinary coronavirus evolution could easily have been condensed into months...it is important to emphasize that our proposal is fully consistent with the underlying principles of viral evolution as understood today.

According to the master's thesis, samples were taken from four of the miners for "scientific research" and blood samples were sent to the WIV to see if they tested positive for SARS antibodies. Some did, which didn't necessarily mean the miners had SARS: there may have been what is known as a cross reaction to something SARS-like. In addition, in late June 2012, the thymus (a gland involved in immune response) was removed from one patient. The master's thesis does not specify why or whether the thymus was sent to the WIV or to any other group for study. According to Latham and Wilson, depending on the kinds of samples the WIV was sent, Shi's group would have tried to extract viral RNA directly from tissue or blood, tried to assemble a sequence

> and/or to generate live infectious clones for which it would be useful (if not imperative) to amplify the virus by placing it in human cell culture. Either technique could have led to accidental infection of a lab researcher.... Any viruses recoverable from the miners would likely have been viewed by them [the WIV] as a unique natural experiment in human passaging

> offering unprecedented and otherwise-impossible-to-obtain insights into how bat coronaviruses can adapt to humans.

In Latham and Wilson's opinion, RaTG13 had mutated in the miners' lungs until it became SARS-CoV-2. And then, it got loose.

But what about the seven-year time lag between when Shi's lab got those samples from the miners and the onset of the pandemic?

Latham and Wilson thought Shi might have waited to do tests on the miners' samples until after the BSL-4 became available—in 2018. As Shi would tell *Science*, after Latham and Wilson's article was published, that was when they assembled the full sequence of RaTG13, though they failed to publish it until 2020.

But it made no sense to me that Shi's group would fail to try and sequence viral genomes immediately from the samples taken from those miners. If they failed to grow virus in culture, they would have tried to extract RNA fragments, tried to make a genome assembly, tried to synthesize whatever they found and to infect human cells or humanized mice with it. It wasn't as if those samples were the result of random bat sampling in the hope of finding something SARS-like. These samples were from patients who had been infected with something not only SARS-like but more deadly than SARS. The money Shi's lab got through EcoHealth Alliance from the PREDICT program was to survey for just such dangerous spillover viruses. On the other hand, maybe after some of the samples tested positive for a SARS-like coronavirus, Shi's lab was instructed to leave further investigation to others, such as the military group in Chengdu, or to work with the samples but not to publish on them.

Shi had told *Science* that her lab knew nothing of SARS-CoV-2 until the patient samples arrived on New Year's Eve. Was she telling the truth?

I had no idea, but I no longer had reason to trust her.

In conclusion, Latham and Wilson presented various facts about SARS-CoV-2 to support their thesis. They pointed out that many papers had shown the SARS-CoV-2 spike has many times greater binding affinity to the human ACE2 receptor than SARS. "Such exceptional affinities..." they said, "do not arise at random, making it very hard to explain in any other way than for the virus to have been strongly selected in the presence of a human ACE2 receptor." The spike and its structure permit the virus to bind to a particular host and find its way to particular kinds of cells: this is known as tropism. Structural analysis of the spike protein, of its receptor binding domain and within that, its binding motif, showed that it was very much like the spike of RaTG13, except that RaTG13 has no furin site. They reminded readers that the furin site in the SARS-CoV-2 spike is only found in one other beta coronavirus—MERS. (The first MERS case was reported in November of 2012, just as the surviving miners were leaving hospital.) Wilson and Latham thought the SARS-CoV-2 furin site could be an example of convergent evolution in which the same problem is solved the same way by unrelated species. Or SARS-CoV-2 may have acquired this site from the miners' lungs. There is a protein (ENaC-a), which is present in some human airway epithelial and lung tissues, which also possesses a furin site. In plants, hosts and viruses sometimes trade positive RNA strands. SARS-CoV-2 is a positive RNA strand virus. Perhaps there was a trade made in the miners' lungs. They reiterated that there had been almost no evolution of SARS-CoV-2 since the pandemic began. "The numerically largest analysis of SARS-CoV-2 genomes...found no evidence at all for adaptive evolution.... It implies that SARS-CoV-2 is highly adapted across its whole set of component proteins and not just at the spike. That is to say, its evolutionary leap to humans was completed before the 2019 pandemic began." They reminded readers that Shi's lab had shown (in *Nature*) that RaTG13 is the closest genome to SARS-CoV-2 across its entire length, which is consistent with the notion that SARS-CoV-2 evolved from RaTG13.

Finally, they appealed to parsimony. The proximal origin paper in *Nature* argued that the creation of SARS-CoV-2 involved a jump from bats to another species, likely a pangolin, and then a jump from a pangolin imported to Guangdong to humans. The paper offered no explanation as to why or where SARS-CoV-2 became so well-adapted to humans and why it exploded in Wuhan, many hundreds of miles from Guangdong. Latham and Wilson insisted their own thesis was more parsimonious. It proposed just one jump from bat to human, already documented by the master's thesis, followed by evolution in the miners' lungs.

How did this evolved entity, SARS-CoV-2, get to Wuhan? In the miners' samples.

I read and reread their thesis. It had merit. But there were problems. It seemed unlikely to me that there could have been over 1,100 nucleotide changes in RaTG13 in just a four-month period in those miners' lungs. I thought the miners might just as easily have inhaled another SARS-like virus closer still to SARS-CoV-2 that was also circulating in bats in the mine. In other words, RaTG13 might be a cousin but not the parent.

But if RaTG13 did evolve into a human-adapted virus in those miners' lungs, how did it get loose? How did a sample held in a freezer at -80°C become a live virus that caused a pandemic?

AFTER HE LEFT office as deputy national security advisor to President Trump, Matthew Pottinger would make the claim on CBS's *Face the Nation* that the People's Liberation Army of China has been doing biowarfare experiments on animals in the Wuhan Institute of Virology since 2017.[307] A Harvard study in the summer of 2020, examining satellite imagery and checking Baidu search queries for words consistent with symptoms of SARS-CoV-2, found that a lot of cars had been parked at hospitals in Wuhan from late August 2019 compared to the same period the year before, and there had been significant search interest in diarrhea (another known symptom of SARS-CoV-2), suggesting that something bad was circulating in Wuhan even then, just before the WIV took its database offline.[308] An American intelligence contractor claimed that by tracing the pings of cellphones characteristically in use at the WIV, and by looking at satellite images of traffic flow, it appeared that from October 7 to October 24, 2019, a high-security area of the WIV had no cellphone use (though one cellphone with a home base in Singapore moved around inside one of the campus buildings for a brief time), and there may have been a "hazardous event" between October 6 and October 11.[309] These assertions could all be true but did not demonstrate that SARS-CoV-2 leaked from the WIV.

I kept going back to Shi's papers and statements—

thinking about the sloppiness, the deception by omission, the contradictions. Maybe Shi *had* waited until 2018, after the BSL-4 was fully open, to sequence RaTG13. She admitted to *Science* that she did experiments with SARS-related coronaviruses in animals in 2018 and 2019 whose results have not yet been published, but she said that none of those viruses were close to SARS-CoV-2. Was that true? Though she told *Science* she had not been able to isolate RaTG13, her use of the world "isolate" was careful. She wasn't asked directly whether she had synthesized the RaTG13 sequence and tried to infect cells with it. Maybe this was a case of didn't ask, won't tell.

Shi had told *Science* that she moved her coronavirus experiments to the BSL-4 in part to train staff on less deadly viruses, but also because the BSL-3, which the government mandated must be used for SARS-CoV-2 studies, did not have the infrastructure to permit work with nonhuman primates, probably macaques. But she did not say *when* she started working with macaques in the BSL-4. Her language suggested this began with SARS-CoV-2, but there was no definitive statement. Xiangguo Qiu had done Ebola experiments on macaques at the NML in concert with researchers from China. Maybe that was why Qiu had been invited to the WIV: not just to help train the WIV's staff on how to work in a BSL-4, but on how to work in a BSL-4 with macaques. Shi did not specify which SARS-related viruses she'd used on animals. What if she'd infected macaques with something from the miners' samples? What if a macaque infected a researcher who had no symptoms but spread it around?

Lab accidents happen for many reasons, including while handling animals who violently object to being subjects of study. Another leading cause is sloppy practice by researchers, especially students and trainees.

The DRASTIC group reported later that it found a transcript from the Second China-US Workshop on the Challenges of Emerging Infections, Laboratory Safety, and Global Health Security. That workshop was held in Wuhan, May 16 and 17, 2017. Among the topics of the meeting were lab safety, gain-of-function experiments, gene editing, targeting.[310] In answer to a question after her talk, Shi said she had infected live bats with Nipah to see how long it took for them to clear the virus. Nipah is a Risk Group 4 pathogen which may only be legally studied in Canada in a BSL-4 lab. Shi had told *Science* that the BSL-4 lab was not permitted to import Ebola at that time, yet Nipah is more lethal than Ebola: Nipah kills up to 70 percent of those infected.[311] Did she do the work in a BSL-3 before the BSL-4 was opened?

In a 2018 story by an online magazine called *Sixth Tone* published in Shanghai, a young post-doc in Shi's lab, Luo Dongsheng, demonstrated for the reporter and the camera how he bagged and swabbed bats in a cave in Hubei. Luo also brought 12 of those tagged live bats back to the WIV for study. He explained to the reporter the value of the extensive WIV online database (taken offline in September 2019), which lists over 2,000 samples (stored at -80°C) as well as genomes and gene sequences derived from bats. One such stored sample had been vital in the SADS study. It had proven to be identical to the SADS virus isolated from those piglets between 2016 and 2018. Sampled from a bat and stored in the lab long before the piglets were infected, this sample demonstrated that bats are the reservoir for SADS.[312] Zhou Peng, one of the most senior researchers in Shi's group, explained to the reporter that the purpose of this sort of research is to learn how bats harbour multiple viruses without becoming ill, and having learned that secret, to apply it to human beings. Luo also told the reporter that the lab's researchers go on sampling expeditions on an annual cycle all over China, from Tibet to the south.

But the images which accompanied this *Sixth Tone* story were more instructive than the interviews. They showed Luo at the entrance to a dark cave at twilight, standing, according to the reporter, on layers of bat dung as he set up a bat capture net. He was dressed in ordinary street clothes. He was not wearing personal protective equipment, no mask, no visor, no goggles, no Tyvek suit. A close-up did show him wearing thick woven gloves as he held a bat and swabbed it. But the glove had a hole in it. If miners could pick up a deadly coronavirus infection while sweeping feces from a bat-ridden mine, why not lab researchers entering bat caves year after year from Tibet to the south without proper protection? Why not researchers bagging wild bats and bringing them back to the lab to breed? One of the known lab accidents at the National Microbiology Laboratory in Winnipeg involved researchers bringing wild birds back to the lab to study, birds that turned out to be infected with avian flu virus. If that could happen in Winnipeg, it could happen at the WIV.

As an associate of the DRASTIC group would later reveal, the WIV applied for two patents dealing with bats. One was for a method to feed insects to bats so they could be held over the winter in the lab and bred there. Another, filed in June 2018, was for interconnected bat-breeding cages. It was granted in April 2019.[313] Bats, though small, can bite. Their feces carry viruses. Rhesus macaques are bigger, stronger, much smarter, and more dangerous than bats.

Their sexual appetites do not stop at the species boundary. One of my colleagues tells the story of how, on her first trip to the jungles of Malaysia, she fell in love with her travelling companion when he saved her from a female macaque who attacked her in a fit of jealousy. That macaque was way too interested in my colleague's male companion: she considered my colleague her rival. He just managed to peel that macaque off her back as it bit down on the top of her skull.

It is noteworthy that in November 2019, the WIV also applied for a patent on a device to close quickly a finger wound while working in a pathogen lab.[314] This demonstrates that accidents are not unknown in the WIV's high containment facility.

It seemed to me that a lab accident/escape at the WIV had to be considered more than likely. It was probable. It was almost inevitable.

12. Following the Money

IT BOTHERED ME more than I can say that I was unable to get a response from Shi Zhengli, Peter Daszak, or even EcoHealth Alliance's communications person. I could understand Shi's silence. She lives beneath the thumb of an aggressive authoritarian government. But Daszak—so quick to attack the lab-leak suggestion as something only a conspiracy theorist would consider, so quick to raise Cain in major media and among Nobel laureates about that retracted NIH grant—runs an American charity almost entirely sustained by US taxpayers. I thought he would—should—be open to questions from just about anyone, even from a curious Canadian.

I hunted around to see if he had answered my questions already while being interviewed by someone else. I found a podcast/video called *This Week in Virology*, "the podcast about viruses, the kind that make you sick."[315] It is hosted by Vincent Racaniello, a virologist at Columbia University. The interview took place at the Nipah Virus International Conference at Singapore's Duke-NUS Medical School, where Linfa Wang is an esteemed leader: Racaniello thanked

Wang for bringing him to the conference. It ran from December 9 to 10, 2019, just as the pandemic took hold in Wuhan.[316] The interview was posted on May 19, 2020.

The two of them sat at right angles to each other on 1960s-ish armchairs in an otherwise empty hall. It must have been connected to a larger auditorium: at one point I could hear faint applause off-camera. Racaniello slumped in his chair with a laptop in his lap, a dark-haired man dressed casually, no jacket, no tie. Daszak was a little more formal—a grey suit jacket, grey pants, sneaker-like lace-ups, blue shirt, no tie. His clothes were those of a senior academic, not a man then earning $410,000 a year from the charity he leads.[317] His skin was very pale, his head large and long and mostly bald, hair short and grey on the sides. From the way he filled his armchair I estimated he is well over six feet, his body thickened by good living.

He planted himself, feet flat on the ground, vaguely moving his right hand for emphasis as he responded to Racaniello's questions. He seemed personable, even affable. He presented himself as someone who loves what he does, which he called applied science. Yet he seldom laughed, and at times there was a whiny edge to his voice. He kept bringing up the One Health concept, which posits that wildlife conservation, good animal husbandry, and human health are one and the same. Apparently to some, the interconnectedness between humans and everything else alive—or half-alive like a virus—is still a new idea.

Racaniello wanted to know about his personal story and asked if he'd pronounced his last name correctly. Daszak said he had and explained that he was brought up in the UK, but his father was Ukrainian, thus the last name. Where he came from in the UK he did not say, but his accent sounded like the Midlands blended with hint of posh and an edge of midtown Manhattan. I was wrong about the Midlands. His brother John, an operatic tenor, has a wiki that says he was born and raised in Manchester, which is farther north. John's website presents a slim version of brother Peter, but also bald, long-headed, and wearing a grey jacket.

Daszak told Racaniello that from the age of eight he'd wanted to be a zoologist studying reptiles on the Amazon. He ended up a parasitologist almost by accident. As an undergraduate in zoology at University College of North Wales, he had to do a research project picked from a list. By the time he made his choice, the only project left was about a parasite found in a lizard's gallbladder. It turned out to be fascinating and hooked him on research. He got his PhD in 1994 at University of East London on coccidiosis, an intestinal parasite that infects many animals, including humans. After that, he did a couple of one-year post-docs (at Kingston University in Surrey), but job prospects were not favourable, so when his wife was offered a biotech job in the US in 1996 in Atlanta he tagged along. While waiting for his work visa, at his wife's suggestion he volunteered at the Centers for Disease Control. He was by then expert in electron microscopy and started work at the CDC just as samples from the first Nipah outbreak arrived for study. Some of the images shown at this conference were images he made at that time, he told Racaniello. By then it was known that Nipah should only be studied in a BSL-4 because its death rate is so high.

"You know," interjected Racaniello, "for years in Germany they worked on Marburg in glove boxes; it was okay, nobody got sick."

Daszak got a job at the Institute of Ecology at the University of Georgia, researching an amphibian disease. He eventually discovered that a new fungal disease was driving the decline of the amphibian population, in particular bullfrogs. CSIRO, then Linfa Wang's employer in Australia, gave him a prize for that work. He got another job in 2001 coordinating a conservation medicine consortium involving five US universities[318] and a conservation charity called the Wildlife Preservation Trust International Inc., the precursor of EcoHealth Alliance. But health got better funding than conservation. "The health projects were so well funded they sort of took over," he told Racaniello. When the Wildlife Trust Inc.'s president left, "I got offered the job about 10 years ago."

My ears perked up when Daszak pointed out that EcoHealth Alliance works closely with W. Ian Lipkin at Columbia. At another point he mentioned that his staff does a lot of late-night video conference calls with their colleagues abroad to keep track of the 30 countries where they have projects. The funding agencies don't like to spend money on travel if it can be avoided, he explained.

How do you pay for all this? Racaniello asked.

"We have a history of being charitable, saving wildlife, we get the public who donate, we have a charity gala, but over 80 percent of our funds comes from federal support," Daszak replied.

Is it difficult to get money for your needs? Racaniello asked.

"Ah...well...it's great when you get it," said Daszak. "...I mean it's pretty brutal, isn't it...it's just hard to spend so much time on the ground and it gets rejected...."

Which made it sound as if EcoHealth Alliance had had a hard time winning grants in the dog-eat-dog competition for US science funding. Which was not the case, as you will see.

Racaniello asked whether EcoHealth Alliance had gotten money from the Gates Foundation.

No, no money from Gates. Gates is interested in orphan diseases, Daszak said. We're not.

That too caught my attention. I expected him to explain that he has a relationship with the Gates Foundation through the Global Virome Project. The Gates Foundation is one of the Global Virome Project's founding supporters, along with Merck, the giant pharmaceutical company.[319] The Project is the brainchild of Dr. Dennis Carroll, the former director of USAID's Pandemic Influenza and Other Emerging Threats Unit, home of the project known as PREDICT. It was Carroll who launched PREDICT, which became EcoHealth Alliance's largest funder. The Global Virome Project's aim is to discover and sequence every virus that might have zoonotic potential much faster than PREDICT was able to do. PREDICT's researchers only found 1,600 new viruses in ten years. The Global Virome Project's supporters figure there are 1.6 million viruses still undiscovered and 800,000 may have the potential to harm humans. PREDICT was just the proof of concept for the Global Virome Project.

While still at USAID, Carroll sought interest from illustrious international parties (and pharmaceutical interests), modelling the needs of the project with $3 million in USAID seed money. He organized a feasibility conference at Bellagio, the Rockefeller Foundation retreat on Lake Como, in 2016. George Gao was invited and was very interested from the start. Daszak is listed as the organization's secretary-treasurer.[320] Carroll is the chair of the project's Leadership Board, which includes a Canadian, Jennifer

Gardy, an epidemiologist formerly with UBC who works for the Gates Foundation. Another board member is a professor at Columbia who was formerly with USAID, the Rockefeller Foundation, and the WHO. Carroll was told by lawyers at USAID that he could not work for USAID and the Global Virome Project at the same time, so he retired in 2019 and is now affiliated with the Scowcroft Institute of International Affairs at the Bush School of Government. Though he told me the Global Virome Project is organized as a 501(c)(3) charity, I could find no record of it. He also astonished me when he insisted that "gain-of-function/dual-use is the biggest biological threat we have.... Dual-use is where we expedite our own demise." Carroll believes that synthesizing viruses has become so simple that the next generation will be doing it in their garages, the way Gates and Jobs first worked on computers. He didn't seem to be aware that PREDICT had funded such experiments through EcoHealth Alliance.

Daszak shared none of this background with Racaniello. Instead, he explained that his funding strategy is to go for government grants, not foundation money, because foundations change their aims every five years, so there's no stability. When his organization changed its focus from conservation to One Health, grant opportunities opened in a very significant way. Daszak explained that the One Health concept (adopted by the UN and various other international entities) posits that animals, humans, and the environment are one, so the health of one affects the health of all (the ecologist's version of the Three Musketeers' oath). He did not mention that this doctrine was invented by EcoHealth Alliance's executive vice president, William Karesh, a veterinarian well-connected to the US Department of Defense. Daszak wanted Racaniello to know that EcoHealth Alliance is helping to build scientific capacity in countries which may have significant viral spillover issues due to land use changes, new roads into former wilderness areas, people living too close to wildlife, and people crowding into wildlife markets where animals that would never meet in the wild are thrown together, raising the threat of dangerous spillovers. EcoHealth Alliance works with the approval of foreign governments, Daszak said. He loves the collaborations EcoHealth Alliance sets up among different types of scientists—mathematicians, economists—people sitting in a room brainstorming, everyone bringing their own perspective to solve a problem. He mentioned that some countries, like China, are getting paranoid about that, but the cure is more collaboration among scientists, more open, transparent collaboration. He told Racaniello he works with labs in about 30 countries through contracts and subcontracts. He likes to have at least one of his technical people in every lab that EcoHealth Alliance funds.

I wondered who that person might be in Shi's lab.

Do you go to the field? Racaniello asked.

Daszak said, "I have two countries that I still, kind of, I'm in charge of, in the organization, China and Malaysia.... I go out there a heck of a lot."

Racaniello wondered about EcoHealth Alliance's surveillance work in Western Asia. But first, he wanted know where Western Asia is. He seemed to think Singapore was part of Western Asia. "Which is about where we are, right?" he asked.

No, Daszak said, it is mainly "the Stans," such as Pakistan and Kirgizstan, where security is a problem. That's why Western Asia is of interest to the DOD (the US Department of Defense), he explained. "There's a lot of pretty unpleasant pathogens out there. Plague is still out there. No one's really done deep surveys of the bats out there...so that's what we're looking at." He said EcoHealth Alliance's grants for surveillance in western Asia come from the Defense Threat Reduction Agency.

Racaniello asked no follow-up questions about that, as if he found it unremarkable that a US charity acts as paymaster for projects in foreign countries that are not good friends of the US, though they are of interest to the US defence establishment. Instead, he moved to EcoHealth Alliance's Rift Valley fever surveillance program in South Africa. Can we get it [Rift Valley fever] in the US? Racaniello wanted to know.

Yes, through meat, said Daszak. The project was trying to work out the seasonal patterns of disease to predict outbreaks.

Racaniello asked about the surveillance and pandemic prediction work EcoHealth Alliance does in other places, and soon Daszak was talking about China. Daszak mentioned his group's discovery of SADS, a disease that affects pigs in a country with an enormous number of pig farms. "That emerged out of bats and just, the capacity for these viruses to pop up seemingly out of nowhere and cause 25,000 pig deaths and for that to go unnoticed for a long time, is incredible...that's the worrying part."

When discussing the problems of wildlife trade and the pandemic risk they represent, Daszak brought up coronaviruses. "SARS coronavirus emerged from a wildlife market, well, the first pandemic of this century, so it's a big event...." He said his group had traced back where SARS came from to wet markets but found it was carried by bats, not civets, so then they went to southern China and did surveillance of bats. "We've now found after six or seven years of doing this over 100 new SARS-related coronaviruses.... Some of them get into human cells in the lab, some can cause SARS disease in humanized mouse models and are untreatable with therapeutic monoclonals, and you can't vaccinate against them.... These are a clear and present danger," he said.

So what do we do about it, asked Racaniello.

"Well," said Daszak, "Coronaviruses...you can manipulate them in the lab pretty easily...[the] spike protein drives a lot of what happens with coronavirus.... You can insert [different spike proteins] into the backbone of another virus...." And off he went describing the gain-of-function experiments planned for the soon-to-be retracted NIAID grant, the kind of work Shi's lab had been doing for years, making coronaviruses with pandemic potential to create a pan-coronavirus vaccine to deal with whatever nature coughs up. He mentioned they were working "with Baric" on these things.

Racaniello didn't raise any questions about the risks of such research.

I waited for Daszak to add a few more details, hoping he would describe his relationship with Shi Zhengli and the WIV.

But he mentioned neither. Not once.

THAT INTERVIEW PROVIDED some information about EcoHealth Alliance's interests, but not much about where its money comes from, where it goes, and nothing at all about Daszak's political reach. The Global Virome Project's website is better on that score. Its page on Daszak claims his research "has been instrumental in identifying and predicting the origins and impact of emerging diseases across the globe." That included "identifying the bat origin of SARS, the drivers of Nipah virus emergence, publishing the first global emerging disease 'hotspots' map, discovering SADS coronavirus...discovering the disease chytridiomycosis as the cause of global amphibian declines." It also listed his appointment to the National Academy of Medicine and as chair of the National Academies of Sciences, Engineering, and Medicine's Forum on Microbial Threats. It mentioned he gives advice to the Director for Medical Preparedness Policy of the White House National Security Staff and is "a regular advisor to WHO on pathogen prioritization for R&D." It even mentioned his membership in the Cosmos Club.[321] Also, it claimed Daszak "authored [not co-authored, please note] over 300 scientific papers" and has been "the focus of extensive media coverage" including: the *New York Times*, the *Wall Street Journal*, *The Economist*, the *Washington Post*, *US News & World Report*, and broadcast appearances on *60 Minutes*, CNN, ABC, NPR's *Talk of the Nation*, *Science Friday*, and *Fresh Air* with Terry Gross.[322]

The Global Virome Project's site was not helpful on where EcoHealth Alliance gets its money. However, as Daszak told Racaniello, EcoHealth Alliance is a 501(c)(3) organization, a tax-exempt charity. These organizations must make publicly available the annual reports they file with the IRS. These reports are supposed to show the origin of the money they take in, who they give grants to and how much they give, how much they spend on lobbying, office expenses, travel, how much they pay their leading executives, and who is on their boards. So I read EcoHealth Alliances' IRS filings. They presented an interesting history.

The charity was formed in 2000 under the name Wildlife Trust Inc. A precursor organization, Wildlife

Preservation Trust International Inc. had changed its name and transferred to this new entity $7.8 million, some of which came from the Gerald Durrell Memorial Fund. The executive director was Mary C. Pearl, a PhD from Yale who had previously worked for another conservation charity and had a relationship to the *New York Times*. Contributions and allocations amounted to $675,000 in the first year. Half went to staff salaries. Mary Pearl got $55,000. The aim of the charity was conservation efforts in various places, particularly in Latin America. A renowned environmentalist, Tom Lovejoy of the Smithsonian, was on the board, as were two Mars candy bar heiresses. Its home office was Prospect Park, Pennsylvania. It got money from and gave money to the Consortium for Conservation Medicine, the group of five universities that hired Peter Daszak as its executive director in 2001. When I looked up the CCM's board of advisors, I found a very distinguished list of scientists including W. Ian Lipkin of Columbia University.

In the report for the year ended 2002, Daszak was listed as paid staff of the Wildlife Trust but was also described as executive director of the Consortium for Conservation Medicine. The trust paid him $80,000 a year plus $20,211 in benefits. Government contributions at $231,921 were only 10 percent of total revenue, most of which came from donations.

The two groups merged in 2002–2003, the year of SARS. Daszak became vice president of the Wildlife Trust. Daszak appears to have brought money with him in the form of a grant he won from the NIH/NSF while he was the consortium's executive director. That grant supported his work on the famous paper on bats as the reservoir of SARS published in *Science* in 2005.[323]

By 2005–2006, the organization had moved to New York, to an office building on West 34th street in Manhattan. Its total assets were $8.9 million. Two-thirds of its new money came from donations, about $1.9 million from government grants.

The year ending in June 2009 marked a big change. At first, Pearl was still president, earning close to $200,000, but Daszak replaced her as president that year. Grants and assets were down from the year before, but health research in 20 countries was part of the organization's new goal to achieve "sustainable solutions to preserve nature and protect human health from wildlife borne diseases." Government grants were $1,307,988. Donations were $3,902,487. Instead of listing the foreign organizations they gave grants to, as shown in earlier returns, now only regions were named as recipients. East Asia and the Pacific got a grand total of $90,000. Why were they hiding the names of recipients? Were they handing money to governments? To individuals? Normally, the names of recipients are listed in returns.

The report for the year ending June 2010 showed Daszak as president and the name of the organization changed to EcoHealth Alliance. Total revenue was over $5 million. The return mentioned that its goals included those of the USAID PREDICT pandemic threat program, to "assess capacity and develop plans for the implementation of wildlife surveillance support. Develop models of disease risk and spread; implement a smart...wildlife surveillance strategy to identify and target high-risk wildlife in the regions most vulnerable to zoonotic disease emergence." There was a grant from NIH/NIAID for Nipah surveillance in Bangladesh, to model bat-to-bat studies, to model hotspots for viral diversity in bats, and to identify new viruses in bats. Grants from the government totalled $3,159,174, more than half the organization's total income. East Asia and the Pacific got $184,572 for emerging disease research. This might have marked the second year the WIV got research money from EcoHealth Alliance, but there was nothing on this return to tell me that.

The next year, 2011, total contributions doubled to over $9.5 million, of which two-thirds were government grants. Daszak's salary and benefits climbed to $252,000. Most of the charity's costs were salaries, bonuses, office, travel, and conference costs. About $1.8 million went to groups outside the US, most to South Asia. William Karesh became vice-president for health and policy. Karesh is on the WHO's International Health Regulations roster of international experts, has appeared at the Aspen Institute, is on the Council on Foreign Relations, and has had something to do with an organization called the Bipartisan Commission on Biodefense. His LinkedIn shows connections to the Department of Defense, the Department of Agriculture, and the World Bank. Daszak's salary and benefits rose to $290,000. South Asia got by far the largest chunk of largesse aimed at foreign shores. Several universities in the US got about $100,000 each, except for one. The Mailman School of Public Health at Columbia got $735,988. The Mailman School is where W. Ian Lipkin hangs his hat.

In 2013, revenue was down, and the board lost its Mars heiresses. Government grants were by far the largest source of income at $7 million. Grants to East Asia and the Pacific climbed to $288,000, but that was still only a third of what was spent in South Asia. Grants given in the US climbed to $1.4 million. The Mailman School got $1,172,415, more than ten times any other US institution, which that year included Princeton, the University of California at Santa Cruz, and Stanford. Despite the revenue decline, Daszak's total pay and benefits rose to about $300,000.

In 2014, a new government agency appeared on the returns. In addition to PREDICT and NIAID's emerging infectious disease programs grants, the Defense Threat Reduction Agency of the Department of Defense provided money to create a tool to rapidly diagnose outbreaks and provide "early warning necessary for countering biological threats." The contract required the creation of a network and a tool that could handle a high volume of data in real time, with feeds connected to "EcoHealth Alliance's emerging infectious disease repository." That was the first and last mention I saw of such a repository. East Asia got $204,630. South America got $553,000. South Asia got $933,000. The Mailman school got over $1 million. That year, Daszak's LinkedIn got an endorsement from Shi Zhengli. She was identified as "one of China's leading virologists. She conducts world class virology at a globally recognized center of excellence. Her work includes identification of bats as the reservoir of SARS and analysis of SARS-CoV receptor binding."[324]

By 2015 total revenues were up to $10 million. PREDICT 1 had been replaced by PREDICT 2, another five years of funding. South American grants were greatly reduced, grants to East Asia and the Pacific were essentially the same, as was the grant to the Mailman School in the US—another $1 million.

In 2016, revenue increased to almost $12 million, but there was an audit. The former well-paid CFO was replaced by a less well-paid CFO.

In 2017, total revenue, almost all from government grants, rose to $14 million. For the first time since the early days of the organization, in this filing, grants made were linked to specific countries, though institutions were not named. Malaysia got $431,055, but China got more—$476,362. In the US, the Mailman School had to make do with $849,802.

In 2018, revenue jumped to $16 million. USAID's contribution was $11 million; the Department of Defense (Defense Threat Reduction Agency) gave $2,358,326. Health and Human Services (Homeland Security) gave $670,988. NIAID was not listed (though it paid about $580,000 as part of a continuing grant). Daszak's salary, bonus, and benefits came to just over $400,000. The total of grants and gifts received over the five-year period ending in June 2018 was shown as $59,247,130, most of it flowing from the US taxpayer and aimed at projects of interest to the US government. Foreign grants made were only recorded by region. East Asia and the Pacific got $1.6 million; East Asia and the Pacific also had the greatest number of offices listed—six.

The last filing available was for the year ended June 30, 2019, just before the pandemic began and the PREDICT 2 program came to an end. Total revenue had reached almost $18 million with $15.5 million from the US government. Grants in the pipeline totalled $3.3 million. The Defense Threat Reduction Agency was paying for the creation of a serological assay to be used in Malaysia to detect antibodies to henipa- or filoviruses in bats, animals, or livestock. It was paying for the Rift Valley fever project in South Africa and "understanding the risk of bat-borne zoonotic disease emergence in Western Asia" for about $740,000.

Once again, the Mailman School at Columbia University got the lion's share of US donations, but it was joined by the Henry Jackman Foundation, a close-to-Congress flow-through charity seeking to improve military medicine. East Asia and the Pacific had seven offices and got $2,365,734. One organization was listed as a recipient of $195,450—the "Institute of Microbiology of the Chinese Institute," likely the Institute of Microbiology of the Chinese Academy of Sciences, one of George Gao's places of employment. Daszak's total compensation had risen to $410,000.[325]

The audited report filed for 2020 showed the organization's total revenue cut in half to a little over $8 million.

GovTribe, a platform which displays all US federal government grants and contracts, showed that EcoHealth Alliance had pulled in $109.7 million since 2013 in grants and contracts from a plethora of US federal agencies, some of which are not listed on its IRS returns. NIAID has never been the main the source of EcoHealth's financial well-being. The bulk of its funding came from PREDICT and from contracts let by the Defense Threat Reduction Agency and the Department of Defense. Freelancer Sam Husseini reported in *Independent Science News* that EcoHealth Alliance's military funding as of December 2020 amounted to almost $40 million.[326] The exact number since 2013 is $38,999,593. Husseini also confirmed with the University of California, Davis, where Professor Jonna Mazet directed PREDICT on behalf of USAID, that EcoHealth Alliance's share of that $207-million program was $64 million.

In other words, EcoHealth Alliance has been deeply involved in dual-use research in foreign labs beyond the reach of US regulatory agencies yet funded by US agencies and departments. Gain-of-function experiments supported by the NIAID can only be done in BSL-3 labs, yet Shi told *Science* that until recently, she had done coronavirus work in BSL-2s as well as BSL-3s. Funds from NIAID and PREDICT sent to the WIV may have been used for work done in conditions considered unsafe under US rules. This is the science equivalent of major brand retailers sourcing the products they sell from Bangladesh, where working conditions are dangerous, but costs are rock bottom.

In effect, EcoHealth Alliance acted as a cut-out, a new definition for charitable purposes. Its filed returns demonstrate that Daszak has been a very efficient political operator, distributing US money through sub-contractees to advance US foreign policy, domestic security, and defence interests in countries not always friendly to the US. In science, those who can distribute research money acquire power. When that money is given to serve a nation's geopolitical interests, the transactions become more complex. When the US government cut the NIH/NIAID grant to EcoHealth Alliance (saying in its first letter to EcoHealth Alliance that the project no longer met NIAID's objectives), Daszak raised a public fuss decrying the intrusion of politics into science funding, which should be decided on merit alone. His voice was amplified by those who saw the retraction as a Trumpian abuse of US science institutions. In response, NIH/NIAID wrote another letter to EcoHealth Alliance. As *Vanity Fair* later reported, that letter, dated July 8, 2020, stated that the agency was concerned that EcoHealth Alliance's sub-contractee, the WIV, was doing experiments that could pose a risk to the health of its workers, the people of China, and the people of the United States, without regard to proper biosafety rules and regulations that all grantees must adhere to. It asserted that EcoHealth had failed to properly monitor its sub-grantee in that regard. The grant was therefore suspended until EcoHealth Alliance complied with a list of demands.

First on the list was the provision of a sample of the "SARS-CoV-2 virus that WIV used to determine the viral sequence." The WIV also needed to explain the disappearance of Huang Yanling, "whose lab web presence has been deleted." (Huang Yanling was the subject of Matthew Tye's YouTube documentary. She was a student at the WIV who was rumoured to be pandemic patient zero. The WIV first denied knowledge of her, then said she'd left in 2015 and that she was alive and well but offered no proof. Information about her had been removed from the WIV's site.) In addition, NIAID wanted responses to the State Department's concerns about safety at the WIV, an explanation for diminished cell phone traffic at the WIV in October 2019, and to know why it appeared roadblocks had been set around the campus for one week during that month. It also needed to know why the WIV "failed to note that the RaTG13 virus...was actually isolated from an abandoned mine where three men died in 2012 with an illness remarkably similar to COVID-19." (Shi would insist to *Science* a week later, that RaTG13 was never isolated, only sequenced.) In addition, the letter said:

> EcoHealth Alliance must arrange for WIV to submit to an outside inspection team charged to review the lab facilities and lab records, with specific attention to addressing the question of whether WIV staff had SARS-CoV-2 in their possession prior to December 2019. The inspection team should be granted full access to review the processes and safety of proce-

> dures of all of the WIV field work (including but not limited to collection of animals and biospecimens in caves, abandoned man-made underground cavities, or outdoor sites.) The inspection team could be organized by NIAID, or, if preferred, by the US National Academy of Sciences.[327]

Daszak did not make this letter public, but he said he could not meet its demands. Aside from the one regarding Huang Yanling, why would that be? When a charity accepts money from a government agency to do things that it subcontracts to a foreign organization—even one owned by a foreign power—it remains accountable for how the money is spent. Daszak had remade a conservation charity into a vehicle for investigating things important to the defence of the US, things a foreign government might like to hide, and in the case of China, has hidden and continues to hide. EcoHealth Alliance's work has never been about scientific merit alone. He told Racaniello that he does "applied science," and that is the case. EcoHealth Alliance has applied American eyes to labs in dangerous parts of the world where experiments are done that might not pass muster under biosafety rules in the US. No wonder, as *Vanity Fair*'s Katherine Eban would later document, some high up in the State Department repeatedly warned colleagues not to investigate the possibility of a lab leak from the WIV because it could come back to haunt them.[328]

Neither the WIV nor Shi Zhengli's names appear as grant recipients on any of EcoHealth Alliance's IRS returns. The IRS might want to ask about that.

13. But Then Again...

FALL 2020. I cannot tell you what the weather was like because I had stopped recording it in my notebook. There was no point. The days had acquired pandemic-induced sameness, an unvarying flatness independent of weather. Every day I rose, read the papers, worked, ate, walked around the neighbourhood for exercise, left the sidewalk for the road to avoid those who wore no masks. I washed my hands. I ordered food online. I greeted like a friend the delivery man who brought the groceries because he was one of the very few people who knocked on my door. Rosh Hashanah and Yom Kippur passed without celebration, without distinction. Every day I wondered whether today would be the day the federal government would send me those 8,000-plus pages recording the number of accidents at the National Microbiology Laboratory in Winnipeg. Every day I was disappointed.

As Canadian case counts began to rise again, despite the care so many took to do what public health officials mandated, anyone could see a second wave was building. Anyone, that is, except officials in Ontario who had failed to take advantage of the summer lull to train and hire more PSWs to work in long-term care. As outbreaks in these homes began again, as deaths among the helpless elderly mounted ever higher, I took to muttering the phrase "criminal negligence causing death" as if saying these words could change the facts.

I had at least made progress exploring the secret of the virus's origin. I had covered two desks with files, files piled on top of files, piles of files threatening to slide to the floor. One desk also held a large Amazon box bursting with newspaper clippings and printouts. HP was doing land-office printer ink business thanks to me. By day I felt confident enough about the narrative coming together in my head that when anyone asked what my research revealed, I said I had almost concluded that SARS-CoV-2 entered the human population due to a lab accident. I wasn't sure which lab; there were so many possibilities—a military lab in Chengdu? In Nanjing? Did it leak from the Wuhan Center for Disease Control and Prevention, carried into the streets by an asymptomatic researcher who trapped wild bats without wearing protective gear? Or did it come from Shi's lab when someone opened a fridge, took out a frozen sample from those miners' lungs, extracted RNA, assembled a genome, synthesized a virus, and passaged it in macaques who fought back?

But by night, certainty shrank, questions ballooned. By night I had to restrain myself from climbing out of bed and going down to my desk to check for the umpteenth time that something important *had* been withheld from an article, that there was a *real* pattern to the way this withholding had been done, not one created by my imagination. By night, an inner voice accused me of failure to consider all alternatives.

You were predisposed to that conclusion from the beginning, it whispered. *Maybe you pushed together all the facts that fit and deep-sixed the rest. Maybe you just don't want to admit that nature is way too vast and mutable to comprehend, never mind capture in a test tube, and forget about predict, which is why you want to find a human hand at work in this. That would be reassuring. It would suggest we can prevent a recurrence, that we can control our circumstances, that we are the great conductors of the symphony of life. And who are you to do this anyway?*

I was sure Stephen, my husband, would have said these things if he'd been alive to challenge me. We would have had a foot-stomping-slam-the-door fight about whether the information I'd gathered about the likelihood of a lab leak and cover-up was sufficient to conclude anything. I would have said, *most scientific hypotheses and assertions, like most journalist's investigations, are built from compilations of evidence; the process is inductive, inferential, additive (or subtractive). In science, peer-reviewed published results are believed until disproved, and in journalism it's the same, except the disproofs come a hell of a lot faster.* He would have said *yes, okay, but you're investigating the cause of a deadly pandemic. You seek to cast blame. For blame you need the equivalent of fulfilling Koch's postulates; you need to find evidence that the virus existed in a sample in a lab before the pandemic.* At least I think that's what he would have said: but maybe that's what I would have said to him if the shoe were on the other foot, if he were the one ploughing through articles written in prose so dull it induces something like narcolepsy. If he had been around, and if he had said it, I would have said *Koch's postulates are all very well, but when Linfa Wang and Peter Daszak and Shi Zhengli and colleagues asserted in 2005 that SARS spilled over from horseshoe bats, Koch's postulates were not fulfilled, but* Science *published anyway. The fact that China has not opened its labs to scrutiny or produced samples of anything means something. It has to mean something.*

Night after night I lay in bed, eyes fixed on the ceiling, trying to reassure myself about my reasoning by listing the indicia of a lab leak. I always started with China's guilty behaviour, because a government's guilty behaviour, unlike a person's guilty behaviour, *does* mean something. You can be sure many clever people in that government debated what to do and what not to do before a path was selected. That's why inferences can be drawn when vital information is withheld.

There was that five-day delay before George Gao at China's CDC started an investigation into the pneumonias in Wuhan, though doctors and three private genome companies had reported that a SARS-like coronavirus was making patients sick. There was the order to destroy all those early samples or turn them in to the correct authorities. There was the failure to accept help from experts willing to come to China, an offer made early in January by Robert Redfield, the head of the US CDC. There was the failure to promptly inform the WHO of a SARS-like disease though China was required to do so by international regulation. There was the attempt to hold back publication of the virus's sequence. There was the failure for weeks to acknowledge human-to-human transmission, though that was obvious from the earliest days. There was the failure to shut down Wuhan until after millions had climbed aboard trains and planes for the annual New Year trek home to friends and family, which allowed the worldwide spread of the virus, which further muddied the waters about its origin. All this suggested the government of China knew exactly who to blame as soon as it became aware of the disease in Wuhan, and it wasn't a wild bat in a southern cave or a pangolin in a wet market.

And what about the way China agreed in May 2020 with the WHO Assembly's resolution that there should be an independent investigation into the virus's origin, but then dragged its feet for another seven months negotiating the terms? All the while, China's officials publicly insisted that SARS-CoV-2 came from somewhere other than China. First, they claimed it was brought to Wuhan by American soldiers who participated in the military games in October 2019. Then it came from Spain, and finally it came on frozen food imported from abroad. Almost as troubling was the way leading Western scientists with relationships in China, or beholden to EcoHealth Alliance or to US science funders looking to avoid trouble, published so quickly in *Nature* that the virus was most likely natural in origin, though they had no evidence of that. And along came that statement in *The Lancet* describing those who even asked about the lab-leak possibility as conspiracy theorists. The *Nature* letter insisted the virus's sequence showed no sign of human manipulation, that furin cleavage sites could be seen in other viruses (though not close relatives to this one), but failed to fully acknowledge that gain-of-function/passaging experiments would leave no trace and that lab-leak incidents are frequent, not rare. Yet due to the authors' stature, and the journal's reputation, that *Nature* letter became like Horatio at the bridge, holding off for almost a year the publication in major peer-reviewed journals of articles making the opposite case. And what about the order from the very top that all scientific publications on SARS-CoV-2 by Chinese scientists had to be approved by the powers that be and their publications coordinated?

Suppression of facts and orchestration of publications were only suggestive, not conclusive.

The miners' story was more important. Their symptoms matched those of a SARS-CoV-2 infection. Yet China did not report to the WHO on their illnesses and deaths, though it was required by international regulation to do so. The miners' story was told only in theses: theses are never widely read. And that too seemed deliberate, possibly coordinated. An article published in a peer-reviewed journal in 2017 by Canping Huang, based in part on his doctoral thesis (both done under George Gao's tutelage), mentioned the mine, but not why it mattered. Yet his thesis contained important details about what happened to the miners, including who investigated their illnesses and what was found. Shi snuck RaTG13, which came from bat feces found in that mine, into the February 2020 *Nature* article on SARS-CoV-2, describing it as the closest sequence to SARS-CoV-2. Her *Nature* article did not mention where RaTG13 came from, when it was sequenced, that her lab had gone to that mine to take samples at least four times in an 18-month span, or why. Shi told a bit of the story to *Scientific American* months later, possibly because Daszak had mentioned the Mojiang County mine as the origin to RaTG13 to Jon Cohen in *Science* at the end of January 2020. But she diverted attention from the connection between the miners' deaths and coronaviruses by claiming falsely that they died of a fungal infection. She said nothing about still having samples from their lungs in her possession. Then came the revelation in her July email to *Science* that she had sequenced RaTG13 in 2018, but not published it, though she had published a portion of its sequence under a different name years before. When asked by *Science* why she did not sequence the whole genome right away, she said she didn't have sufficient sequencing resources and the RdRp was not close enough to SARS to be of interest. That made no sense: it was a coronavirus from a mine where people got sick and died. As the PhD thesis revealed, those samples had tested positive for a coronavirus. At minimum, you'd think she would have sequenced RaTG13 immediately to compare it to whatever she found in the miners' samples.

But none of that was conclusive either.

The most convincing evidence for a lab leak was the virus itself. Most of the epidemiological papers agreed that sometime in November 2019, SARS-CoV-2 exploded into the human population in Wuhan, home to the Wuhan Institute of Virology, the world's largest repository of SARS-like coronaviruses, and to one other lab studying bat-borne viruses, the Wuhan CDC. From the beginning, SARS-CoV-2 appeared to be extremely well adapted to humans. It entered cells through the ACE2 receptor, but it could also enter human cells in two other ways, aided by its unusual furin cleavage site that enabled the merger between human host and virus. Structural studies showed SARS-CoV-2 has a much stronger binding affinity for the human ACE2 receptor than SARS, but only a weak affinity for the bat ACE2 receptor. We know that it was well adapted to humans from the start because hundreds of scientists around the world posted full genome sequences as the disease appeared in their communities. These sequences barely changed for four months, though SARS-CoV-2 infected many millions around the world. It didn't produce a more infectious variant until nine months after it first appeared. If it had not been so well adapted, the pressure to fit itself to its human hosts would have produced many mutations very quickly. That is what happened with SARS: it changed rapidly, adapting itself to humans as it spread among us, though only 8,098 of us got it.

Isn't that enough, Stephen? I asked the ceiling.

ONE DAY, I downloaded the list of people associated with *Virologica Sinica*, the WIV's journal. I had not quite given up all hope of speaking with Shi. I hunted there for someone who might be willing to plead my case to her. The board of advisors was primarily composed of leading China-based scholars, including: Hualan Chen, who heads the second BSL-4 in Harbin; George Gao, listed as director of China's Center for Disease Control and Prevention and professor at the Key Laboratory of Pathogenic Microbiology and Immunology, Institute of Microbiology, Chinese Academy of Sciences; Linfa Wang, honorary professor at University of Melbourne, professor at Duke-Nus Medical School in Singapore, and member of the Chinese Academy of Sciences. I thought a Canadian connection might be more helpful.

A significant number of people on *Virologica Sinica*'s editorial board had major jobs in the West, including a fellow with the Institute Pasteur in France, a group leader at the Pirbright Institute in the UK, the chief scientific director with the US Army Medical Research Institute of Infectious Diseases, and the virology lead at the NIAID/NIH. The list also included several professors at US schools of note: University of North Carolina at Chapel Hill, UCLA, UC Santa Barbara, University of Florida. Many had Chinese names, suggesting they might have taken advantage of Deng's policy, gone to the US for higher education, and stayed. There were only two Canadian-based people on that list. One was Xiangguo Qiu, who listed her workplace as University of Manitoba.

A more likely prospect appeared higher up on the masthead, on the list of associate editors. Basil M. Arif was identified there as Scientist Emeritus with the Great Lakes Forestry Centre, Canadian Forest Service, in Sault Ste. Marie.

I called the forestry centre. A security guard answered the main number. He said almost no one was working in the centre's labs thanks to the pandemic. When I asked after Dr. Arif, he said, "He's a world traveller, that guy." He had no idea when he would be in, but he promised to pass a message if he saw him.

I found a home phone number for him. No answer. I left a message. Then I looked him up on ResearchGate. One paper jumped out from the many he'd done on viruses that infect insects. He'd coauthored a 2017 paper describing the synthesis and rescue of a baculovirus, at that time the largest virus synthesized. His co-authors, except for Just M. Vlak of Wageningen University in the Netherlands (also on the editorial board of *Virologica Sinica*), were all researchers at the WIV. Both the genome synthesis and the rescue had been done in the WIV's labs, and all funding came from agencies of the government of China. The last author on the paper was Zhihong Hu. She was a co-author on the 2005 *Science* paper on bats as the probable reservoir of SARS, and the former director general of the WIV who opened it to the study of bat-borne coronaviruses. No Zhihong Hu, no Linfa Wang, no EcoHealth Alliance relationship with Shi Zhengli, and the rest would not have become history.

Okay, I thought. If I can't get to Shi, I'll settle for Zhihong Hu. Maybe Arif can talk her into speaking with me.

Reading that rescue paper, any lingering doubt I had about whether researchers at the WIV could synthesize a coronavirus and insert a furin cleavage site without leaving a trace vanished. Baculoviruses have a double-stranded DNA genome with six times as many base pairs as SARS-CoV-2. The paper detailed the significant number of steps necessary to synthesize a baculovirus from a genome sequence alone. The known sequence of a baculovirus, held in the WIV's repository of microorganisms, was used as the template. Small pieces of the genome were fabricated with overlaps so they could be aligned using polymerase chain reaction techniques. These small fragments were joined into larger ones. The larger ones were introduced into yeast cells, where they recombined to form a whole DNA sequence. That sequence was checked for accuracy against the original sequence. Then this synthetic genome was amplified in *E. coli* and introduced into insect cells, which produced active baculovirus. The article included a full-page illustration that laid out all the steps required to get the job done. It was published in the American Chemical Society's synthetic biology journal. Yes, there is a whole journal dedicated to synthetic biology—that's how many labs are doing this work.

The abstract opened with the following statement:

> Synthetic viruses provide a powerful platform to delve deeper into the nature and function of viruses as well as to engineer viruses with novel properties.

It ended with:

> We validated a proof of concept that a *bona fide* baculovirus can be synthesized. The new platform allows manipulation at any or multiple loci and will facilitate future studies such as identifying the min-

imal baculovirus genome and construction of better expression vectors...[329]

As the article made clear, baculoviruses have long been used for the biocontrol of insect pests like spruce budworm, which can destroy softwood forests. But after countless interactions with their insect hosts, baculoviruses became more innocuous. The main point of the project was to make a synthetic baculovirus to test which genome changes might make it more effective. They aimed at making a better baculovirus vector to carry the genetic instructions to make certain proteins into insect cells. In other words, they made a platform for gain-of-function experiments.[330]

I looked for anything I could find on Basil Arif. The *Virologica Sinica* website showed that he'd been an associate editor for at least twenty years, since *Virologica Sinica* first became available in English in 2000. The Sault Ste. Marie Forestry Centre mentioned him in a small article on the 75th anniversary of its founding. It was set up to deal with a massive outbreak of spruce budworm in Ontario forests. Scientists from around the world were hired to develop biological control methods. Basil Arif arrived in 1972 and "became a world leader in the molecular biology and genomics of insect baculoviruses and entomopoxviruses. Dr. Arif and his team developed a genetically modified spruce budworm virus that has a faster mode of action."[331]

He was also mentioned in a Sault Ste. Marie news story titled "Baghdad-born Saultite pleads for peace." There was a picture of a man in a mauve tuque and white lab coat smeared with ketchup. On a cold day in March 2003, Arif and his wife (also wearing a lab coat smeared with ketchup) had taken to the streets to protest the sanctions against

Iraq and the impending American invasion.[332] Arif explained to the reporter that he had more reason to hate Saddam Hussein than "anyone in North America" because after Saddam's invasion of Kuwait, which provoked the 1991 Gulf War, his father, his mother, his younger brother, aunt, cousins, and friends were all killed by the US cruise missile bombardment. "I owe it to my family to say, for God's sake, don't do it again," he told *SooToday News*.[333]

I found another article by Arif published in *Virologica Sinica* in 2016. Co-authors included Hedi Zhou and Zhihong Hu of the WIV, as well as Bryan Eaton, a retired Australian scientist formerly with CSIRO and a colleague of Linfa Wang's. He'd also been a co-author on the 2005 *Science* paper on bats as the probable reservoir of SARS. This 2016 article described the history of the Zika virus then sweeping through the Americas. In 1947, British epidemiologist George Dick was studying yellow fever virus at a research station set up by the Rockefeller Foundation in Uganda. The station was in a forest called Zika, which is near Lake Victoria between Kampala and Entebbe. Dick placed some "sentinel" macaques onto rickety platforms high in the forest canopy, hoping they'd be bitten by mosquitoes carrying yellow fever. One of the macaques got sick with something else. Dick isolated the virus from the monkey, and then isolated it again from the mosquitoes that carried it. While it paralyzed mice, which meant it could attack neurons, in monkeys the disease was mild. Dick determined it was new and named it Zika, after the forest. Three people in Nigeria got sick from it in 1954. A researcher named Simpson caught it at the same research station in Uganda in 1964, carefully noting his symptoms, extracting his own serum, infecting mice with the isolate. Eventually, 12 strains were isolated, but Zika remained innocuous for the next 43 years. Then, in 2007, it caused a human epidemic in Micronesia, which is in the western Pacific, more than halfway around the world from Zika Forest in Uganda. There were more outbreaks in Tahiti and in Moorea a few years later. In 2016 it appeared in Brazil (and Mexico and eventually 50 more countries). Pregnant women bitten by infected mosquitoes gave birth to babies with tiny heads. Somehow, an innocuous virus discovered in an East African forest 70 years before had made its way to Brazil, entered human fetal cells, and wreaked neurological havoc. Why had it changed? No one knew.

The article related a similar story of the discovery and transformation of a virus called chikungunya. It was first isolated at the same East African research station in Uganda where Dick had discovered Zika. Initially it was carried only by the *Aedes aegypti* mosquito and was also "innocuous." But in 2005 it caused an epidemic in Reunion Island in the Indian Ocean, killing 260 people. The only difference between the killer strain and the original found in Uganda was one mutation in the virus's envelope protein. This single mutation gave chikungunya the capacity to enter new mosquito hosts that like to bite people and can be found everywhere, not just in Africa. "This one mutation allowed the virus to be carried by and multiply in the Asian tiger mosquito...which is known for its anthropophylic bias and exists in many countries in Europe and North America." By 2007 chikungunya had expanded its range far beyond that East African forest and sickened 1.4 million people.[334]

At the end, there were a few paragraphs memorializing George Dick, who had taught both Arif and Eaton microbiology—just the two of them in the class—at Queen's University in Belfast.

So now I knew Basil Arif came from Baghdad. Most of his family had been killed in the first Gulf War. He had studied in Belfast under a great epidemiologist. He was the classmate of an Australian scientist who had worked with Linfa Wang, Peter Daszak, Shi Zhengli, and Zhihong Hu on that groundbreaking 2005 *Science* paper. Arif got his PhD at Queens University in Kingston, Ontario, but somehow became an assistant editor on the WIV's *Virologica Sinica* journal in 2000. Before SARS. Long before Shi began to work on coronaviruses. Why and how?

To my chagrin, his article also forced me to think again about the lab-leak narrative. Zika had started out in Africa as a virus innocuous to humans and stayed that way until 2007, but by 2016, it was everywhere and could break into human fetal cells. A single mutation had transformed the equally innocuous chikungunya to something deadly. Could SARS-CoV-2 have a similar history, innocuous until suddenly it wasn't? Was it possible that the authors of the proximal origin piece in *Nature* had it right?

ARIF CALLED ME back. He had a nice voice, a friendly manner and was willing to be interviewed. Unfortunately, the day before he called, I'd hurt my back. I had to put off the interview for a week. He said he understood about back pain: herniated discs had bedeviled him for several years, but he'd recovered, and two years ago he'd swum the Sea of Marmara from Europe to Asia.

As I waited to heal, unable to do much more than swallow painkillers, I played with all kinds of story lines about how Arif connected with the WIV. They all turned out to be crap. He was born in Baghdad when the Hashemite king was still in power. Then came the revolution of 1958, and the king was shot dead. This was followed by another coup in 1963, involving one of two brothers named Arif (no relation). When the Baathists came to power in 1968, the second Arif brother was taken to the airport at gunpoint by Saddam Hussein and sent into exile in the UK. As Basil Arif was growing up, Baghdad was still a lovely city with night clubs, gardens. After the Baathists came to power things "went down." He was only 16 when he left Iraq to go to school in the UK. Because he didn't have much money, Northern Ireland, being inexpensive, was attractive. Studying with Dick in Belfast had been a wonderful experience. But he didn't know Dick was the first to isolate and characterize Zika until he researched that 2016 article: "I almost fell off my chair."

He graduated from Belfast in 1967, before the Troubles. He had scholarship offers to do his doctorate from various schools in various places, but a post-doc he knew kept telling him he should do his PhD in Canada. It was "the best decision I ever made," he said. He studied at Queen's University in Kingston (which he liked because it's on a lake) under Peter Faulkner, who also specialized in yellow fever. When he was doing his post-doc with Faulkner, a molecular biologist named Tom Angus came to the lab and ended up inviting him to come to Sault Ste. Marie. He loved the work there and had what he called "a fabulous career."

So how did you end up connected to the WIV in China? I asked.

He took two sabbaticals. For the first he went to Cologne in Germany because he spoke a bit of German and thought he'd improve it. To his surprise, all the seminars were given in English, so his German did not improve. His next sabbatical was at Wageningen University in the Netherlands.

There he met Zhihong Hu. She'd come from China in 1995 to do what he called her "sandwich PhD." That meant she was expected to do work in two labs at the same time—one in the Netherlands, the other in China. He wasn't sure whether this agreement was between the WIV and Wageningen University, or between the Chinese Academy of Sciences and the Netherlands. As in Germany, no one spoke Dutch; all seminars were in English. But Zhihong Hu's English was bad, so "I helped her with papers." In 1998, she invited him to the WIV to give seminars. She was director general of the WIV for eight years, he said, and the "science programmes flourished under her tenure."

After the invitation to give seminars, "they invited me as visiting scientist professor, they nominated me for a special prize, we collaborated on a number of projects, and I helped some PhD students. We became close friends with Hu and her gorgeous daughter, who is at Queen's now."

Well, what about Shi Zhengli, I asked, holding my breath. Did you get to know her?

"Zhengli is a very dear friend of mine," he said.

So I told him what I had been learning and what I'd been thinking. I explained I had tried to reach Shi for comment but got no response. What did he think: was she in trouble with the government over the accusations of a leak from her lab? When NBC went to the WIV to do an interview about it, she wasn't interviewed. Why would that be?

"She's not in trouble," he said. "Not in China."

He explained that the way the Chinese system works, the Chinese Academy of Sciences would stipulate that the boss of the institution would give the interview. "When you're the head of the WIV, you must be a member of the Communist Party," he said. "Some scientists I know couldn't give a hoot about that...within the institution, they have official members of the Communist Party. They are always there. Having said that, the research there is totally independent of political pressures."

Are you sure about that? I asked. I mentioned the withholding of important facts from journal articles, the obfuscation about human-to-human transmission of SARS-CoV-2.

"That's my experience," he said. "After COVID-19, maybe. Before that...."

So how often, I asked, have you been there?

More than 20 times, he said, "and no political pressure, I can still talk freely...I would think right now that Zhengli wants to bury her head in the lab and not talk about this coronavirus at all....Zhengli went back four years to see if this virus was in the lab. It's probably from Southeast Asia, and then an intermediate animal and from that to people."

That suggested he'd either spoken to or corresponded with Shi Zhengli or Zhihong Hu about this. Shi hadn't specified in *Scientific American* that she'd gone back only four years to see if she had a sequence that matched SARS-CoV-2, at least not in anything I'd read. She'd told *Scientific American* she'd searched everything. If she'd only gone back four years, that might not necessarily take her to whatever had been extracted from the miners' samples.

And there was nothing about the coronavirus sequence that suggested changes made in a lab, he added.

But you guys made a baculovirus from scratch. What's to say SARS-CoV-2 wasn't done that way too? I asked.

He paused to educate me. The baculovirus had been made according to the known sequence of a virus held in their repository. "If you synthesize, you delete unnecessary genes to arrive at a minimal genome that is more efficient.... We tried to make a more efficient expression vector.... If you make a virus, to synthesize it and introduce certain serious mutations you will *see* that—[you will see] there's a big insertion here."

Well SARS-CoV-2 has insertions, there's the furin cleavage site...a furin cleavage site is unusual in a coronavirus, I said.

You've got me thinking, he said.

I told him about the miners' story, about the samples sent to Shi's lab that she'd never published on or mentioned though she had the opportunity. He began to talk about how politics takes science to another level, how China is an "autocratic system."

But science has to be transparent to be science, I said.

"You're thinking like a Canadian," he said. Scientists are transparent unless it's taken to another level, and then scientists do what they're told to do, he said. "Hu told me, 'we have no choice.'"

Although he said he never had political discussions with any scientist in China and had no idea about Shi's political leanings, if any, one particular story Hu told him years ago still stuck in his head. She had mentioned a concert held in a square in the city. All around the square there were apartments where people lived who had not paid for tickets. "The police went around to people's doors and said they were not allowed to sit on their balconies because they didn't pay. And people did what they were told to do." No one went out on their balconies to listen.

He said he would ask Hu about the miners' samples. But he thought they could have been set aside under orders from the top. "Just like Trump giving orders to his doctor so the doctor will not say when he last had a negative test," he said.

But doesn't withholding information reduce trust in what is published? I asked.

Of course, he said. There had been a scientist at the forestry centre who was intentionally inaccurate in one of his documents. Arif went to his managers to say this guy lied. "With lies, credibility disappears forever. If you're hiding something, what can we trust?"

Why would Shi say the miners died from a fungal infection when the doctors said a virus?

"I can't see Zhengli saying that. She has published in the best journals in science."

I assured him that she had said it in *Scientific American*. And what about failing to sequence those samples taken from them? Would she have done that?

"Her curiosity [would] be killing her," he said.

I asked him about her character. I told him about her email responses to *Science*, in particular what she said about Trump owing them an apology. I said that didn't sound like a careful scientist talking. Would she say something like that? Was that in character?

"I can't see her saying that. Come on. For Heaven's sake. She is far from that. Not aggressive.... I can't see her saying, 'He owes me an apology.' I know she's been bombarded and chased to the point of taking a jump. Her science is no longer important. I don't know how she can work. When I asked Hu, she'd say she's very tired, very busy. I hope to hell they did not destroy that fantastic scientist."

I told him Shi told *Science* that she'd not been able to isolate RaTG13, the closest known sequence to SARS-CoV-2, that she'd only managed to put together the sequence. I told him that when she was asked why she hadn't sequenced the whole genome right away, instead just doing an incomplete RdRp, she'd told *Science* she didn't have financial or staff resources to sequence everything, not until next-generation sequencing became available, and the RdRp wasn't close enough to SARS to be of interest.

"They have more money than you can shake a stick at," he said, sounding like Kobinger. "If she said 'no money to sequence,' I do not believe her."

He said researchers at the WIV could go wherever they want to study. Graduate students can go to any meeting anywhere. He would tell me later, when I sent him a query about next-generation sequencing capacity, that the WIV did a lot of their sequencing at the Beijing Genomics Centre. As far as he knew, "that institute has all the genomics and proteomics capacities that it needs."

So let's talk about the journal, I said. When did you get involved?

He thought it was 1998 or the 2000s. "At that time, it was Chinese-only, with an abstract in English. No one took it seriously. It was the decision of the CAS to make it English-only with the abstract in Chinese."

Did you read or speak Mandarin?

Neither, he said. And still doesn't. "You can function there in English," he said. "They want you to give seminars in English."

Did you ever meet Xiangguo Qiu?

He had not.

Was he aware that Xiangguo Qiu sent 15 strains of class 4 pathogens to China, allegedly to the WIV's BSL-4 where Shi works, apparently without a material transfer agreement?

"That boggles my mind," he said. "How do you send this without an agreement? What can I say? The scientists I dealt with in Wuhan were totally trustworthy. Under political pressure?"

I could almost hear the who-knows shrug over the phone.

He said he would write to Hu and maybe to Shi Zhengli to see if they would talk to me.

ARIF SENT AN email a few days later to say that he had an answer back from Zhihong Hu. "She would like to decline to be interviewed. She indicated that the scientists at the WIV have done many interviews and I get the feeling they are getting tired of them and want to concentrate on their research instead. She said that Zhengli is very tired of all the politics associated with this virus. I can see that because she is a very gentle soul and her commitments have always been science and research."

ON OCTOBER 20, 2020, a peer-reviewed journal, *Frontiers of Public Health,* finally published an article that had spurred the Latham/Wilson thesis when it appeared as a preprint. The authors, Monali Rahalkar and Rahul Bahulikar, went through the facts reported in the master's thesis and the PhD dissertation. They asked: where are those miners' samples now? Why had the WIV failed to mention them in any published paper? Why had the miners' illnesses and deaths in 2012 from a SARS-like coronavirus infection not been reported to the WHO? Could the miners' samples or another sample from the mine have carried a virus that leaked from the WIV and resulted in this pandemic?[335]

A peer-reviewed article is harder to ignore than a preprint. On December 7, 2020, almost exactly a year after the first case of SARS-CoV-2 was documented in Wuhan, an addendum (as distinct from a correction), appeared in *Nature* attached to the original article by Shi's lab claiming RaTG13 is the closest sequence to SARS-CoV-2.

Jonathan Latham sent me the link to the addendum.[336] I read it carefully. I had learned to read everything published by Shi's lab very carefully. The addendum purported to provide more information about the "bat SARS-related coronavirus (SARSr-CoV) strain RaTG13 reported in our Article." But it wasn't only about RaTG13. It was also a non-denial denial of the Latham/Wilson thesis and a rebuttal of the allegations in the July NIAID letter to EcoHealth Alliance concerning what must be done to get that grant money flowing.

The authors wrote that between July 1 and October 1, 2012, "we" (the Shi lab) received 13 serum samples from four patients, one of whom had died, who "showed severe respiratory disease." The patients were miners who'd been clearing bat feces from a mine preparatory to mining copper, in Tongguan Town, Mojiang County, Yunnan. They were admitted to hospital "April 26-27," 2012. The samples "we received" were gathered by hospital staff over the months of June, July, August, and September 2012. In an attempt to learn what the cause of the disease might be, using PCR (polymerase chain reaction) methods "developed in our laboratory," they had tested the samples for the RdRps of "Ebola virus, Nipah virus and bat SARSr-CoV Rp3, and all of the samples were negative for the presence of these viruses."

I knew I'd read about SARSr-CovRp3 somewhere but had to look it up again. It's the name for a SARS-like genome sequence described in the 2005 *Science* article about bats as the natural reservoirs of SARS. It had been sampled but not isolated from bat feces. Although the 2005 article claimed this genome was "essentially identical" to SARS, in fact it was only 92 percent identical to the Toronto sample of SARS to which it was compared. It was more distant from SARS than RaTG13 from SARS-CoV-2.

The addendum said they also tested the serum samples for antibodies against the nucleocapsid proteins of these three viruses. "None of the samples gave a positive result."

I sat there, staring at the addendum, questions exploding in my head. Is that all they did? They must have tried to isolate or at least sequence whatever they could get from those miners' serum samples. Searching for specific antibodies wouldn't necessarily produce information about an unknown virus. Half the miners had died from something that produced SARS-like symptoms. The lab was supposed to be looking for SARS-like viruses that might spill over. This one *had* spilled over.

The addendum didn't say whether they'd tried other things and failed. Instead, it said the lab had "recently" revisited the samples. That at least answered one question: Shi still had the samples. They used a different method (called ELISA) to see if they could get a match for the SARS-CoV-2 nucleocapsid protein, "which has greater than 90% amino acid sequence identity with bat SARSr-CoV Rp3." This "confirmed these patients were not infected with SARS-CoV-2."

Did it? It was not credible that the WIV was unable to extract full sequences from those samples. Recently, sequences sufficient to identify mammals, plants, and bacteria have been assembled from sediments more than 16,000 years old gathered from a Mexican cave.[337] Neandertal DNA genomes have also been recovered from cave sediments—not from bones, from sediments—dating back

50,000 to 150,000 years.[338] Getting viral sequences from those miners' samples should have been easy. So why hadn't they published a word about what they found? Why did this addendum only say what they didn't find, not what they did?

"We suspected that the patients had been infected by an unknown virus," the addendum continued. "Therefore, we and other groups sampled animals including bats, rats and musk shrews in or around the cave."

But why would they go after samples from bats and other animals before exhausting what could be learned from the samples taken from those miners?

"Between 2012 and 2015, our group sampled bats once or twice a year in this cave and collected a total of 1,322 samples." But Shi had told *Science* that her team visited the mine four times in 18 months. Which was correct? From these samples "we detected 293 highly diverse coronaviruses, of which 284 were designated alphacoronaviruses and 9 were designated betacoronaviruses on the basis of partial RdRp sequences." All nine were SARS-related, "one of which (sample ID4991; renamed RaTG13 in our Article to reflect the bat species, the location and the sampling year) was described in a 2016 publication." The addendum said they only sequenced part of its RdRp, which they uploaded to a public site in 2016.

> In 2018, as the next-generation sequencing technology and capability in our laboratory had improved, we performed further sequencing of these bat viruses and obtained almost the full-length genome sequence (without the 5' and 3' ends) of RaTG13. In 2020, we compared the sequences of SARS-CoV-2 with our

> unpublished bat coronavirus sequences and found that it [SARS-CoV-2] shared a 96.2% identity with RaTG13.

This addendum left gaping holes. Did they apply next-generation sequencing techniques to the miners' samples? And if not, why not? And what about the other eight beta coronavirus sequences discovered in the mine/cave? And why was there no explanation of how RaTG13's close propinquity to SARS-CoV-2 was unknown when that other paper with Shi's name on it was submitted the same day, claiming two other sequences were closest? And what did she mean by "when next-generation sequencing in our lab improved"? Next-generation sequencing technology has been available since 2007. It was used by Shi's lab for a paper published in early 2012.[339] There was no need to wait until 2018 to use that method to sequence the sample containing RaTG13 or to sequence viruses found in the miners' samples. Nowhere did the addendum explain why all the sequences and samples the lab gathered over its many visits to that cave/mine had not been uploaded to a public site. It didn't even say how many remained unpublished.

I WROTE TO Wilson and Latham to see what they thought of it.

Wilson wrote back immediately.

> ...we found the addendum significant for the absence of the following tests—tests that would better answer the question of what pathogens were or were not in the miners' samples.

> 1. Viral metagenomics—something the Shi lab published on in 2012...and that other Chinese labs were also doing at the time—and which would be even easier and more effective now, since metagenomic/nextgen sequencing techniques have advanced significantly since these earlier publications.
>
> 2. PCR using primers that would identify any SARS-related coronavirus that was present in the samples and /or PCR to identify any coronavirus present in the samples (using "pan-coronavirus" primers). And now, PCR could be done to look for any SARS/2 related COVs specifically.
>
> 3. Test for histoplasmosis to support Shi's claim that the miners were killed by a fungus.
>
> Also absent from the addendum were the actual data and controls from the experiments, and descriptions of the exact methodologies (e.g. PCR primer sequences) used—information that might allow independent researchers to determine whether or not Shi's tests could in fact identify a SARS-2 related Coronavirus were it indeed present in the miners.
>
> We are expected to take all the addendum's claims at face value, clearly inappropriate at this point.

I wrote to *Nature* about the addendum. Answers to my questions came by email from Dr. Francesca Cesari, Chief Biological Sciences Editor. I asked if the authors had supplied the addendum unbidden or if *Nature* had asked for it. She replied that the editors at *Nature* asked for it because "comments were raised with us about the paper, and after careful consideration, we concluded that the original paper was insufficiently clear about when the full-length sequencing of the bat virus was performed and that an addendum was necessary for readers' understanding of the article." Cesari denied that the addendum was asked for because of the controversy over the miners and their samples, in particular the Latham/Wilson article. *Nature's* only concern was to clarify the provenance of RaTG13. Cesari confirmed that the RaTG13 sequence had been submitted with the original paper, which confirmed it arrived the same day the other paper was submitted to *EMI* with Shi's name attached, which did not mention RaTG13. I also asked whether the editors at *Nature* had asked for information about the samples from the miners' lungs or if the authors had provided it unbidden. The authors had provided it without a request, I was told. Though I asked if the authors provided any tables or data about the miners' samples or the tests done on them, Cesari did not answer directly but referred me back to an earlier answer which said the addendum was only meant to clarify the original paper's reference to RaTG13. I asked who the peers on the original paper were. Cesari explained that *Nature* gives peer reviewers a choice about whether they are to be thanked by name in the publication or not. In this case the peers chose to remain anonymous. "Our editors are confident that the peer reviewers for this paper were well qualified, independent and rigorous," she wrote.

IF SHI OR the editors at *Nature* thought this addendum clarified matters and would quell concerns, wrong again.

14. Brass Nerve!

In July 2020, *The Lancet* announced the creation of a commission on the pandemic. *The Lancet* was founded in Britain in 1823—before Darwin toured the world on the Beagle, before Victoria became Queen. It is natural that such a prestigious medical journal would be interested in the worst pandemic since 1918. (Especially since it had been forced to retract an article on chloroquine only the month before, leading the *New York Times* to ask whether peer review is "broken.")[340] The 1918 pandemic had an impact so enormous that it brought an end to World War I, which led to the rise and fall of empires, the enfranchisement of women, and the Roaring Twenties, which led to the Great Depression and World War II, which led to the funding of big science by governments, the creation of institutions of global governance, and so much more. It was clear already that the health, economic, social, and geopolitical impacts of SARS-CoV-2 would be far greater than those of the 1918 flu.

This genius of a virus had already brought several Western countries to their knees. It had upended the basic principles of neoliberal "free" market globalism, such as: that manufacturing should be done at the lowest price by the lowest-paid workers with the least protections in the poorest countries; that these goods should flow from the developing countries to the developed in just-in-time delivery systems; that borders should permit the flow of goods and money yet be impervious to desperate refugees; that science should be a global phenomenon involving collaborations between scientists living under the thumbs of autocratic governments and those in democracies supported by taxpayers who have no idea what's going on in the lab down the street. By July 2020, hundreds of thousands had died and millions had been sickened; refrigeration trucks had been turned into morgues in the beating heart of global capitalism, New York City; borders had been shut to all but essential travellers (though the rich could still find a way around the rules); small businesses were collapsing and unemployment was headed for Depression-era heights; government deficits were exploding; bureaucrats were trying to lure back vaccine production while throwing huge money at Big and small Pharma for any vaccine that looked promising; and a generation of children were learning to fear contact with their friends and their grandparents, spending their days staring at screens. The leaders of some governments—China, Russia—were taking full advantage of the crisis to crush political opposition and push out their borders at the expense of their neighbours.

If SARS-CoV-2 was shown to be the product of gain-of-function experiments by scientists in China supported by the taxpayers of the European Union and the US, the rage against globalized science would be considerable. Mobs with pitchforks came to mind. Thus, the *Lancet's* pandemic commission.

JEFFREY SACHS WAS appointed the commission's chairman. Sachs is an American economist. He was once associated with the rapid transformation of the inflation-ridden nationalist economies of Latin America and the planned economies of the former Soviet Union and its empire, which led to their inclusion in the global economic system. In the 1980s, at the beginning of his academic career at Harvard (tenured at age 28), he was invited to advise Bolivia on how to stop its 14,000 percent inflation without destroying its democracy. Though he'd never been to Bolivia, he said he could do it in a day. His advice was taken seriously. Two years later, Bolivia was a neoliberal economy with only 10 percent inflation but a vast array of people living in abject poverty (except for those growing coca for the global cocaine market, which expanded exponentially). He provided advice to Solidarity after it won control of Poland, which resulted in the Polish unemployment rate soaring from just above 0 to 35 percent. After the collapse of the Soviet Union, he offered advice to Russia. In the 1990s, he became an advocate to end world poverty through sustainable development.[341] He is director of the Earth Institute's Center for Sustainable Development at Columbia University.

The phrase *sustainable development* is credited to Gro Harlem Brundtland, a former prime minister of Denmark, who headed a UN commission on environment and development in 1983. (Later, she became director-general of the WHO: she dealt with the SARS pandemic.) The phrase was relentlessly promoted by Canadian oil entrepreneur/UN environment impresario Maurice Strong, who sent one of his close business associates, a Texas oil lawyer named Chip Lindner, to help organize the commission for Brundtland. The phrase is meant to convey the notion that the world can prevent the worst of climate change while still enjoying economic growth—growth being necessary to alleviate poverty. The method proposed in 1992 at the UN Earth Summit in Rio de Janeiro, organized by Strong, was to let governments of developing countries, such as China, pollute freely as they industrialized, while advanced Western countries radically reduced their carbon dioxide and methane emissions by investing in green innovation and systems for trading pollution credits. When I interviewed Strong for a book I was working on,[342] he also argued for a global tax to support a global governance system. He insisted that climate change, a threat to the entire globe, can't be addressed by individual nation states acting on their own. When I asked if such a global system would be democratic, Strong's colleague Lindner explained that democracy might not be tenable. Strong also told me that working through the UN gave him much more power than any cabinet minister in a democratic government like Canada's. He could arrange his own funding, create his own relationships with multinational business interests, fund NGOs to influence public policy in many countries. Years later, when he was accused of having played a role in the oil-for-food scandal to get around sanctions against Iraq (he was exonerated), he took himself to China until things cooled down. He had important friends there.

Sachs is also familiar with the power of the UN to spread important ideas and is a friend to China. He has been a special advisor to Secretaries General Kofi Annan, Ban Ki-moon, and António Guterres. He was appointed an advocate on sustainable development goals by Guterres.[343] In 2020, as Trump began to call SARS-CoV-2 *the China virus* and the US intelligence community announced it was revisiting the lab-leak theory, Sachs gave interviews explaining why it was important for the US to expand its relationship with China and not slip back into a dangerous Cold War stance.[344] In an opinion piece he wrote with William Schabas, a Canadian expert on human rights law, he argued that there is no evidence of a Uyghur genocide and that China's actions toward the Uyghurs need to be understood in the context of terrorist actions years before in Xinjiang and Beijing. In an interview given in January 2021,[345] Sachs also compared the treatment of Uyghurs to US racism and war-mongering. He argued China's internment of Uyghurs in re-education facilities was comparable with American actions against terrorists taken after 9/11.[346] To be fair, he gave that interview before Amnesty International published a report detailing China's appalling abuses against its Muslim minorities: pervasive surveillance, forced labour, imprisonment without trial or cause, an ongoing plan to cut down the Uyghur population by millions, and if these allegations were not bad enough, systematic torture. Amnesty compiled testimony of 50 former internees in the Xinjian re-education camps who spoke of torture done to them and witnessed by them. The report mentioned something called the tiger chair, in which people are forced to sit, their limbs clamped in painful positions for days on end, not released to urinate or defecate.[347] Sachs still teaches at Columbia, including at the Mailman School of Public Health.

As chair of the *Lancet* commission, Sachs created a task force to investigate the origin and spread of SARS-CoV-2. Guess who he asked to lead it? If your first thought was Peter Daszak, 500 brownie points to you.[348]

By July 2020, despite Daszak organizing that February statement to *The Lancet* decrying the lab-leak idea as conspiracy theory, despite his tweets and media interviews reiterating that view,[349] it was obvious that any investigation into the pandemic's origin would have to consider the possibility of a leak from Shi's lab. In other words, Daszak would have to investigate a lab he had funded and overseen. So, when his appointment to lead this task force became public, knowledgeable scientists were amazed. Even Kristian Andersen, lead author of the proximal origin paper in *Nature*, who had corresponded with Daszak in early February about how to counter the lab-leak theory,[350] tweeted: "Not the right person for the job...that's ridiculous." Dan Sirotkin, co-author of the first peer-reviewed paper to suggest a possible lab origin,[351] said: "having Peter Daszak lead this is like having Goebbels chase war criminals."[352]

On November 5, two days after Biden defeated Trump in the US election, the WHO made an announcement of its own. It presented the terms of reference for a "global study" of the origins of SARS-CoV-2[353] and its list of "independent" experts who would execute it.[354] The WHO stressed that it was not conducting this study itself but was acting as the convenor for independent experts. It had been working toward this goal since February 2020, when it convened an expert group on the pandemic in Geneva. (Daszak was at that meeting.) In May, all 140 member states of the WHO, including China, had voted in favour of a resolution that the WHO should try to determine the origin of the virus to prevent continuing re-infection from that source and the spread of the disease to other reservoirs. But it wasn't until July that the WHO was allowed to send two people to China to negotiate the terms for such a study. (A journalist with the *South China Morning Post* asked at a WHO press conference if the names of the negotiators would be forthcoming. The answer was *yes*. But when I emailed the WHO press office and asked for those names, the answer was *no*.)

The terms of reference for the study revealed that it would not be independent. Unlike the aftermath of the SARS pandemic, when renowned experts such as W. Ian Lipkin were invited to China, and Linfa Wang went in as the member of a WHO mission, this time only scientists in China would do the fieldwork, gather the data, interpret the data, and present it to the experts commissioned by the WHO. There would be 17 international experts and 17 Chinese experts conducting the joint study. The WHO presented this as an equal partnership, though the terms of reference showed China would decide which tests would be done (and could veto any expert it didn't like on the WHO's expert list simply by refusing a visa). In addition, seven of the 17 members of the international team were not independent academics but were employees of the WHO or the World Organization for Animal Health.[355] The names of these employees and the Chinese scientists were not included on the list of experts.

One of the ten named international experts was Dr. Marion Koopmans, a veterinarian and chief of viroscience at Erasmus Medical Center in the Netherlands. She is a close colleague of Ron Fouchier's, whose gain-of-function experiment with ferrets and avian flu virus had raised a storm. According to the WHO, people were chosen for their expertise, but also with an eye to geographical and gender representation. Yet there were only two women on the list and no one from Latin America. There was a Russian, a

German, a Dane, someone from the UK, someone from Qatar, someone from Japan, someone from North Vietnam, someone from Australia, and one lone American.

If you guessed the lone American was Peter Daszak, 1,000 brownie points to you.

These experts had been selected by the WHO from names submitted by member states and from applications sent in by individuals in response to the WHO's call for volunteers. *Vanity Fair* reports that the US submitted three names, but not Peter Daszak's. He must have applied for this job himself.[356]

In other words, by November 2020, Daszak had managed to get himself appointed to *two* groups of scientists commissioned by important institutions to inquire into the origin of SARS-CoV-2. Both would have to examine his work to be credible, which meant neither would be credible.

I tried hard to understand this behaviour, to characterize it. Did the word *shameless* apply? Even Maurice Strong, who made a practice of positioning his people on all sides of a negotiation, wouldn't have tried it. I had an aunt-by-marriage who was raised in small-town Alberta. She knew how to turn a phrase. When describing a person with poor manners she'd say: "he wasn't brought up, he was *dragged*." She'd have said of Daszak: "*he's got brass nerve!*"

Daszak's brass nerve was almost equaled by the WHO's. In its first report on the results of phase one of the joint study, it would publish the following declaration regarding interests:

> The WHO international team was finalized with the completion of administrative procedures, including a declaration of interest and a confidentiality

> undertaking. All declared interests were assessed and found not to interfere with the independence and transparency of the work. The declared interests were shared with all team members and were managed by the WHO Secretariat.[357]

The WHO was not going to share this information, and thanks to the confidentiality agreements, none of the participants would be able to talk about it either. Peer-reviewed journals like *The Lancet* and the *New England Journal of Medicine* have learned from hard experience that scientists' findings may be biased in favour of their own and their funders' interests, which is why pharmaceutical companies offer gifts of various kinds to physicians/researchers as well as funding their studies.[358] All serious peer-reviewed journals demand that authors publicly declare who funded their work and list any personal interests that might affect their findings.

The report would also explain that:

> All members of the team served in their personal scientific capacity and not in that of any institution or government with which they were associated.[359]

Insisting that the scientists from China had been shorn of their government's interests by virtue of their appointments to this team was like peeling the outer layers off an onion, painting it red, and calling it an apple.

Maybe *shameless* was the best word after all.

In December, just before *Nature* published Shi's slippery addendum, Daszak's *Lancet* task force issued a statement insisting its members would keep an open mind about all theories of origin, including various versions of a lab leak.[360]

JANUARY 6, 2021. The first AstraZeneca vaccines were sliding into ancient arms. Canadians, enduring another lockdown to bend the curve of the second wave, were head down over screens as Trump supporters stormed the Capitol in a failed attempt to prevent the certification of Biden's election. The rioters erected a gibbet to hang Vice President Mike Pence. They ransacked Nancy Pelosi's office. People died. That same day, two members of the WHO's international study team set off for China. But at the last moment, there was a glitch. China refused to let them into the country. The director-general of the WHO, Tedros Adhanom Ghebreyesus, said he was disappointed with this behaviour,[361] which was unusual as he had almost always bent like grass in the wind before China's demands, though China is not among the WHO's largest funders. In fact, it's far down the list.[362] In fact, Canada contributes much more.[363] So does Pakistan. The US is by far the largest funder, followed by the Bill & Melinda Gates Foundation.

More negotiations between the WHO and China ensued, which were not explained. A week later, the team was at last on its way.

By then, some major media had begun to take the lab-leak thesis seriously. AP had sent reporters to get pictures of the mine in Mojiang County where the six miners got sick in 2012. They were prevented from getting close by men in plain clothes whose cars blocked the road.[364] Samples had even been confiscated from Chinese scientists who went to the mine. The AP story referred to its earlier reports about the clampdown in China on all science having anything to do with the origin of SARS-CoV-2, and about the government giving hundreds of thousands of dollars in grants for military scientists working in southern China, but controlling what they publish, "while actively promoting fringe theories that it could have come from outside China." It also reported that China's side of the WHO joint study had failed to examine human samples from surveillance hospitals dating back before 2020 to help pinpoint when the pandemic began. That didn't bode well.

On January 14, as the WHO's international team finally arrived in Wuhan, a horde of reporters lay in wait. In a video that made the news, Daszak was easy to spot. He was the large bald man wearing a medical mask and waving at the reporters.

As BBC correspondent Robin Brant put it, the "investigators arrived here as a propaganda effort, lead [*sic*] by China's state media, is in full swing to question whether the pandemic originated here in the first place." He quoted Dale Fisher, chair of the global outbreak and response unit at the WHO, who said he "hoped the world would consider this a scientific visit. It's not about politics or blame but getting to the bottom of a scientific question." Then he added that most scientists "believed that the virus was a 'natural event.'"[365]

The next day, just five days before President-Elect Biden took over the government of the US, the US State Department put out a statement. It was called "Fact Sheet: Activity at the Wuhan Institute of Virology." It said the Chinese Communist Party had "systematically prevented a transparent and thorough investigation of the COVID-19 pandemic's origin, choosing instead to devote enormous resources to deceit and disinformation." It argued that millions had died and their families "deserve to know the truth." It said the US government had not determined whether the pandemic had been caused by a spillover from an animal or "was the result of an accident at a laboratory in Wuhan, China." It said scientists in China "have researched animal-derived coronaviruses under conditions that increased the risk for accidental and potentially unwitting exposure." Then it listed new "previously undisclosed information" and information from open sources. It claimed:

> The US government has reason to believe that several researchers inside the WIV became sick in autumn 2019, before the first identified case of the outbreak, with symptoms consistent with both COVID-19 and common seasonal illnesses. This raises questions about the credibility of WIV senior researcher Shi Zhengli's public claim that there was "zero infection" among the WIV's staff and students of SARS-CoV-2 or SARS-related viruses.

The fact sheet accused the CCP of preventing independent journalists and global health authorities from interviewing researchers at the WIV, including those who became ill. It argued any credible inquiry into the origin of the virus must include interviews with these researchers. It raised the gain-of-function research done at the WIV. It raised the issue of the miners, and the fact the WIV had not been transparent about its research on RaTG13. It said WHO investigators should get access to lab records and find out why the WIV had altered and then removed online records "of its work with RaTG13 and other viruses."

The fact sheet raised the specter of secret military work

done at the WIV and acknowledged that the US might have inadvertently paid for that too.

> Despite the WIV presenting itself as a civilian institution, the United States has determined that the WIV has collaborated on publications and secret projects with China's military. The WIV has engaged in classified research, including laboratory animal experiments, on behalf of the Chinese military since at least 2017. The United States and other donors who funded or collaborated on civilian research at the WIV have a right and obligation to determine whether any of our research funding was diverted to secret Chinese military projects at the WIV.

It concluded:

> ...as WHO investigators begin their work, after more than a year of delays—the virus's origin remains uncertain. The United States will continue to do everything it can to support a credible and thorough investigation, including by continuing to demand transparency on the part of Chinese authorities.[366]

Those at the top of the State Department must have decided that there was no point anymore in trying to protect the department from accusations that its own agency, USAID, might have funded experiments that created the pandemic. Its analysts seemed to have been reading the preprint literature on the lab-leak possibility, including: the DRASTIC report about the WIV taking its virus databases offline in September 2019; the works of Rahalkar and Bahulikar, Deigin and Segreto; and the Wilson-Latham thesis. It did not speak well for the independence and objectivity of establishment science publications like *Nature*, *The Lancet*, and even *Science* that most of this important new information had not been published by them.

THE INTERNATIONAL TEAM members were put into quarantine for the first 14 days of their stay in Wuhan. They worked with their Chinese teammates online. When they were finally let out of stir, reporters trailed them as they went from meeting to meeting, filing stories about their visit to the Wuhan Institute of Virology and to the new museum nearby, the shrine to China's "victory" over SARS-CoV-2. Apparently, one room in that museum has an entire wall devoted to a giant photo of President Xi, China's viral Napoleon.

At the end of the visit, the WHO-convened joint team held a press conference. Thomson-Reuters Shanghai, *Agence France-Press*, *Wall Street Journal*, Hubei Broadcasting & Television Network, CGTV, and Sky News attended. Mi Feng, spokesperson of the China National Health Commission, presided, spoke, and turned it over to Professor Liang Wannian, team leader for the Chinese side, who spoke at length, and then to Peter Karim Ben Embarek, a Dane described as the chairman of the international team. Wikipedia says he is a program manager with the WHO. His entire career has been either with the FAO or the WHO. His speciality is foodborne zoonoses, though he has also worked on MERS. He spent several years in the WHO's China office advising the government of China on food safety issues. In 2017 the Chinese Institute of Food Science and Technology gave him a prize—its Scientific Spirit Award. Dr. Marion Koopmans, the only foreign expert member of the international team at the press conference, was introduced as well, but only to answer questions. She didn't get to make a speech.

It was Embarek's opening comments that made the news. He insisted that "all the work that has been done on the virus and trying to identify its origin continue to point towards a natural reservoir of this virus and similar viruses in the bat population. But since Wuhan is not a city or an environment close to these bat environments, a direct jump from bats to the city of Wuhan is not very likely." The team, he asserted, had worked in a rational manner through the evidence in three main areas of inquiry. Then they applied a risk analysis method to consider the likelihood of four possible scenarios of origin. Though no evidence had been found of SARS-CoV-2 in any animal sampled, the team believed one possible scenario of the origin of SARS-CoV-2 was a spillover from one animal direct to humans, likely via a bat. A more likely one was a spillover from a bat to an intermediary animal that passed it to humans. Another possibility was that the virus had arrived in China on frozen foods. The fourth was a lab leak. "However, the findings suggest that the laboratory incident hypothesis is extremely unlikely to explain the introduction of the virus into the human population and therefore is not a hypothesis that will imply to suggest future studies into our work to support our future work into the understanding of the origin of the virus."[367] His phrasing was convoluted: what he seemed to mean was that this scenario was so unlikely no further work would be done on it.

A reporter asked for the numbers that led to this risk ranking.

There were no numbers.

THREE DAYS LATER, at a press conference in Geneva, with Peter Embarek standing beside him, the director-general of the WHO insisted that all theories of origin remained on the table, maybe not to be explored in this report or in the phase two report to follow, but still on the table.[368] He also insisted that this wasn't a WHO study, it was a WHO-convened independent study, as if that made any difference at all to the price of tea. Meanwhile, the *New York Times* carried a story that the Chinese team had refused to hand over important epidemiological data relating to the earliest days of the virus, the so-called line data, which includes who was interviewed and what they said.[369] Apparently, discussion had grown heated between some international team members and their Chinese counterparts. The *New York Times* was then accused on Twitter by WHO team member Thea Fisher of twisting her words: it was just a passionate exchange of views that had led to raised voices and that happens all the time in academic circles, she explained. Peter Daszak tweeted about the trust and openness the team shared: "We DID get access to critical new data...."[370] In another tweet, Daszak added: "It's disappointing to spend time w/ journalists explaining key findings of our exhausting month-long work in China, to see our colleagues selectively misquoted to fit a narrative that was prescribed before the work began. Shame on you @nytimes!" These tweets were featured by China's media.

By the middle of February, it was obvious a war to control the narrative about the WHO-China joint study had begun though the phase one study report had not yet been published.

On February 13, 2021, Biden's new national security advisor, Jake Sullivan, put out a statement. He pointed out that the US had great respect for the WHO's experts and its work fighting COVID-19 and that the Biden administration had reversed the Trump administration's withdrawal from the WHO.

> But re-engaging the WHO also means holding it to the highest standards. And at this critical moment, protecting the WHO's credibility is a paramount priority. We have deep concerns about the way in which the early findings of the COVID-19 investigation were communicated and questions about the process used to reach them. It is imperative that this report be independent, with expert findings free from intervention or alteration by the Chinese government. To better understand this pandemic and prepare for the next one, China must make available its data from the earliest days of the outbreak.[371]

ON MARCH 4, 2021, though the WHO joint study report had still not been made public, a group of scientists and interested parties put out an open letter calling for a "full and unrestricted international forensic investigation into the origins of COVID-19." They argued that the terms of reference, the lack of correct skill sets among the international members of the team, the duress under which China's scientists operate, the inability for individual patients or researchers to speak to the team privately and in confidence, the control of all information by China, its veto over who was on the team and its veto over what was published, all militated against a real investigation of anything. They insisted they supported the WHO, but that reporters had gotten the wrong idea that what happened in Wuhan in January had been a real WHO investigation into the origin of the virus. They produced a long list of requirements for a proper investigation. The letter was organized by Jamie Metzl, a well-known Washington insider. He is a senior fellow with the Atlantic Council and an advisor to the WHO. Formerly, he was the executive vice president of the Asia Society. Before that, he worked with the Clinton administration on the National Security Council. Before that he was deputy staff director of the Senate Foreign Relations Committee when its chairman was Senator Joe Biden. In other words: he has good contacts with the Biden administration. The statement was signed by Alina Chan of the Broad Institute of MIT and Harvard, Monali Rahalkar, Rossana Segreto, Gilles Demaneuf, Francisco de Asis de Ribera Martin, Rodolphe de Maistre—all members of the DRASTIC group using their own names—as well as the chief US critic of gain-of-function research, Richard Ebright. A significant number of researchers from Australia signed, including Nikolai Petrovsky, a professor of immunology at Flinders University, who very early on raised the possibility of a lab leak. The largest number of researchers who signed were affiliated with major science institutions in France.[372]

MARCH 26, 2021, four days before the WHO report was released, CNN carried a two-hour documentary by Sanjay Gupta on the way the pandemic had unfolded in the US. Gupta interviewed the US doctors who'd managed the pandemic response during the Trump administration. Robert Redfield, former head of the US Centers for Disease Control and Prevention, was interviewed at length. Redfield was asked if he thought the virus had a natural origin.

While giving his answer, Redfield was careful to explain that he is a virologist by training, he's spent his life in virology. He said that, in his opinion, this virus was not the product of nature, based on the way it behaves. "I am of the point of view that, I still think that the most likely aetiology of this pathogen in Wuhan was from a laboratory. You know, escaped...." He didn't believe that a virus jumping from a bat to a human could so quickly become so efficient in transmission. That sort of efficiency is what lab experiments engender. "I just don't think this makes biological sense," he told Gupta.

Dr. Fauci, head of NIAID, one of Shi's funders through EcoHealth Alliance, did not concur.[373]

ON MARCH 29, 2021, one day before the WHO joint study report was to be released, NPR carried an exclusive interview with Peter Daszak. According to NPR's reporter, Suzette Lohmeyer, Daszak, "part of an investigative team that did two weeks of research in China, said it found evidence that wildlife farms may be a potential source of the main spillover event."

"Additional research is needed on farms in southern China that breed exotic animals for food, the report is expected to say," Lohmeyer wrote.

Daszak explained that 20 years earlier, to alleviate poverty, the Chinese government had encouraged people in the south—provinces like Yunnan—to farm the kinds of wild animals popular with Chinese epicures and sell them in wild animal markets. Ever since, these farms had been breeding and selling animals susceptible to SARS-CoV-2 such as civets, porcupines, pangolins, raccoon dogs, and bamboo rats. Farmers had sold these products to the Huanan Seafood Market, where environmental samples of SARS-CoV-2 were found and where it was first surmised the pandemic began. Daszak said it was important to note that the government of China, on February 24, 2020, had ordered all these farms to close, though in 2016 the wild animal farms already employed 14 million people in a $70-billion industry. He told the reporter that these farms "could be the spot of spillover, where the coronavirus jumped from a bat into another animal and then into people." The "team" had found new evidence while in China that these farms "were supplying vendors at the Huanan Seafood Wholesale Market in Wuhan, where an early outbreak of COVID-19 occurred."

The reporter checked Daszak's theory with Linfa Wang, described as a virus expert who studies bat-borne viruses at Duke-NUS Medical School in Singapore. (The reporter also identified Wang as part of the WHO-convened joint study team, though that was not the case.) Wang told the reporter that there were lots of positive samples taken from the market, and two strains had been isolated and recovered.

I read this story with interest because the wild animal farm origin theory had not been mentioned at the team's press conference six weeks earlier. I wondered if things had been discovered since. Surely Daszak wouldn't tell a reporter that something would be in the report when it wasn't.[374]

THE 120-PAGE REPORT of the "China Part" of the WHO-convened joint study was made public the next day, March 30, 2021. The epidemiology section was fascinating.[375] The Chinese scientists had pulled together a significant amount of data from China's national disease surveillance systems. In China, one network of hospitals and clinics reports influenza-like illness plus one more symptom, known as SARI, and another keeps watch for flus. The team was told that Wuhan has no hospital in the SARI reporting system. The closest SARI sentinel hospital to Wuhan is in Jingzhou which is two and half hours away by car and has a population half the size. There are only two flu sentinel hospitals in Wuhan.

Graphs displaying data from those two flu sentinel hospitals in Wuhan showed that there was a sudden jump in cases—from under 500 to almost 4,000—in the forty-seventh week of 2019, at first mostly among children, then in adults. Laboratory confirmed cases of flu showed the same pattern, first in children, then in adults. Similar graphs were presented for Hubei province as a whole. The SARI surveillance hospital in Jingzhou showed some SARI type cases in children, but no increase among adults in the final weeks of 2019. In addition, "respiratory tract samples collected as part of...surveillance in Wuhan, elsewhere in Hubei Province and in Shaanxi Province in 2019 were tested retrospectively for SARS-CoV-2 by nucleic acid tests. All were negative."

Retrospective tests were done on samples taken from probable flu cases that appeared in Wuhan sentinel flu hospitals in the first three weeks of January. Nine samples were positive for SARS-CoV-2: these patients came from six different districts in Wuhan, indicating the virus had already spread widely by the first week in January. Unfortunately: "no samples from adults were available for testing in the last three weeks of December 2019 so conclusions about SARS-CoV-2 causing...[flu like symptoms] in adults in December cannot be made."

That seemed extremely odd. Where had those samples gone?

As I reread that section, I finally figured out what had been bothering me about China's authorities insisting that samples tested by commercial labs in December 2019 as positive for a SARS-like coronavirus be destroyed or handed over to authorities. Here was the issue again, only this time it was missing samples from Wuhan's flu sentinel hospitals. COVID-19 symptoms can be flu-like. People who caught it in the early days of the pandemic might have been misdiagnosed. Why would there be no adult flu samples available for December 2019? Get rid of those samples and no one would be able to prove that patients with flu-like symptoms in December and earlier actually had SARS-CoV-2. The earliest patients—where they lived, worked—might pinpoint the place of origin for the virus within the city of Wuhan. The missing flu samples were such clear evidence of a cover-up, I could almost smell the gun smoke.

The epidemiology team did collect information about purchases of drugs from pharmacies to see if there had been a sudden surge in requests for drugs to fight colds, flus, or coughs. They found that demand for these products had gone up year by year since 2016, and so there was nothing untoward in the fact that these purchases in 2019 were higher than 2018.

They looked at mass gatherings in Wuhan that might have aided spread. A table showed there had been several events in the last four months of 2019. The largest took place in November. Except for the military games in October, the number of foreigners involved was small. The team asked for follow-up of any reported sicknesses among visitors to the military games, again pushing one of China's origin theories.

They looked at death data. Was there a surge in

unexpected deaths? If so, when? The team members explained that deaths and their causes are recorded in hospitals by doctors treating patients, but deaths outside hospitals are recorded by local health workers. However, the data are online, so it was possible to make year by year, week by week comparisons. The graphs showed deaths rose to a sharp peak beginning in the third week of January 2020 in Wuhan. Excluding confirmed and suspected COVID-19 deaths, there was still a peak between weeks four and seven of 2020. When deaths in people over 65 were separated out, it was clear that deaths in that age group had been going down in 2019 when compared to earlier years, but deaths in this group rose sharply in weeks four to seven in early 2020. In Hubei province outside Wuhan, there was no such peak, which suggested that the pandemic had not yet spread outside the city.

They looked at deaths in Wuhan from pneumonia, because earliest cases of COVID-19 might have been listed as caused by pneumonia. There was a huge jump starting the first week in January 2020, peaking in the seventh week. Then it fell sharply. Among those over 65 who were classed as having died from pneumonia, the shape of the graph was identical. But Hubei province showed only a moderate increase in pneumonia deaths over the same period. The conclusion:

> During the period of August-December 2019, review of all-cause as well as pneumonia-specific mortality surveillance data provided little evidence of any unexpected fluctuations in mortality that might suggest the occurrence of transmission of SARS-CoV-2 in the population in the period before December 2019.

> This does not exclude, however, the possibility that some SARS-CoV-2 circulation was occurring in the population at a low level, as changes in mortality at the population level would be unlikely to be sufficiently sensitive to detect this possibility.

There had been a grand total of four deaths beyond what would have been expected: two from pneumonia in week 40 and one in week 44 of 2019. These deaths occurred in two different districts of Wuhan. "Given the time lag from onset of disease to COVID-19 associated death of a median of 17...days in Wuhan, the documented rapid increase in all-cause mortality in week 3 of 2020 and pneumonia-specific deaths in week 3 of 2020 suggests that virus transmission was widespread among the population of Wuhan by the first week...of 2020." Two weeks later there were excess deaths in other parts of the province which suggested: "the *epidemic in Wuhan predated the spread in the rest of Hubei Province*" (italics mine).

Then they did clinical review of files. There is a national reporting system for SARS and for other notifiable diseases in China. The cases "that were identified with the earliest onset occurred in December 2019 and were reported to the national Notifiable Disease Reporting System and published." Reporting is online and goes direct from hospitals, townships, and primary care clinics to the China CDC. But SARS-CoV-2 was only added to the list of reportable diseases on January 20, 2021 (the same day Shi and Gao and others submitted their papers on SARS-CoV-2 to various journals). At first, patients with SARS-CoV-2 diagnoses were asked about all close contacts they had two weeks prior to the onset of their illness and about their contact with the Huanan Seafood Market. But case definitions changed as the symptom list expanded and the focus on the Huanan Seafood Market was dropped when it became clear that some early cases had no contact with that market. The test developed for SARS-CoV-2 was not available to make firm diagnoses until mid-January.

All SARS-CoV-2 cases reported with onset in December had been reviewed. The Chinese team said 174 had been officially reported. Of these, 100 were confirmed by laboratory test and 74 were diagnosed by their symptoms. Other "cases" were identified as part of the search for potential cases with onset in December 2019 (including some that were included in early publications). After clinical review by the Chinese team, none of these other cases were considered compatible with COVID-19 disease, "leaving only the 174 notified cases."

The team moved the date of the first official case of SARS-CoV-2 from December 1 to December 8, 2019. The report said: "The case with the earliest onset date reported to the NNDRS became ill on 8 December 2019. The clinically diagnosed cases were generally reported in the second half of December with the first clinically-diagnosed case having onset of illness on 16 December."

In other words, the team had disappeared from this study the first three cases identified in peer-reviewed articles though they had tested positive for SARS-CoV-2. The smell of gun smoke was growing stronger.

"The earliest cases were mostly resident in the central districts of Wuhan, but cases began to appear in all districts of Wuhan in mid- to late December 2019," the report said. Of the early cases, 44.6 percent had no history of market exposure, either to the Huanan Seafood Market or to another market in the city. Further, "The case reported with the earliest onset date (8 December) had no history of exposure to the Huanan market."

Anticipating objections to the removal of the first three cases from the record, explanations for their exclusion followed. The report said that the man previously considered to be the first case, who became ill on December 1 and is described only as a 62-year-old, had had a minor respiratory illness that responded to antibiotics. But he did have COVID-19 as of December 26. His wife was admitted to hospital the same day and turned out to have COVID-19 as well. She did have contact with the market. "This couple, together with their son, became part of the first recognized family cluster of COVID-19." Case two, a younger woman who got sick on December 2, had pneumonia but tested negative for SARS-CoV-2 throughout her hospitalization, the report said. The third case also responded to antibiotics. Blood collected in April—four months later—proved to be negative for SARS-CoV-2 antibodies.

All of this was suspicious. More suspicious was the admission that the supporting material for this section, which was considerable, had not actually been reviewed by the international team members. Why not? The report said there was too little time. The international team had recommended that early patients should be re-interviewed and that the cases thrown out should be re-examined, which was a weak form of protest.

China's team also made a retrospective file search for possible COVID-19 cases in all health institutions of Wuhan. Because the symptoms varied and many people remained asymptomatic, there could have been a lot of transmission going on that nobody noticed until the "explosive outbreak" in Wuhan "from the middle of December 2019 onwards." The search went back to October 2019. But the international team was only presented with information that went back to December 10, 2019. The team had reviewed the files. When cases looked likely, if the patients could be found, blood was obtained from them "where possible." There had been 73,253 cases with fever, flu symptoms, respiratory distress, or pneumonia. But the team's review of those files produced only 92 cases for follow-up. Several of the patients had died, so the Chinese team had only been able to get blood from 67 patients. All had tested negative for SARS-CoV-2 specific antibodies. But antibodies to SARS-CoV-2 may not last in the blood for over a year, so the international team asked for review of the review. And they asked for tests of stored blood samples.

China's team had acquired throat swabs collected between October and December 2019 from Wuhan Union Hospital as well as 106 samples from three hospitals outside the city and more from Henan. They were all tested for SARS-CoV-2 nucleic acids. All results were negative.

The international team wondered about blood samples of convenience—meaning blood donated to blood banks or stored in research biobanks. When Shi Zhengli and her colleagues searched for evidence of SARS-CoV-2 circulating in Wuhan cats, she got blood samples from stray cats and house cats, but also cat blood stored in veterinary blood banks. But what had been done with stored cat blood had *not* been done with stored human blood, though the international team had asked for it. They had asked that blood centres in China test for SARS-CoV-2 antibodies in blood donated from October to December 2019. Representatives from the China's blood centres came to talk to the international team: they explained this sort of work would require several kinds of regulatory and ethical approvals, which they did not have.

To summarize: the first three cases had been scrubbed, so the first confirmed case of COVID-19 now occurred on December 8, 2019, though the first three cases had been reported in peer-reviewed journals as having tested positive for SARS-CoV-2 using PCR—a proper test. There was an explanation, but it was wonky. Out of more than 73,000 possible cases of illness that might have been caused by SARS-CoV-2, only 92 cases had looked interesting enough for the Chinese team to test: 67 had been tested, but all were negative. No samples from flu hospitals were available to test for the critical last three weeks of December. No rise in disease or deaths had appeared either inside or outside Wuhan until later in January 2020. No point of origin in the city had been established because the confirmed cases in January were from all over the city. While blood stored in various transfusion and research centres might answer questions as to where and when SARS-CoV-2 began circulating, that work had not been done. And never would be done.

THE NEXT SECTION dealt with molecular epidemiology. Assuming, as this study did, that the virus originated in an animal and spilled over to humans through an intermediary, and then adapted itself, the question addressed was which animal was host and which was the intermediary. The team was also concerned about other animals that might have become infected (like mink, cats, etc.) and could become new reservoirs for infection. The team was interested in any progenitor genomic strains found in animals that might have led to SARS-CoV-2. They thought spillovers might have happened repeatedly.

The most likely host, the report asserted, was a bat, but no virus with a genome close to SARS-CoV-2 had been found in bats. The team examined the huge number of genome sequences of SARS-CoV-2 that had been sampled from human patients around the world and posted to public websites. Their mutations could pinpoint emergence, chains "of resurgence," clusters of infection, the rate of spread, adaptations, and even the number of founders. The first sequence publicly available had been uploaded by Zhang and Holmes on January 10 to Virological.org and also to GISAID, the public platform created to track flu genomes. Between January 10, 2020, and February 16, 2021, GISAID had acquired a total of 487,487 SARS-CoV-2 genome sequences posted from 238 countries.

Various methods were used to try to trace back to the first genome sequence of SARS-CoV-2. This involved trying to reconcile different genome sequence systems. There are three public platforms to which sequences are posted, the earliest dating back to the very beginning of gene sequencing in the 1980s. One is in Europe. Another, founded in the US, is at the National Center for Biotechnology Information. The third is called the International Nucleotide Sequence Database Collaboration. The Americans and Europeans established portals specifically for SARS-CoV-2. There are different naming systems in use. The first graphic representation of SARS-CoV-2's mutation and spread was created by nextstrain.org, a site maintained by a group in Washington State led by Trevor Bedford at the Fred Hutchinson Cancer Center. It presents sequences found on GISAID. China has three systems of its own that record

SARS-CoV-2 sequences, but these are only partly available to the public, and don't record the same information. However, the Chinese systems also include the data from foreign centres. Do they share their own collections and analyses? Not so much. "As at 4 February 2021, the database has integrated 437,808 non-redundant sequences, of which 2,089 are released from China," the report said.

There are 768 early SARS-CoV-2 genome sequences (defined as assembled before January 31, 2020) from 26 countries, of which 514 are Chinese. Of the 2,089 total sequences from China, 2,028 are from humans, 28 are from the environment, 33 sequences are from possible animal hosts (pangolin and bat), from pets (cats and dogs), or from animal experiments (mouse and hamster). The number of confirmed patient cases sequenced between December and the end of January was 494.

A table showed which database these sequences came from, when and where collection took place, and the name assigned to each sequence. I scanned it quickly. There were 25 complete, and three partial, human genome sequences that dated from December 2019. No sequences were collected from outside Wuhan before January 26. However, the first sequences from samples taken from seafood packaging and cold chain products were not collected until August 2020. In other words, these cold-chain sequences were found so late in the game they could not support the idea that the virus originated abroad, hitched a ride on frozen seafood to China, and produced a pandemic in Wuhan.

The team found two distinct lineages in the early sequences. Lineage A has mutations that also show up in the closest known bat-derived genomes, RaTG13 and RmYN02. Lineage B was found in the first human sequence collected in Wuhan on December 16, 2019. Some thought Lineage A might be the ancestor of Lineage B but there was no consensus on that.

One table showed the dates of early case and environmental sequences that had been collected around the world. On that table, the first sequence shown is from Italy. It was collected in week 49 of 2019, with three more collected on week 51, but these are marked with asterisks as "partial" sequences only. The next sequences were sampled in China, two in week 52 and 26 in week 53 of 2019. China produced 12 more in the first week of January. Mexico had three, Thailand had nine, and Spain had one.

Another table showed the "publicly available" sequences gathered in China by province. In week 52, there were two, in week 53, there were 26, and all samples were from Hubei. None were shown for other provinces until week two of 2020. The earliest samples belonged mainly to Lineage B. There were 29 sequences (because all samples were sequenced by more than one research group to ensure accuracy) from the earliest batch of 13 samples. These were assembled from scratch, so there were differences between them. Samples from two patients were identical. "The number of mutations of these 13 early cases ranged from zero to three relative to the reference genome." So the virus had not been circulating in Wuhan for very long, or there would have been a lot more mutations.

But these samples were drawn many days into patients' illnesses, so it was possible that even these few mutations occurred in the patients. Most of the early samples came from people working in the Huanan Seafood Market or who had visited there. Two sequences taken from environmental samples in the market had almost no mutations.

But when the early samples were graphed for location, they were scattered all around the central area of Wuhan. There were no obvious clusters.

The team wanted to know the time to the most recent common ancestor of 66 early sequences. Ten academic groups had tried to answer this question using several different methods of calculation, which produced different times ranging from September to mid-December. The preponderance pointed to a common ancestor between the middle of November and the middle of December.

The team followed up on those early sequences found outside China. They listed the sewage sample from Barcelona, Spain, that showed bits of RNA that might have been from SARS-CoV-2, which was collected on March 12, 2019. (They did not mention that even the authors of that paper were skeptical as to whether they had found something significant: all other sewage samples collected between January and December 2019 were negative.) In Italy, the first reported case dated to February 21, 2020, but there was a positive sewage sample from Northern Italy that dated to mid-December. A throat swab taken from a child in early December had tested positive. Skin biopsies taken from a woman in November were positive for a number of SARS-CoV-2 probes, but the woman tested negative with a PCR test, though antibodies were found in her blood taken in June 2020. Similar early finds showed up in France in late December, including neutralizing antibodies found in blood sampled in mid-December. A sewage sample taken on November 27, 2019, in Brazil showed positive results. Tests on banked donated blood in the US (the kind of survey China had not done) turned up 100 positive tests between December 13, 2019, and January 17, 2020, long before the first US patient tested positive. The report acknowledged: "Collectively, these studies from different countries suggest that SARS-CoV-2 preceded the initial detection of cases by several weeks. Some of the suspected positive samples were detected even earlier than the first case in Wuhan, suggesting that circulation of the virus in other regions had been missed." But then the authors backtracked to say that none of these study findings had been confirmed "and serological assays may suffer from non-specific signals." Meaning such tests are not reliable.

They turned again to the genomes, the closest and next closest to SARS-CoV-2. They discussed RmYN02, pulled out of bat feces collected in Mengla County, Yunnan, between May and October 2019. The sequence was found by doing a wide metagenomic scan of all sequences found in the collected bat feces. RmYN02 was not as close to SARS-CoV-2 as RaTG13 but it was deemed to be important because it had amino acid inserts similar, but not identical, to those found in SARS-CoV-2, so SARS-CoV-2 was not unique among beta coronaviruses in that regard. They also mentioned a sequence that is 92 percent identical to SARS-CoV-2 that was found in bats in Cambodia in 2010. But these sequences do not bind to human receptors. The report observed: "These findings do show that the ongoing search for the origins of SARS-CoV-2 should consider wider geographical ranges, multiple potentially susceptible species, and a sampling design that includes knowledge on number and densities of colonies."

The team found 72 "reliable genomes" of coronaviruses from 13 species. The vast majority came from animals that had been collected in China but their origin was not guaranteed. For example, pangolins sequenced in China may have been smuggled from another country. They attached a long list of animals infected with SARS-CoV-2 found around the world, such as mink, which had been culled in Denmark and the Netherlands.

The team concluded that it is necessary to link molecular information to epidemiological information to trace origin. Since viruses found in the Huanan Seafood Market were essentially identical, "that suggested a spreading event." The time to most recent common ancestor suggested "that virus transmission or circulation date might be recent, in late 2019." The most closely related genomes had been found in bats, and reports of early finds from other parts of the world "require follow-up."

THE NEXT SECTION dealt with samples taken from various places, including the markets, and from animals around China. It opened with the usual assertion that 75 percent of "emerging human infectious diseases have animal reservoirs, including wildlife."

"Analyses show that these spillover events are driven by factors that include large-scale environmental and socioeconomic changes, including land use change, deforestation, agricultural expansion and intensification, trade in wildlife, and expansion of human settlements."

I could almost hear Daszak's voice intoning this argument.

"...SARS-CoV-2 is also thought to have its ecological niche in an animal reservoir."

The team looked at animals and animal products as well as other potential routes for the emergence of SARS-CoV-2 in people associated with the Huanan Seafood Market, including "exposure to contaminated animal meat or food products that are refrigerated or frozen, or the introduction of the virus by people infected elsewhere."

They looked at outbreaks in Beijing and in Qingdao, where the only link between patients was frozen food packaging, or at least the only link they discovered. They looked at environmental samples from markets, sewage, and animals alive and dead. When the Huanan Seafood Market was closed January 1, workers went in 30 times over the next two months, sampling and isolating viruses where they could. They went after sewage, surfaces, individuals, animals for sale, animal parts in storage, stray cats, mice, rats living in the market, commodities in cold storage, etc. Some 923 samples were collected and tested but only 73 were positive and most were environmental. Only four came from sewage or even from other markets in Wuhan. They isolated SARS-CoV-2 strains from three samples. A vendor selling wildlife products tested positive. But animals on the farms of suppliers to the vendors in the market all tested negative. No animals sold in the market came from abroad: "No living or dead animals of foreign origin were identified from the sales records in late December 2019." And all the wildlife sold in the market came from licensed wildlife farms: "No illegal trade in wildlife has been found."

Some 457 animal samples were tested, including unsold samples found in freezers and fridges, stray animals trapped in the market, animal feces, and animal products sold in other markets in Wuhan. They tested rabbit, snake, bamboo rat, rat, chicken, salamander, etc. All were negative. Animals from farms that supplied the markets were tested: all were negative. Samples were taken from animals in other provinces such as Yunnan, Guangdong, and Guangxi.

Animals tested included pangolin, civet cat, *Rhinolophus affinis* bat, *Miniopterus schreibersii* bats, bamboo rat, macaque, bear monkey, porcupine, and fox. Of 1,287 samples taken, all tested negative. Between May and September 2020, 27,000 samples of wild animals were collected and sampled. "All SARS-CoV-2 NAT were negative."

They checked thousands of domestic animals across the country, food animals like pigs, cattle, sheep and chicken. All negative.

In 2019, routine animal surveillance for animal diseases took place across China. Samples were gathered from 222 large farms, 130 small farms, 67 scattered households in small towns and villages, and 25 slaughterhouses. These samples were retroactively tested for antibodies to SARS-CoV-2. This was done again in 2020 with a larger number of samples. Some of the 2020 tests included samples done at the Xinjiang Production and Construction Corps. All tests were negative for SARS-CoV-2.

Stored animal tissue samples collected in 2018 and 2019 were tested using new methods. All told, 26,807 samples from 24 provinces and autonomous regions were tested using a pan-coronavirus method and then for SARS-CoV-2. The pan-coronavirus tests produced 1,711 positive samples. The SARS-CoV-2 specific tests were all negative.

Tests of animals in Wuhan between November 2019 and March 2020, including dogs and cats, were all negative. (This was strange considering Shi's paper had found cats that tested positive.)

They looked at the countries of origin of cold-chain (frozen) products imported into the Huanan Seafood Market. Apparently, epidemiological analysis showed that the first three cases of COVID-19 in the market all had a history of exposure to cold-chain products. All tests done on cold-chain products in the market were negative. But:

> after China successfully controlled the COVID-19 epidemic in Wuhan in April 2020, a series of clustered epidemics occurred in various places. According to the experience of prevention and control of these epidemics, especially the successful traceability results of Xinfadi in June, Dalian in July and Qingdao in October 2020, it is confirmed that SARS-CoV-2 can survive and maintain infection activity in cold-chain products and packaging for a long time which provides a scientific basis for the possibility of introduction of SARS-CoV-2 through cold chain products.

Yet all the environmental sampling done in the market was consistent with infected humans shedding the virus into the environment, not humans being infected by something in the market. No evidence of wild animal infection was found there. No evidence of SARS-CoV-2 circulating among "domestic livestock, poultry and wild animals before and after the SARS-CoV-2 outbreak in China" was found either. Yet recommendations were made for further study involving more of the same kinds of tests, especially tracing back cold-chain products to their places of origin outside China.

There was nothing in this report that supported what Daszak had told NPR about wild animal farms being a possible point of origin for the virus.

THE LAST PORTION of the report was styled "Possible Pathways of Emergence." So far, 111 pages had been devoted to every possible pathway except for an accidental leak from a lab. The time to the most recent common ancestor pointed to mid-November 2019, in Wuhan, with an explosion of cases all over Wuhan by early January. Crucial epidemiological evidence had been withheld or scrubbed or not sought, such as: the samples from patients in flu sentinel hospitals for the last three weeks of December; the first known cases of SARS-CoV-2 that had been scrubbed although previously published after peer-review; no tests of blood from donations made in the last quarter of 2019. The blood tests might have helped establish when the first cases occurred if any positive samples were found, and from there, who the first people to get SARS-CoV-2 might have been, where they lived and worked. The earliest sequence evidence only dated from the end of December, by which point there were mutations sufficient to define two lineages, but not to establish which was first. The reason given as to why the retroactive blood study had not been done was that it would require difficult-to-get ethical permissions—a strange excuse from a country with no qualms about using the tiger chair to torture Uyghurs in Xinjiang.

In this situation, the absence of evidence had become evidence. Nothing was found suggesting SARS-CoV-2 spread from a particular animal, wild or domestic, to humans. The very extensive studies done on wild and domestic animals across the country were all negative. Even the rats and cats hanging around the Huanan Seafood Market had not been infected with SARS-CoV-2. There was no evidence for any early human infections caused by cold-chain samples at the market, and in any case more than 40 percent of the earliest cases had no ties to any market. These negative findings reinforced what had always been obvious: this pandemic started in Wuhan in humans, and it spread so quickly because it was already adapted to humans—much better adapted than SARS.

So where was the section on the studies done concerning a lab-leak hypothesis? I flipped to the end. Nothing. I flipped back to the table of contents to see if I had dropped pages when I printed the report. There was a list of site visits that showed the team went to the Wuhan Institute of Virology, but nothing in the body of the report described what happened there. There was also a visit to the CDC Hubei, but not to the Wuhan CDC where bats have also been sampled and tested for viruses, and whose researchers had been videoed entering bat caves without protective equipment. But I found no section detailing studies done on the lab-leak hypothesis.

The last section of the report listed four hypotheses on how the virus might have been introduced into humans, including arguments for and against. These were:

1) direct zoonotic transmission, or spillover

2) introduction through an intermediate host followed by spillover

3) introduction through the cold/food chain

4) introduction through a laboratory incident.

The joint team used a risk assessment scale to assign a level of likelihood to each of these scenarios "considering the available scientific evidence and findings." They arrived at a scale stretching from very likely to extremely unlikely.

The authors repeated the arguments for and against

direct zoonotic transmission and found that scenario to be "possible to likely."

The second scenario, Daszak's favourite, for which there was a great deal of speculation but no evidence at all, was assessed as "likely to very likely."

The third, cold-chain transmission, official China's theory of choice, was assessed as "possible."

And the fourth? A lab incident? The report said: "We did not consider the hypothesis of deliberate release or deliberate bioengineering of SARS-CoV-2 for release, the latter has been ruled out by other scientists following analyses of the genome." This was a sly reference to a theory brought to public attention with the help of Steve Bannon, President Trump's former éminence grise, and Guo Wengui, a Chinese billionaire and founder of Lude Media, living in exile in the US. Li-Meng Yan, a former post-doc from a leading lab in Hong Kong, had produced two preprints claiming SARS-CoV-2 is an engineered bioweapon unleashed to destroy the economies of China's enemies. No one except Fox News took these preprints seriously.[376] This report relied on the proximal origin article in *Nature* as the refutation of that thesis.

The report then acknowledged that laboratory accidents do happen and "different laboratories around the world are working with bat COVs." It acknowledged that working with virus cultures and inoculating animals could lead to an infection of a researcher in labs "with limited biosafety, poor laboratory management practice, or following negligence." It acknowledged that the closest "known" coronavirus to SARS-CoV-2 had been sequenced at the Wuhan Institute of Virology. It also acknowledged that the "Wuhan Center for Disease Control and Prevention moved on 2 December 2019 to a new location near the Huanan market. Such moves can be disruptive for the operations of any laboratory."

Wait a minute, I yelled as I read that line of text. The Wuhan CDC's move on the second of December was news to me. Tian Jun-Hua works for the Wuhan CDC. For years he has trapped and sampled bats, hunting for viruses. He has been a co-author with E.C. Holmes on several papers, including the first report of the SARS-CoV-2 genome. Why was the relocation of the Wuhan CDC dropped into this report like it meant nothing? And why had the authors failed to point out that both CDC and WIV researchers bring live bats back to their labs from the wild, often failing to wear proper protective equipment while doing so?

Accidents in containment labs are not rare anywhere. Lab leaks of the SARS virus happened in Singapore and Taipei and twice in China during and after the pandemic. SARS was supposed to be handled only in BSL-3 labs, but at the CDC in Beijing good biosafety practices were not adhered to. One of the members of the DRASTIC group published a preprint about those events. The Beijing CDC's BSL-3 was connected to two labs with lesser containment. A graduate student who worked in one of the labs prepared attenuated SARS improperly. Two people were exposed but did not get sick. Months later, another grad student working in the lab complex, who swore she never entered the BSL-3 lab, left town on a train. She got sick enough that she went right back to Beijing to a hospital. Her mother arrived in town to help her. In the hospital, the grad student was diagnosed with SARS. Her mother got SARS. Her mother died. Nurses who treated the student became ill. Hundreds of people had to be isolated. The WHO was informed and sent experts in to help trace the origin of the leak. It turned out that the three interconnected labs, directed by the same eminent authority who had declared SARS to be caused by chlamydia, were very crowded, the students ill-supervised, and the SARS virus was kept in a fridge out in the hall. Nothing about that case, nothing specific about any previous lab accidents in China, was mentioned in this report.[377]

The authors turned to the arguments against the lab-leak hypothesis. The closest known relatives of SARS-CoV-2 are relatively distant, the report said. "There has been speculation regarding the presence of human ACE2 receptor binding and a furin-cleavage site in SARS-CoV-2." The report did not state the nature of that speculation but said "both have been found in animal viruses as well, and elements of the furin-cleavage site are present in RmYN02...." The team's next disproof was that there "is no record of viruses closely related to SARS-CoV-2 in any laboratory before December 2019, or genomes that in combination could provide a SARS-CoV-2 genome." This was a peculiar statement: Shi had admitted that the WIV has many sequences which it has not published (thus no record) and that she retained samples taken from the miners who became ill in 2012 (without publishing what she found in those samples). There was no evidence, the report continued, that SARS-CoV-2 was circulating "among people globally" and the surveillance program in place was limited regarding the number of samples processed. Therefore, the risk of the accidental culturing of SARS-CoV-2 in the laboratory was also extremely low.

> The three laboratories in Wuhan working with either CoVs diagnostics and/or CoV's isolation and vaccine

> development all had high quality biosafety level (BSL3 or 4) facilities that were well-managed, with a staff health monitoring programme with no reporting of COVID-19 compatible respiratory illness during the weeks/months prior to December 2019, and no serological evidence of infection in workers through SARS-CoV-2 specific serology-screening.

The problem with that statement was that Shi had told *Science* her work on coronaviruses had also been done in BSL-2 labs. The report insisted that though the Wuhan CDC lab moved on December 2, 2019, it "reported no disruptions or incidents caused by the move. They also reported no storage nor laboratory activities on CoVs or other bat viruses preceding the outbreak." Thus, the assessment of likelihood of a lab leak was "extremely unlikely."

The report's annexes gave more detail about what happened at the meetings with the Wuhan CDC and the Wuhan Institute of Virology. Annex Four described the meeting between representatives of the Beijing CDC, the Hubei CDC, and the Wuhan CDC and the joint team. The annex made no reference to a discussion about the Wuhan CDC moving its offices close to the Huanan Seafood Market on December 2. But it revealed that one of its staffers had tested positive for SARS-CoV-2. This was explained as part of a "family cluster." No other information was given. Annex Five described the meeting between the joint team and senior staff at the Wuhan Institute of Virology. The team got a tour (shorter in duration than NBC's). But no lab records were shown. The leadership of the WIV explained that all staff had tested negative for SARS-CoV-2 from January 2020 forward. No staff had reported respiratory illnesses in 2019 either. Stored samples taken from staff in April 2019 and in January 2020 had tested negative. Most of the meeting involved discussion on how the WIV could counter the conspiracy theories. The joint team was told that the young woman rumoured to be patient zero, who'd left the WIV in 2015, had refused to speak with the media. They were told that all researchers wear PPE when sampling bats in caves. They were told that the miners in Mojiang died of a fungal disease. They were told that recombinant experiments had only been done at the lab since 2018, and that other SARS-like viruses discovered at the Mojiang mine (visited seven times) were too distant from SARS-CoV-2 to be its ancestor.[378]

The team made recommendations for further study, which included "regular administrative and internal review of high-level biosafety laboratories worldwide" and follow-up "of new evidence supplied around possible laboratory leaks," implying SARS-CoV-2 might have leaked from a lab in another country.

I read the report's lab-leak refutation twice. It had a grand total of three citations, though the other sections of the report had pages of endnotes. The first referred to the proximal origin letter from *Nature*. Though there is an extensive literature on lab leaks, no such papers were cited. The second citation referred to a paper published in 2017 about how host and viral traits predict zoonotic spillover from mammals. It had nothing whatever to do with lab leaks. The third was for a psychology paper on "a technique for the measurement of attitudes."

This was a breathtaking display of arrogance. The report's authors had treated reasonable concerns raised by other scientists with contempt. They had treated the families of the sick and the dead with contempt.

They seemed to think they could get away with it.

THE SAME DAY the WHO published this joint study, several nations issued a statement about it, 14 to be exact, including the US and, for a change, Canada. Notably the European Union did not sign, nor did France, Germany, or Italy, but Denmark and Norway did. How did 14 countries negotiate, write, distribute, and sign this statement on the same day the report came out? Their diplomats had gotten their hands on advanced copies.

The statement went on at length about the WHO's praiseworthy work fighting the pandemic (with nothing said about its appalling tardiness to declare it). Everybody, the statement said, wants to work with the WHO and the global community to discover the origin of the pandemic "in order to improve our collective global health security and response." But "we join in expressing shared concerns regarding the recent WHO-convened study in China, while at the same time reinforcing the importance of working together toward the development and use of a swift, effective transparent, science-based independent process for international evaluations of such outbreaks of unknown origin in the future."

I waded through a swamp of similar phrases until I finally beat my way to the statement's point:

> ...the international expert study on the source of the SARS-CoV-2 virus was significantly delayed and lacked access to complete, original data and samples. Scientific missions like these should be able to do their work under conditions that produce indepen-

> dent and objective recommendations and findings. We share these concerns not only for the benefit of learning all we can about the origins of this pandemic, but also to lay a pathway to a timely, transparent, evidence-based process for the next phase of this study as well as for the next health crises.[379]

The likelihood of that? Unlike the joint study team, I could put a number to it.

Zero.

15. DRASTIC Measures

JOURNALISTS NEED TO know where their information comes from, including the areas of expertise, the histories, and something of the characters of the sources who give it to them. Information from anonymous sources or from documents delivered in brown envelopes or attached to encrypted emails can be dangerous. Relying on it can lead to the publication of something that is untrue and designed to hurt someone. On the other hand, if a journalist has known a source for years, if they've gossiped together at dinner parties, have mutual friends, drunk from the same wine bottles too often, too much trust can also lead to a bad end. Dealing with sources in the intelligence community is especially fraught. Sometimes they offer the straight goods, sometimes they hold vital things back. It is up to journalists (and non-fiction authors) to try and check the facts independently, but if they are derived from top-secret files, that may be impossible, so once again it comes down to trust and judgment. Sometimes lawyers are consulted. Once, the lawyers for a magazine I was writing for took me down to a basement room (were they afraid of being bugged upstairs?) to grill me about a story I had written about a secret trial involving the transfer of classified information to a foreign country. The lawyers asked me why I hadn't come to them first. Why would I? I replied. They muttered about how, under the Official Secrets Act (since repealed), I could get ten years if the government came after me. The editor thought that was unlikely as the government had already gone after another journalist over the same matter, which had created a cause célèbre. We published and held our breaths, which is a lot like jumping off a cliff and hoping to land in deep water instead of on the rocks.[380]

The Edward Snowden story is another case in point. This is how it went: a man who said he once worked for the CIA and was employed by a contractor to the NSA (National Security Agency) approached a filmmaker known for documentaries about American surveillance misbehaviours. The source did not use his own name and insisted on communication via encrypted emails. He also approached two journalists working for competing publications, one of whom was well known for taking the NSA and the secret FISA court system to task, the other for the big prize he had on his shelf. One of the journalists, a freelancer, took the story to the UK headquartered *Guardian*, which had opened a new digital outlet in the US. The other worked for the *Washington Post*. The source offered to show classified documents he'd smuggled from his workplace, documents that demonstrated how electronically intrusive in the lives of their own citizens the British GCHQ and the American NSA had become, which flew in the face of constitutional and legal protections. The source asked to meet the journalists and the filmmaker in Hong Kong. The *Washington Post* reporter didn't go but got the story anyway. In Hong Kong, the source—Edward

Snowden—went through serious rigamarole to protect from electronic prying eyes his computer, the external thumb drives he'd loaded with classified downloads, and his conversations with the journalists from the *Guardian*—Glenn Greenwald and Ewen MacAskill—and filmmaker Laura Poitras. He showed them how the world's internet data, most of which passes through undersea cables between the US and UK, is tapped and surveilled, how no email or cell phone call (not even the cell phone calls of Angela Merkel) is safe from this electronic hoovering, and how items of interest are shared among the Western allies known as Five Eyes (Australia, New Zealand, the UK, the US, and Canada). He explained that major communications corporations and other tech giants had enabled these intrusions through backdoors in their systems. And that was just for openers.

He handed over a treasure trove of data for the journalists to study: it was up to them to determine what could be safely published. He didn't want to compromise the security of an agent or harm his country, he told them, so he was leaving those judgments to them. At the last moment, he announced he would explain on camera who he was and why he'd done this, though the journalists had been willing to keep his identity secret. They went home and fact-checked their stories as best they could, calling up US security agencies and the White House for comment on what they were about to print. The video in which Snowden outed himself was uploaded to the *Guardian* website right after the journalists' first story appeared. After the reporters and filmmaker left Hong Kong, he'd hidden in the home of refugees living in the city, conveyed there by the Canadian human rights lawyer handling their case. When the stories ran, Snowden ran too. There was a public drama at the airport in Hong Kong after the US pulled his passport and he tried to find a country to keep him safe from American wrath. He ended up in Russia. The *Guardian*, not protected in Britain by constitutional rights regarding freedom of the press, submitted to British government officials who descended on its offices and destroyed the hard drives carrying the documents. (But there were copies held in the US, so why did they bother?) The documents were real. The publications got prizes. The filmmaker won an Oscar. Snowden remained stuck in Russia, his punishment for trying to stop the NSA from doing unconstitutional things. Everybody involved got famous.[381]

This is the best illustration I know of the weird and dangerous dance that goes on between sources who won't give their names and journalists who want a great story. It's a little like the relationship between a virus and a host. There is risk for both. There will be change for both, which may be wonderful or dire.

Fear of that dance is why I was so slow to contact members of the DRASTIC group. DRASTIC is an acronym that stands for Decentralized Radical Autonomous Search Team Investigating COVID-19. I first came across the name in the summer of 2020. I didn't follow up because I thought it was safer to rely for information on peer-reviewed articles, on preprints that looked as if they would survive peer review, on documents filed with the IRS, on video and print interviews with people using their own names. In the beginning, DRASTIC members communicated their finds on Twitter, though they soon began to publish preprint articles as well. But anybody can get themselves a Twitter handle and put up anything they like. So I ignored DRASTIC, even after I spoke at length with Jonathan Latham.

In retrospect, I am amazed at how much time I spent

worrying about whether Wilson and Latham were as independent as they appeared to be. Latham is from the UK but did his PhD in the US, whereas Wilson is American. I wondered about the possibility that they relied on some unknown foreign funder with some sort of hidden agenda. Their organization, bioscienceresource.org, is a 501(c)(3), just like EcoHealth Alliance. But according to its filings, it didn't seem to have enough money to support two researchers and a publication. In some years, the organization had filed no IRS reports. Latham explained that if a charity earns a negligible amount it need not file a report that year, which only made the problem worse. So where does your money come from, I asked? Latham explained, very reluctantly. He asked me not to publish what he told me, which also made me nervous, yet in the end I decided they were trustworthy because their work speaks for itself and they put their names on it.

DRASTIC certainly came up when I asked Latham how he and Allison Wilson learned about the provenance issue, and about the master's and PhD theses so fundamental to their argument. He mentioned Yuri Deigin and Rossana Segreto. I told him I'd looked up Deigin, and that his Russian connections made me wary, and that Segreto didn't have a lot of publications to her name. We also discussed the preprint (later a peer-reviewed publication which has had at least 46,000 views) by Monali Rahalkar and her partner Rahul A. Bahulikar, which described the theses. I'd been pleased to see that Rahalkar and Bahulikar are legitimate scientists working at good institutions in India. I didn't look further at how they found the theses because I was being wilfully blind. When Rahalkar and Bahulikar's paper was finally published in a peer-reviewed journal in October 2020, they credited a "twitter user @TheSeeker268" with finding them. Theseeker is part of the DRASTIC group, as are Deigin and Segreto and Rahalkar and Bahulikar.

Who was this "seeker" who found such important documents yet hid behind a pseudonym? Was he/she a whistle-blower in China? I thought that might be the case. Surely he/she would have to read Mandarin to search a Chinese database for those theses and figure out why they mattered. Not knowing one way or the other gnawed at me. The more I leaned toward the lab-leak theory, the more finding out about DRASTIC and theseeker seemed important. If those theses hadn't been made public in the West, all the interested parties (Daszak, Baric, Lipkin, Andersen, Holmes, Fauci, Collins, Shi Zhengli, George Gao, and of course Xi Jinping and the WHO) would have succeeded in shutting down any serious public discussion about, let alone investigation of, a possible lab leak. Shi Zhengli would have gotten away with telling *Scientific American* just enough about those miners to give RaTG13 a thin provenance. I wouldn't have wondered why several important labs in China all rushed to sample bats in the same mine, or why peer-reviewed papers by Gao and Canping Huang, who wrote the doctoral thesis, did not say exactly why they went to that mine, only that they discovered interesting viruses when they did. Without knowledge of those theses, there would have been less pressure on *Nature* to publish an addendum on the origin of RaTG13. Shi would not have been forced to admit that she retained samples from the miners and had tested them "recently" for presence of SARS-CoV-2, failing to say what she *did* find. All of that led back to one person—theseeker.

One day, long after I began to write this book, I found myself muttering: what if theseeker is a spook for some government with a bone to pick with China?

I GOT IN touch with DRASTIC at the end of April 2021. A person who uses the name Billy Bostickson—a pseudonym—seemed to be the organizer, but I wanted to speak with someone in DRASTIC who uses his/her real name. I asked Jonathan Latham how I could check that the WIV's samples-and-sequences databases remained offline. He suggested I get in touch with Monali Rahalkar in India. Monali Rahalkar handed me off to Gilles Demaneuf, a DRASTIC member who uses his own name.

Demaneuf lives in New Zealand and works as the data science technical lead for the Bank of New Zealand according to his LinkedIn bio. He was educated at the Lycée Sainte Geneviève, Versailles, in math and physics, followed by an MSC in applied mathematics, sciences, and management from CentraleSupélec in 1991, a school with relationships to big universities around the world, including Tsinghua University and Peking University. Then he got a certificate in financial risk management in 1999. He speaks French, English, and German. There is one three-year gap in his bio, after he graduated from CentraleSupélec. Then, in 1994, he went to work at Lehman Brothers on various forms of derivatives, staying for eight years. From 2002 to 2008 he worked for Calyon Investment Bank doing derivative structuring (hotshot investment bankers around the world traded derivatives with abandon, which helped bring on the near collapse of the global economy in 2008). By 2009 he was an investor and partner and head of quantitative analysis for a company called Sentient Technologies Holdings Limited, located in Hong Kong. He stayed there for six years, and then moved to Auckland and a bank called ASB and, in 2018, to the Bank of New Zealand. On the side, he became a co-founder and managing director of a food company called Paleo Choice. It is clear from his biography that he is more than adept at dealing with data.

He responded to my email right away to tell me that the databases were still offline, giving me links to prove it. He also told me that information about these databases had also recently disappeared from the website called China Science Data. He sent me a link to the original article on the databases and their removal.[382]

A typical supersmart person, I thought. Not content to answer only the question asked. Later he would tell me he has Asperger's.

I asked if someone could speak with me about how DRASTIC came together. He thought I meant Billy Bostickson. "Billy won't be able to talk to you," he wrote. He said I could communicate with Billy via email, or chat with him if I preferred. But Demaneuf is in New Zealand, a huge time zone distance from Toronto, so I contented myself with emails to him. I asked if he could tell me about the group's history. He said it started with some scientists who were upset "by the absence of scientific method...and the categorical dismissal of a possible lab pathway." I thought he was referring to the *Nature* proximal origin letter, but that February *Lancet* statement seemed to be more important. "The *Lancet* letter looked to us very much like fake moral outrage parading as science."

These researchers "found each other" while doing their own research and "the conversations quickly converged in private forums on Twitter and became loosely organized around Billy Bostickson...in an informal group called DRASTIC." That was followed by the first peer-reviewed paper discussing the possibility of a lab leak and some unexpected features of SARS-CoV-2 published by members of DRASTIC, “who also quickly found out about the Mojiang mine accident and its link to RaTG13.” DRASTIC is an interdisciplinary group, Demaneuf wrote, with relevant skills, continuously collecting data from open-source means. Listing the skill sets involved, he included these: “lab-building and operations...Chinese language and culture, OSINT (open-source intelligence)...DRASTIC can sound a bit militant or irreverent at times,” he wrote. “That partly comes from having done a lot of hard work when the ‘expert opinions’ based on very little were so dismissive.”

To answer my next set of questions he switched to a ProtonMail account, which is encrypted. He told me more about himself, including that he first worked in the City of London, and thus Peter Embarek, chairman of the WHO joint study group into the origin of the virus, had referred to him as “the banker.”

How did he know about you at all, I asked?

Members of the DRASTIC group had written to members of the joint study group asking questions, he explained.

He added more detail on DRASTIC’s origin. It formed in February 2020, he wrote, after Daszak’s statement appeared in *The Lancet*. But Demaneuf himself did not connect with Billy Bostickson until June 2020: up until that time, he and a few others had been using another site, but it was attacked by so many “Chinese trolls” they decided to abandon it. He provided me a list of original members according to when they got involved: most, he said, had started wondering about the virus’s origin on January 20, 2020 (the day all those papers from scientists in China were submitted to various leading journals, the day Xi admitted person-to-person transmission). Dan Sirotkin and his father Karl published a paper on the lab-leak possibility, he wrote.

(Actually, it was Dan Sirotkin alone who published that early paper on January 31, 2020). "One can say DRASTIC came together in February. Then Rossana (Segreto), Yuri (Deigin), Mona, theseeker, etc. joined them."

I asked about Billy Bostickson and why he remained anonymous. I had by then found Billy Bostickson's LinkedIn account. There Bostickson describes himself as coordinator of DRASTIC, but also as being affiliated with the Pedro Kourí Institute, while also doing public relations for RAGE University. I'd looked these institutions up. Pedro Kourí Institute is in Havana, Cuba, and is a WHO collaborating centre, a serious research institution. Rage University teaches social justice organizing tactics from an anarchist perspective. There is an image of a stylized bear on its home page. An *x* crosses out one of its eyes. Contact information reveals no location.

In the About section on Bostickson's LinkedIn I read:

> I am passionate about Anonymity, Subversion, Revolution, Resistance, Freedom, Jutice [*sic*], Equality and Solidarity. I desire the overthrow of all tyrannical governments, organised religions and military juntas by whatever means necessary. Adelante hasta la Libertad!

Under the heading "experience," Bostickson listed himself as coordinator of DRASTIC since January 2020, and claimed to be located in Wuhan, Hubei, China. Yet he was also working for Rage University in Belize. In addition, he said he had been an assistant professor at Universidad ORT in Uruguay from 2012 until 2020, doing an investigation of government and corporate corruption in Asia. He claimed as well to have been a teacher of investigative techniques at the IPEP School of Journalism, also in Uruguay, for a little over a year. He said he had a master's degree in virology, a diploma in computer science, and a school certificate in "Political Persuasion and Subversion." He also said he had full proficiency in several languages: Norwegian, Swedish, English, Spanish, German, and Romanian. Under voluntary experience he listed: "Investigator, Bureau of Public Secrets." I thought that was a joke. But it turns out there *is* a Bureau of Public Secrets. It has an internet address with lots of references to revolutionary literature in many languages and a box office in Berkeley, California. Ken Knabb's writings are featured.

There is also a Billy Bostickson listing on ResearchGate. In the first months of 2021, Billy co-authored several preprint papers: on the WIV databases taken offline; on a proposed forensic investigation of Wuhan Laboratories; on the Wuhan Institute of Biological Products Company; on RaTG13 and the clade of related coronavirus sequences collected from that mine but not published by the Shi lab. That preprint may have driven the Shi lab to produce in a big hurry a preprint of their own on the clade of eight coronaviruses collected in the Mojiang mine. According to Shi's preprint, they are not at all like SARS-CoV-2. But it reads as if written in such haste no one had time to edit it before it was posted on May 21, 2021, on bioRxiv.[383]

Bostickson also listed on ResearchGate that one of his areas of expertise is propaganda.

This combination of science publications plus claimed expertise in propaganda did not inspire confidence. I asked Demaneuf to explain.

Demaneuf wrote that "for perfectly valid reasons Billy has to hide his identity...yes he was involved in some previous project called RAGE. The rule is that we don't ask and we don't tell—we trust him and leave it at that. We have seen enough of his personal interactions and his character, sometimes foibles, to know that he is real and genuine."

I wasn't so willing to trust. But I decided to get in touch by email with Billy Bostickson, whoever that is, to ask questions directly.

He claimed to have started DRASTIC in January 2020, which seemed awfully quick off the mark considering the *Nature* letter and the *Lancet* statement appeared later. Even Shi's preprint of her *Nature* article mentioning RaTG13 didn't appear until January 23. He sent me a link to a DRASTIC research website. It had an opening page with listings such as: *Home*, *The Team*, *Our Works*, *Where to Start*, and *Resources*. Thumbnail images of team members were included under the Team section. Some were photos of real people, some were illustrations. One anonymous person uses a photo of Xi Jinping to represent himself. Billy Bostickson uses the bear image from Rage University. Of the 20 images shown, nine were visual versions of pseudonyms. The Team page had a nice little magazine-style title and deck: "The Origin of SARS-CoV-2 is a Riddle: Meet the Twitter Detectives Who Aim To Solve It."

Billy Bostickson seemed to have acquired PR training at some point, because the way he responded to questions—answer some, ignore others—reminded me of Eric Morrissette of PHAC.

Why the anonymity? I asked.

He claimed to have been working anonymously for ten years "due to locations I work in, for security reasons."

I asked if he is an anarchist.

He said that his political philosophy "is not related to

my work at DRASTIC. We have all sorts working in DRASTIC so we avoid topics like Football, Sex, Politics and Religion." He wrote that he is based in Asia, has no PhD, but "specializes in Biosafety and Biosecurity investigations." He provided links to the five papers he's published. He said he had six more in the pipeline.

"I smelled a rat like many people did and decided to dedicate myself to uncovering the truth," he wrote. "I felt it was a duty to do this out of respect for so many old folks who died and the terrible effect on local economies. That is what kept me going 15 hours a day, every day for more than a year." He wrote that he researched papers by Zhengli Shi and George Gao (using the Western order for their names, which is odd if he is in Asia, unless he is in South Asia.) He shared his concerns on Twitter when Dan Sirotkin posted his article on SARS-CoV-2 on January 31, 2021.

Dan Sirotkin's father Karl Sirotkin has significant medical and scientific credentials. Dan does not appear to have any, though he may have attended Harvard as an undergraduate and seems to have an acquaintance with prison.[384] The younger Sirotkin's paper made accusations against numerous parties and took dead aim at the extreme danger of gain-of-function experiments.[385] Billy Bostickson began corresponding with Sirotkin via Twitter and soon there was a group of five. The group (by May 2021) numbered 26, loosely organized in various working groups, including: investigating conflicts of interest; specific labs in Wuhan; biosafety; the cover-up; missing databases; infected researchers; communication with the media; translating Chinese documents; tracking scrubbed documents; and investigating vaccine development before 2020 in China. "We also have our own counter-intelligence group," he wrote. He included a series of links to media stories about the group (no doubt because major media reporters, when writing about a group as marginal as DRASTIC, feel more comfortable if other media have done so already).

He also included a list of achievements. First on the list was the discovery of the theses and the miners' story, which he attributed to theseeker.

DEMANEUF GOT ME an email address for theseeker, who he described as "a retired teacher in India," which made him sound like a person in his/her 60s. As to why theseeker chose to be anonymous, he wrote: "Again, anonymity is a personal choice based on circumstances. One member has also lost his job and has seen his career set back by a few years at least for not being anonymous. Others like Rossana have been harassed and have changed jobs. I am not anonymous. I have had anonymous phone calls, my emails were read. I had to switch computer and mobile, etc. By the way I am answering you with my Proton Mail [*sic*] account for this exact reason. If you are a journalist working on China you will know what I mean."

I wrote to theseeker. It took a few days to get a response from him. I can use the word *him* because when I asked for his gender in a follow-up email, he admitted he is male. His first reply also came via ProtonMail. In answer to my questions about who he is and where he lives, he wrote that he is currently living in the city of Bhubaneswar. I looked it up. It is the capitol of the Indian state of Odisha, which is on India's east coast near the Bay of Bengal. It has about 700 temples and many sacred landscapes but it was designed as a modern city after Partition in 1949. It's a major business/high tech centre: one of its sister cities is Cupertino, California. It is only 25 kilometres away from Charbatia Air Base, an aerial reconnaissance post of the Indian intelligence agency's aviation unit, the Aviation Research Centre. Wikipedia records that Charbatia was built with the aid of the CIA in the 1960s and "used for aerial surveillance and intelligence gathering of China's strategic forces."

Theseeker wrote that he is in his late 20s, that he studied architecture first, then switched to filmmaking. After that, he taught science to secondary students for two years—until 2019. He said nothing about what he has done since 2019 to earn a living. (Bostickson had also made it clear he has been working on DRASTIC nonstop since January 2020.) As to why he began this investigation, he wrote: "I think I got into it as scepticism started emerging on the mainstream scientific and media narrative on the origins of Covid-19."

"And yup," he was the one who found the theses, about ten days apart. It took a lot of effort "and maybe a bit of luck. Back then I had just discovered this Chinese journal database—CNKI, and I started exploring the database before I got obsessed." He searched it using trial and error and "Chinese keywords, I was looking for titles by particular keywords, authors, institutes, year of publication; using every keyword combination I could think of as relevant, and one day it was bam!"

I found it hard to understand how searching known keywords like Shi Zhengli and the Wuhan Institute of Virology or George Gao or RaTG13 would have brought him to those theses. Gao's name might have led to the PhD thesis eventually, and from there he might have gotten to the master's thesis, but the chase would have been very long. Shi Zhengli's mention of the miners and their illness in *Scientific American* might have reminded him that Peter Daszak had mentioned

Mojiang County, the location of the mine RaTG13 had come from, in the interview he gave *Science* in January 2020. But Shi said in the *Scientific American* story that the miners died of a fungal infection, not from a coronavirus like RaTG13. To search, you need some idea of what you're looking for.

I asked why he withheld his name and country of origin.

Not a conscious decision, he replied. He thought his pseudonym fit, it described what he excels at. "That name is also acknowledged in Monali [Rahalkar] and Yuri [Deigin]'s peer-reviewed papers, so I'm sticking to it." He directed me to two media articles written about him as if to say, *see, these other journalists are writing about me, so what's your problem?* One of the articles was by Ian Birrell. "The COVID dissidents taking on China" was published in UnHerd.com.[386] Birrell used to be a speechwriter for UK prime minister David Cameron, and for a time was deputy editor of *The Independent*.

I asked theseeker if he'd noticed that two papers came out of the PhD thesis, all of which were supervised by George Gao, but that the papers did not mention why the authors went to that mine looking for viruses.

He misunderstood me. He thought I meant George Gao supervised both theses. He replied:

> Both? On miners, there is only one dissertation supervised by Gao Fu as far as I know. I think Li Xu, the Kunming Medical University doctor was best placed to give the details. But yeah, it's surprising that Huang Canping's thesis didn't refer to Li Xu's thesis. In fact, it somewhat surprised me that Canping's thesis would dedicate a chapter on the miners' illness and it went on to record some extra detail

> such as the IgG [a type of antibody found in blood serum which binds pathogens such as viruses, bacteria and fungi], perhaps because KMU doctors only had access to their own clinical records and IgG was analysed outside. Btw, I found some more theses a few weeks back, one mentioning the miners. The translation part is almost done...It fills in a lot of gaps and reveals new details on the sequencing of CoV's from the mine—including Ra4991, 7896 clade and more...

That was the end of the first response. But it raised new questions. His language was American—"yup" and "yeah"—but he used the British spelling for the word *skepticism*. He used the correct form of Chinese names, unlike Billy Bostickson. George Gao's birth name is Gao Fu. He signed off this email in British style with the word *cheers*.

I wrote him back. This time I asked if he'd grown up in North America, because he wrote like one of us. No, he said, never been to the Western Hemisphere. He thought his language might be influenced by his deep interest in American "flicks, TV series, and novels." As to when he started on this SARS-CoV-2 origin adventure, "it must have been somewhere around April, 2020 because up until March 27, I still believed in the wet market narrative. As someone recently pointed out to me, this is what I posted on twitter that day. 'Nobody wants to see their parents or grandma and grandpa die over a stupid virus from an exotic animal market.'"

That was interesting: it suggested he hadn't read the early journal publications that said the first cases had no link to the market.

He explained that his interest was piqued by reading "Yuri's [Deigin's] *Medium* piece about that time and I began educating myself on the topic."[387] He had questions by late April or early May. "On May 13 (perhaps influenced by reading Botao Xiao's paper) I did my first deep dive on Wuhan CDC and I posted a thread on my findings, which included among other things, a tender for disposing of 2.5 tons of laboratory biowaste. (These links were later scrubbed by the Wuhan CDC, but luckily they have been archived.)" And after that, he started looking into the WIV webpages "and of course the CNKI database." He thought that was when the serious discussions about origin arose on Twitter and "when drastic arose and came together."

I asked him what he searched for on the CNKI database, adding that obviously he reads Mandarin.

> Yea, I was looking at all sorts of things SARS related: Mojiang, spike coronavirus, pneumonia, Kunming, Shi Zhengli, Gao Fu, Wuhan CDC, Yunnan CDC, etc., etc. But no I don't speak or understand Mandarin. I would write the text in english, run it through Google Translate, and then copy the text in the CNKI search engine and hit enter. (As I previously mentioned, it's a time-consuming process and it involves a lot of trial and error.) Usually 100s and sometimes 1000s of results would show up, but cnki allows you to filter it by year/authors/institutes, and one can directly translate the search results in english by using Chrome's auto translate option.

His DRASTIC origin story was different from Bostickson's but similar to Demaneuf's:

> No one founded/organized it really. It just happened, like organically. It started from a Twitter chat room called "Daszak's Fan Club" and it's still just a Twitter chat room, but now it's called "drastic" and yea, Bill named it and he's probably the most enthusiastic of the lot. But I see drastic as a recognition of certain facts we found: like the Mojiang theses, official notices, grants, missing databases, tenders, patents, awards and such. Facts that may have a direct bearing on the origins of SARS-CoV-2.

He sent me a link to his most recent Twitter offering.

I wrote him again to ask if he is a chain smoker: Demaneuf had mentioned that. No, he said, he only does the occasional chai cigarette. I wrote him once more to ask directly how he's made his living these past few years. He asked if he could please not answer that question. I said the problem with not answering is it made one wonder if he worked for an intelligence agency. He did not respond to that email.

His story seemed plausible, on one hand, but on the other, not plausible at all. His curiosity and drive to find out more about the origin of SARS-CoV-2 seemed entirely familiar. I'd been that curious myself. But when I put myself in his shoes, trying to search a database in a foreign language, written with unfamiliar symbols, through Google Translate, I threw up my hands. I am a determined reporter, but even if I knew about the CNKI database, I would likely have missed the theses because I would not have trusted Google Translate or Chrome. I'd have had to know that the name *Mojiang* was important, and that something happened there. It seemed to me he must have known what he was looking for—a dissertation on people who got sick in that mine in

Yunnan. Until the *Scientific American* article appeared online on March 11, 2020, no one outside Shi's lab—other than Daszak—knew that RaTG13 came from a mine where miners got sick. About the mine, yes. The sick miners, no. Yuri Deigin's *Medium* post of April 22 had a lot of information about Shi Zhengli, the mine, and the RdRp sequence pulled from a bat sample that was 100 percent identical to the same section of RaTG13, etc. The word *Mojiang* appeared, but no mention of sick miners. If theseeker had put the two articles together, would that have gotten him to the master's thesis? I thought it was possible, but not likely. I thought he'd have needed an insider nudge to get him to the PhD dissertation. And then he'd have had to do the translations. Either he is a much better researcher than I am, which is more than possible, or somebody gave him that nudge. And speaking of Yuri Deigin's *Medium* post: it provides such a wide swath of information about Shi's work since 2004, about Baric's gain-of-function experiments, about EcoHealth Alliance's grant application to NIAID, about the pangolin sequences, it made me suspect that a group had researched it—not an individual who runs a start-up company with no street address and two Russian colleagues.

For a day or two, the idea that theseeker may have had help to find those theses shook my faith in what was derived from them. But while provenance is everything in archaeology and art history, in journalism and non-fiction what matters more is whether the facts offered can be independently proven to be true. In the *Nature* addendum, Shi had confirmed the salient facts when she revealed that the samples from those miners are still in her possession. The questions that remain are: what did she extract from them, and did it lead to SARS-CoV-2?

16. It's Not Over Until the Fat Lady Sings

MAY 2021. THE magnolia in my garden flaunted huge magenta blooms, calling all pollinators. The scent of lilac drifted everywhere, fat bees bumbled in the grass, tiny birds—woodpeckers?—swooped in and out of the hedge, shrieking at competitors to be off. Hawks hovered over the pond in the park nearby, stooping to kill, then dropping the remains of their day—severed rat heads and feet—on well-mannered driveways. One pair of Canada geese stood guard over their last surviving gosling on the pond's muddy bank. This was an uncertain spring: full of hope that we'd beat the pandemic, that we could vaccinate it out of our lives, yet full of fear that hope is for mugs.

Ontario was under yet another stay-at-home order due to the pandemic's third wave. It was worse than the first two, thanks to the incompetence of political leaders. They had failed to grasp that putting business before public health ends in disaster for both, though their science advisors had explained and explained. Maintaining proper social distance, I walked every day around the neighbourhood for my permitted outdoor exercise, thrilled at the unrivalled beauty of a Canadian spring, an explosion of colour and sound and scent triumphant over the sterility of a perilous winter. Most of my neighbours and friends, like me, had been given one shot of a vaccine, so I was less afraid of getting too close to anyone. Most had received Pfizer-BioNTech, made, like Moderna, from insights and technologies derived from genomic science, synthetic biology, and gain-of-function experiments. A few had taken clever AstraZeneca, which is delivered by a traditional viral vector instead of swathed in a nanoparticle. This seemed to be a source of problems: AstraZeneca is a little less effective than Pfizer and Moderna and can induce the formation of a rare kind of blood clot called VITT (Vaccine-Induced Immune Thrombotic Thrombocytopenia), which can kill if not treated quickly. But delivery of the Moderna vaccine had been peripatetic. Pfizer had been forced to reorganize its Belgian production facility to meet demand. Canada's deliveries were supposed to come from that facility. There'd been no Pfizer or Moderna deliveries for many weeks in the winter. Our political leaders, following advice from their advisors at PHAC, kept saying the best vaccine is the first one you're offered. There were lots of pictures of them rolling up their sleeves for AstraZeneca.

When my eldest daughter and son-in-law became eligible to receive it, we conferred about the risk of COVID-19 versus blood clots. The numbers said the danger of COVID-19 is much higher than any risk from AstraZeneca. So my daughter and her husband went to a local drugstore and got it. Then the risk ratios tumbled. First the clots were said to happen to one person in a million, then to one in

500,000, then to one in 60,000. The National Advisory Committee on Immunization (NACI), a group of scholar/volunteers without much in the way of support staff, issued a statement saying that people should carefully consider their actual risk of getting COVID-19 before determining whether to take AstraZeneca or wait for Pfizer or Moderna. People who must leave home to work, people living in hotspots, people travelling on crowded buses, people labouring shoulder-to-shoulder in meat-packing plants or other places susceptible to outbreaks such as crowded warehouses, people sharing too few rooms in high-rise apartments or living in multiple-generation homes—the urban poor, in other words—should go for it. People working at home in neighbourhoods where vaccine rates were high, but infections low, could choose otherwise. A doctor on the NACI explained further that if her sister asked if she should get AstraZeneca, and she said yes, and her sister developed a life-threatening blood clot, she'd never forgive herself. That statement made it utterly clear that there were good vaccines and lesser ones, and the lesser ones were just fine for lesser folks. Which meant we weren't all in this together. Which caused outrage. The government of Ontario ordered that AstraZeneca vaccinations be stopped. But then the story changed again. AstraZeneca vaccinations resumed for those who'd had it as their first dose, then stopped once more. Meanwhile, PHAC kept right on saying the best vaccine is the first you're offered. Did this reflect differing views on the science? Confusion? Or none of the above?

PHAC WAS CERTAINLY not confused when it came to protecting itself from being held to account. For months, the

House of Commons Special Committee on Canada-China Relations had been demanding that PHAC turn over documents showing why Xiangguo Qiu and Keding Cheng had been fired from the National Microbiology Laboratory and all documents related to the shipments of pathogens to the WIV. At the end of March 2021, the committee had issued a compliance order. PHAC delivered documents—heavily redacted documents. The committee demanded unredacted documents. At a hearing on May 10, Iain Stewart, the new president of PHAC (replacing the former president, who'd been shown the door), and Dr. Guillaume Poliquin, the acting head of the NML (replacing Matthew Gilmour, who'd gotten out while the getting was good), testified before the committee. They had a lawyer from the Department of Justice to advise them. Under questioning, Stewart explained that the documents had to be redacted because PHAC had to follow the law. Which law? The Privacy Act. He couldn't talk about the reasons for firing anyone because that's personal information. Stewart also insisted that the shipment of pathogens had nothing to do with the firings. No material transfer agreement was necessary for the shipment of pathogens to be sent to the Wuhan Institute of Virology. Under further questioning, he said that the firings may have had something to do with a possible violation of security protocols, which marked a sudden switch from earlier statements about administrative or policy issues. But he could give no specifics. He insisted, when one of the MPs tried to link that shipment and the pandemic, that it did not pertain to coronaviruses. The committee wanted to know more about the ongoing RCMP investigation. Stewart said he had "no line of sight" on what the RCMP was looking at, so the committee should call the RCMP.

Certain committee members such as Michael Chong (Conservative, Wellington-Halton Hills) and Garnett Genuis (Conservative, Sherwood Park-Fort Saskatchewan) were not pleased. Genuis made a passionate statement about parliamentary committees' rights to see any documents in the government's possession in unredacted form. Stewart simply repeated himself—due to the Privacy Act, we cannot comply. Stewart was reminded that the speaker of the House of Commons had ruled in 2010 that under the Constitution, Parliament is supreme, and its voted orders take precedence over any of its acts. If something needs to be kept secret due to national security concerns, arrangements can be made to review documents in secret session, or even to have them redacted by the clerk of Parliament. The Justice Department lawyer, when asked if he was aware of that ruling, said he was, but repeated PHAC's position: no unredacted documents could be handed over due to the Privacy Act. The committee voted in favour of a resolution to give PHAC several more days to produce them all intact and clear, and if it failed to do so, to take the matter to the House as a whole. Clearly the minister of health, Patty Hajdu, hoped PHAC would be able to keep its dirty laundry in the hamper until after Parliament rose for the summer on June 23. Perhaps she and her cabinet colleagues had decided that there would be an election called before Parliament resumed in mid-September, which could put off the day of reckoning even further.

But two days later, May 12, the *Globe and Mail* carried a front-page report by Robert Fife and Steven Chase that exposed some of that laundry. The story said that CSIS had warned PHAC about Xiangguo Qiu and Keding Cheng and recommended that their security clearances be withdrawn:

CSIS had become concerned about information they were transferring to China. Stewart had admitted in March that PHAC became concerned some time in 2018. The story also recounted what had gone on in the committee hearing two days before. It read as if a group in CSIS had grown so angry about PHAC's reluctance to acknowledge, let alone fix, what had gone on in its high security lab that they decided to give tidbits to reporters who could get them on the front page.[388] One had to wonder: did CSIS raise the alarm before or after the governor general handed out the Innovation Award?

On May 20, another front-page story by the same reporters appeared in the *Globe and Mail*. This one proclaimed that Xiangguo Qiu and Keding Cheng had collaborated with scientists of the People's Liberation Army on six papers, starting in 2016 and ending in 2020.[389] This collaboration was not exactly news: Amir Attaran had made that allegation in 2019 to Karen Pauls, who included it in her CBC report on the shipment of pathogens to the Wuhan Institute of Virology. But Attaran had only mentioned one paper, not six. This story also pointed out that one of those PLA scientists had also worked for the NML. The person named was Feihu Yan. The *Globe* still didn't appear to know the most damaging facts, such as that PLA Major General Chen Wei had worked with Xiangguo Qiu and Cheng on papers even after their removal from the NML. And George Gao's name didn't appear in the story either. Reading the newspaper over breakfast that morning, I wondered if CSIS was holding these things in reserve, or if it didn't know that China's leading civilian and military virologists (not that there's a clear distinction between civilian and military research in China anymore) had used the NML's BSL-4 as if it were their own for years.

When asked for comment, PHAC's spokesperson, Eric Morrissette, explained that the individual named in the story, Feihu Yan, was not in fact an employee of the NML. He pointed out that everyone working at the NML must have a security clearance. He did not explain how Feihu Yan got into the lab and why that NML affiliation appeared on the published paper. Morrissette said more papers done with the PLA would be forthcoming. PHAC also gave a statement to the *Globe* that read: "collaborations with institutions outside of Canada is critical to advancing public-health research and science aimed at improving public health on a global scale." This suggested that PHAC, like Stefan Wagener, makes no distinction between the military researchers of a friendly power, such as the US, and those working for an unfriendly power, such as China. It also suggested that PHAC approved of the relationship with the PLA. If true, I muttered to myself, heads should roll.

Andy Ellis, a former assistant director of operations for CSIS, told the *Globe* it was "madness" for PHAC to be "co-operating with the PLA.... 'It is just incredible naiveté on their part.'" The *Globe* reporters also got comment from a professor at Georgetown, James Giordano, "a senior fellow in biowarfare and biosecurity at the U.S. Naval War College." Giordano said what everyone knowledgeable about China had been saying for some time—that work done with academics in China is available for use by the Chinese military. He explained that China is interested in the soft power aspects of biotechnologies so that "Beijing is able to 'ride in like a white knight' and offer other countries treatments or cures for diseases that threaten their population and increase its international influence." Giordano also pointed to something else: "China is particularly interested in modifying pathogens to create a novel organism that is not listed—and therefore not regulated—in the international Biological and Toxin Weapons Convention."[390]

Coronaviruses are not listed bio-agents like tularemia or smallpox.

In Question Period on May 26 deputy Conservative leader Candice Bergen asked the prime minister: "Communist China cannot be trusted. Will the prime minister commit to ending this research and this co-operation with the regime that...actually wants to hurt Canada?"[391]

The prime minister replied: "The rise in anti-Asian racism we have been seeing over the past number of months should be of concern to everyone. I would recommend that the members of the Conservative Party, in their zeal to make personal attacks, not start to push too far into intolerance toward Canadians of diverse origins."[392] In other words, the prime minister intended to carry on as if nothing of importance had been revealed. Instead of dealing with the facts and offering an explanation, he accused the Conservatives of launching a racist attack by asking the question. Which was the same tactic Daszak and his colleagues had been using since the possibility of a lab leak was first mentioned in *Science*.

On June 2, a majority in the House of Commons voted in favour of a resolution demanding that PHAC produce without redactions all documents in its possession regarding the firing of Xiangguo Qiu and Keding Cheng and the pathogen shipment to the WIV, to be delivered to the speaker of the House within 48 hours. The resolution said the documents would be handed first to the parliamentary law clerk and parliamentary counsel for redaction of information relating to national security or any criminal investigation, and that members of the Canada-China committee would then be briefed in private about what the documents revealed. The minister of health would be called before the committee to testify. Members would be free to describe what had been redacted, or to rely upon that information to write a committee report.

The minister of health's spokesperson said that PHAC would not give unredacted documents to the Canada-China committee but would consider turning them over to the National Security and Intelligence Committee of Parliamentarians (NSICOP), "the best forum for sharing this information in a transparent and accountable manner that protects the security of Canadians." But the only Canadians whose security would be protected if these documents went to NSICOP were the minister and top bureaucrats at PHAC. The NSICOP committee is neither transparent nor accountable. It is not a committee of the House. Its members, while representing all parties in the House and groups in the Senate, are selected and appointed by the prime minister who, in the Canadian system, is in charge when it comes to national security and intelligence. Members all have the highest level of security clearance. They must keep secret everything classified that is shown to them and they cannot plead parliamentary privilege if they leak information and are prosecuted. They are required to keep all classified information secret until they die. The prime minister has the right to review and demand revisions to any of the committee's reports before they become public. The chairman of the committee is a Liberal MP, David McGuinty.[393]

In other words, NSICOP is like Las Vegas: secrets revealed to NSICOP stay with NSICOP.

And so it didn't end there. On June 16, the speaker of the House, Anthony Rota, ruled that handing the documents over to NSICOP was a breach of parliamentary privilege and would not stand: "the government cannot ignore an order of parliamentarians, even on grounds of national security."[394] The unredacted documents had to be handed over immediately as per the House resolution. A majority in the House passed another resolution the next day, finding the Liberal government in contempt of Parliament for failing to produce the documents. The new resolution required PHAC president Iain Stewart to be brought before the bar of Parliament on June 21 to be admonished and to bring the unredacted documents with him. No civil servant or private citizen had been brought before the bar for 110 years. However, Parliament's only recourse if he failed to appear and to hand over the documents would be a vote of non-confidence, resulting in the fall of the government and possibly an immediate election—but not necessarily the turnover of unredacted documents.[395]

Meanwhile, the *Globe and Mail* and the CBC tried to get comment from Xiangguo Qiu and Keding Cheng. They both learned the couple owns two homes in Winnipeg and a property at Gimli on Lake Winnipeg. The picture of the house that accompanied Karen Pauls' CBC story was much bigger and newer than the one my friend had photographed. Their property holdings in Winnipeg, according to CBC, are worth about $1.7 million. A source told Pauls that the couple had also bragged of owning a mansion in China. Their neighbours at the bigger, newer house—the style suggested it was built in the last decade—said they had not been seen for as many as six months, though someone was taking care of the grounds. The *Globe* said the students who rent the smaller house didn't know where Cheng and Qiu are and pay their rent via WeChat. Pauls spoke to Scott Newark, a former Alberta crown prosecutor and Canadian Police Association officer, who wondered about the security clearance procedures they'd undergone and whether their combined incomes of about $250,000 were sufficient to support the property they own, which raised the possibility of another source of income—a red flag for those investigating espionage. The government's extreme reluctance to turn over documents, to explain why the two were fired, to suddenly hide behind national security as a reason for withholding information, reminded Newark of Project Sidewinder, a joint operation of CSIS and the RCMP in the 1990s, in which "the Chinese government and Asian criminal gangs had been working together in drug smuggling, nuclear espionage and other criminal activities that constituted a threat to Canadian security." The Sidewinder report went nowhere: according to Newark, the Liberal government of the day didn't want to antagonize China because it was intent on encouraging trade. A security expert, Christian Leuprecht, of Royal Military College and Queen's University, said China has an extremely dangerous and aggressive bioweapons program, but that the government might not want to lay charges because charges lead to court proceedings and court proceedings drag secrets into the light that may prove embarrassing and have implications for allies. "This needs to be a wake-up call for Canada about how aggressive the Chinese have become at infiltrating Western institutions for their political, economic and national security benefits," he said.[396]

Iain Stewart appeared at the bar of Parliament on the appointed day at the appointed time. I watched some of the event online. He reminded me a bit of Peter Daszak: tall, a bit heavy, grey hair, grey suit, rumpled white shirt, unbuttoned jacket, though he wore a tie for the occasion. He looked bewildered to be there. And he was right to be bewildered. It should have been his political master, Minister of Health Patty Hajdu, standing there and listening to the speaker's admonishment that no one can say no to an order of Parliament. Stewart did not produce the documents.

The speaker had planned to rule on the last day of the parliamentary session, June 23, on a motion to send the sergeant-at-arms to Stewart's office to search for the documents, seize them, and bring them to the House. But on June 21, the government filed suit against the speaker in federal court. The government argued that disclosure of this information could jeopardize national security and harm Canada's international relations. Turning to the court to settle this was so unprecedented, so outrageous, it caught the Opposition flat-footed. Canada's parliamentary system is not built, US style, on separate but equal powers. Canadian governments retain power only so long as they enjoy the confidence of the House of Commons. The speaker said he would "defend the rights of the House.... The legal system does not have any jurisdiction over the operations of the House. We are our own jurisdiction. That is something we will fight tooth and nail to protect and we will continue to do that." But the speaker said he needed more time to consider whether to order the sergeant-at-arms to search Stewart's office, and that it might be up to the next speaker to make that ruling if an election is called before the House returns in September.

The House rose as planned on June 23. Knowledgeable pundits said an election would be called before the summer came to an end.[397]

On the same day, Tom Blackwell reported for the Postmedia chain that Xiangguo Qiu had filed patent applications in China with several co-inventors in 2017 and 2019. One patent was for a method to inhibit Ebola, the other for a detection system for Marburg. Anything invented by federal civil servants is owned by Canada. Federal civil servants can only legally file for patents outside the country with the permission of their minister. Did the minister grant permission? PHAC would not comment. Did PHAC know about the patent applications? PHAC would not comment.[398]

Another front-page story appeared in the *Globe* on July 1, Canada Day. Fife and Chase reported that they had learned that "the RCMP have been informed that Xiangguo Qiu and her husband, Keding Cheng, recently relocated to China after they were fired in January." (This raised obvious questions: When did they leave? How did they leave while under RCMP investigation, with borders closed and few planes flying?) The police were said to be inquiring as to how "two People's Liberation Army scientists gained access to the Winnipeg lab...." The *Globe* was not told the second scientist's name. It seemed clear that the RCMP had decided to leak. Why? Probably because Gary Kobinger, Qiu's former boss, spoke to Fife and Chase. He told them that the RCMP had come to see him in March 2021, 20 months after Qiu and Cheng were suspended, two months after they were fired, which seemed incredibly slow off the mark, even for the RCMP. They'd asked Kobinger about the transfer to China of intellectual property belonging to PHAC. They had not been interested in the shipment of viruses to the Wuhan Institute of Virology, or so Kobinger told the *Globe*. He thought Qiu would have given permission to have the reason for her firing made public if anyone had asked her. Qiu had spoken to him last in December. She'd told him she'd lost weight from all the stress. She'd told him that she hadn't done anything wrong, that it was all a misunderstanding. "They [the RCMP] asked if I thought it was likely, possible that she could have removed material," he told the reporters, suggesting that national security had nothing to do with a terrific female scientist losing her job. He thought MPs deserved to know that.

IN MAY 2021, there was an abrupt shift in the way major American media reported on the question of SARS-CoV-2's origin. The proposition that there might have been a lab leak in Wuhan suddenly went from tin-foil-hat conspiracy theory to a possibility worthy of investigation. The *Washington Post*, the *New York Times*, CNN, the *New Yorker*, *Vanity Fair*, etc., all began to sing from the same songbook. Then they reported on each other singing from the same songbook.[399] It did not surprise me that when major American media finally moved, they moved in unison. Once, when I was covering the Dubin Commission of Inquiry into the Ben Johnson/amateur sport doping scandal, I found myself sitting in the hearing room near two American reporters, one from the *New York Times*, the other from the *Los Angeles Times*. In Canada, reporters from competing news organizations generally do not check with each other about what they're going to report. Yet each time a witness said something of interest, these two kept asking each other: "What can we say about that?"

It's not as if the lab-leak possibility had not been raised by responsible reporters and publications before May 2021. In January 2021, author Nicholson Baker published a very interesting piece exploring the idea in *New York Magazine*.[400] It caused a stir. But *New York Magazine*, like *Boston Magazine*, does not have the clout or the global reach of a front-page story in the *Washington Post* or the *New York Times*. Most major media had continued treating the lab-leak notion as barely worth consideration. And then, all at once, this flip turn.

Why?

First, a long piece by Nicholas Wade that took the possibility very seriously appeared in the *Bulletin of Atomic Scientists*; then leading figures in virology signed a letter to *Science* that said a lab leak had to be investigated; finally, an American intelligence report leaked to the *Wall Street Journal* said three researchers at the Wuhan Institute of Virology had become sick enough to seek hospital treatment in November 2019 (though nobody knew exactly what ailed them).[401]

Nicholas Wade is a respected science journalist and author. He once worked as an editor at *Nature* and at *Science*. For several years, he also reported on science subjects for the *New York Times*. The *New York Times*' editors have the irritating habit of assuming that if something has not appeared in their paper, it either didn't happen or wasn't worth reporting. Once, when I pitched a science story to the *New York Times Magazine*, the editor said he couldn't take it because there was nothing in the paper's archives on the subject. Because it's news, I said, naively. Not to the *New York Times*, he said. However, former *Times* reporters *are* listened to by the editors.

Wade's piece, which appeared on May 5, started this way: "From early on, public and media perceptions were shaped in favor of the natural emergence scenario by strong statements from two scientific groups. These statements were not at first examined as critically as they should have been." He went through the problems with the *Nature* proximal origin paper and the *Lancet* statement. He pointed out that Peter Daszak organized the *Lancet* statement, drafted it, signed it, and declared in print that he had no competing interests. However, Wade wrote, "Virologists like Daszak had much at stake in the assigning of blame for the pandemic. For 20 years, mostly beneath the public's attention, they had been playing a dangerous game. In their laboratories they routinely created viruses more dangerous than those that exist in nature."[402]

And on it went from there, dealing with various issues raised in this book. There was little that was new to anyone who'd read Latham and Wilson or DRASTIC's publications. But for those who hadn't, it was revelatory. Wade explained in detail the importance of the furin cleavage site in making SARS-CoV-2 so infectious. He explained that there is a particular sequence of amino acids where a furin on human cells always makes its cut. The sequence is: proline, arginine, arginine, alanine, or PRRA. The furin cuts between the two arginines. He said the PRRA sequence that is found in SARS-CoV-2's receptor binding domain is unique among beta coronaviruses. He argued it might have evolved naturally, or it might have been picked up from people, but he found both highly unlikely. He thought too many successful mutations, one after the other and in the correct order, would have had to happen for nature to produce such a cleavage site in a coronavirus harboured by a bat. Besides, the particular codons in the SARS-CoV-2 genome (codons are triplets of nucleotides that call up a particular amino acid) are most often found in humans. "The human-preferred codon is routinely used in labs," he wrote. "So anyone who wanted to insert a furin cleavage site into the virus's genome would synthesize the PRRA-making sequence in the lab...." Then he quoted David Baltimore, a Nobel laureate molecular biologist/virologist and former president of California Institute of Technology: "When I first saw the furin cleavage site in the viral sequence, with its arginine codons, I said to my wife it was the smoking gun for the origin of the virus. These features make a powerful challenge to the idea of a natural origin...."

As Rowan Jacobson had done in his *Boston Magazine* profile of Alina Chan, Wade also wrote about the politics involved. He pointed out that certain scientists had moved fast to protect their interests, while others kept quiet out of fear of anonymous colleagues sitting on grant committees who might take revenge on them for outspokenness by denying them money. He quoted a former State Department consultant, David Asher, now at the Hudson Institute, who had explained at a seminar[403] in March the basis for the State Department's fact sheet issued just before Mike Pompeo left office. The fact sheet had mentioned the US had reason to believe that researchers at the WIV had become sick in November 2019. Asher explained to the seminar that the State Department's knowledge of this incident came from "a mix of public information and 'some high end information collected by our intelligence community.'" Asher told the seminar that three people working in the WIV's BSL-3 became sick within a week of each other "with severe symptoms that required hospitalization." According to Wade, Asher characterized this as "the first known cluster that we're aware of, of victims of what we believe to be COVID-19."

Wade's piece was well-reasoned and comprehensive, but the prose fairly thrummed with rage. He wasn't nice about the failures of the media, especially the science media, to ask tough questions. He went so far as to list everyone he thought was to blame for the pandemic. He started with Chinese virologists, naming Shi Zhengli as their public face; China's officials, who withheld or scrubbed vital information while redirecting attention and giving the WHO joint study group the run around; virologists in general, who knew all about the dangers of gain-of-function experiments and nevertheless lobbied for them and got away with almost no oversight:

> Virologists knew better than anyone the dangers of gain-of-function research. But the power to create new viruses, and the research funding obtainable by doing so, was too tempting. They pushed ahead with gain-of-function experiments. They lobbied against the moratorium imposed on Federal funding for gain-of-function research in 2014, and it was raised in 2017.
>
> The benefits of the research in preventing future epidemics have so far been nil, the risks vast. If research on the SARS1 and MERS viruses could only be done at the BSL3 safety level, it was surely illogical to allow any work with novel coronaviruses at the lesser level of BSL2. Whether or not SARS2 escaped from a lab, virologists around the world have been playing with fire....

> ...You might think the SARS2 pandemic would spur virologists to re-evaluate the benefits of gain-of-function research, even to engage the public in their deliberations. But no. Many virologists deride lab escape as a conspiracy theory, and others say nothing.
>
> They have barricaded themselves behind a Chinese wall of silence which so far is working well to allay, or at least postpone, journalists' curiosity and the public's wrath. Professions that cannot regulate themselves deserve to get regulated by others, and this would seem to be the future that virologists are choosing for themselves.

ONE WEEK AFTER Wade's article appeared in the *Bulletin* a new letter appeared in *Science*. After first congratulating science for a job well done in learning how to handle SARS-CoV-2, it ambled into the question of nature versus lab. At the end of the first paragraph, the letter said: "Theories of accidental release from a lab and zoonotic spillover both remain viable." This was not exactly an earth-shattering statement, nor was it surprising that the letter went on to say that the WHO study did not give the two theories "balanced consideration." The co-authors agreed with the director-general of the WHO when he said "consideration of evidence supporting a laboratory accident was insufficient...." They asked for a proper investigation, one that is data-driven, transparent, and objective, involving independent oversight and minimized conflicts of interest. This was a polite way of saying the so-called independent joint study convened by the WHO had failed in all respects.

The real importance of this letter was who signed it. I was not surprised to see Alina Chan's name. She was one of the better-credentialed independent investigators and, as *Vanity Fair*[404] revealed, had even been invited to make a presentation on a panel before some officials in the State Department to discuss the lab-leak possibility as Trump's term ended. I was not surprised to see Marc Lipsitch's name either. Lipsitch had warned for years of leaks from high containment labs. I was relieved to see the name of one Canadian co-author—Dr. David Fisman of the University of Toronto—because Canadian scientists had said next to nothing in public about gain-of-function experiments, synthetic biology, dual-use technology, and the possible lab origin of this virus. The real shocker was the third name on the list—Ralph S. Baric.

Baric had been involved in the drafting of the *Lancet* statement that accused those who spoke of a lab leak of being conspiracy theorists. Baric had worked closely with Shi Zhengli on coronavirus gain-of-function experiments. Baric well knew how to make changes to a coronavirus genome without leaving a trace: no one knew better. Baric also knew how things are done in Shi Zhengli's lab. A member of DRASTIC had pulled from a Wuhan Institute of Virology archive a group photo showing Baric, a big man with a white mustache, sitting front and centre with Shi at an international conference at the WIV.[405] Because Baric put his name to this *Science* letter asking for a proper investigation, the lab-leak hypothesis moved from highly unlikely to highly likely in one jump.

Then, in the last week in May, US intelligence officials confirmed to *Wall Street Journal* reporters that what David Asher had said at a Hudson Institute seminar in March was correct. The *Wall Street Journal* reported that the United States had been informed by an "international partner" that three researchers at the WIV had become ill and gone to hospital in November 2019. One source told the *WSJ* he/she wasn't convinced by the evidence. Another told the *WSJ* that the information came from various sources and was "of exquisite quality. It was very precise. What it didn't tell you was exactly why they got sick." Still others said that in China people often go straight to hospital if they become ill because general practitioners are hard to find. But the *Wall Street Journal* also quoted Marion Koopmans, who'd told NBC in March that she was told three staffers at the WIV got sick in November. "Maybe one or two. It's certainly not a big, big thing," she'd said. No such information had appeared in the main body of the WHO joint study report. It wasn't in the annexes either.

ON MAY 24, 2021, the *Wall Street Journal* declared the lab-leak theory to be mainstream. The next day, so did the *Washington Post*.

READING THE STORIES online as expert after expert reversed their previous pronouncements on how a lab-leak origin was conspiratorial nonsense was like watching a synchronized swimming team rise in perfect unison from the depths of the pool, legs together, toes pointed to the sky. W. Ian Lipkin of the Mailman School at Columbia, a close collaborator with Daszak, co-author of the proximal origin letter in *Nature*, let it be known that he had changed his opinion. He told the *Wall Street Journal* he was concerned that the WIV was doing coronavirus experiments at lower levels of containment than required in the US.[406] Lipkin also told NPR: “It’s possible that the virus was brought into the laboratory, that it was grown inside of a cell line, that somebody became infected and left the laboratory inadvertently and carried the virus with them.”[407] James Le Duc, the now retired director of the Galveston National Laboratory, which had helped train WIV staff for work in the new BSL-4, and had said suggestions of a possible lab leak were malicious, said he too thought a proper investigation necessary.[408] Anthony Fauci told Fox News that he was now “not convinced” that SARS-CoV-2 developed naturally and wanted an investigation into its origin.[409] Francis Collins, director of the NIH, called for an investigation too, though denying (as had Fauci when interrogated by Rand Paul in a Senate hearing), that his organization had ever funded dangerous gain-of-function experiments at the WIV. He claimed the NIH only funds gain-of-function experiments on *bat viruses*, not human pathogens.[410]

As would become apparent in June, thanks to Freedom of Information Act applications, these statements represented the emergence of early opinions kept secret since the start of the pandemic. Fauci’s emails, obtained by *BuzzFeed*, showed he had participated in two early teleconferences on the possibility that the virus’s origin was not natural. The first was on January 31, the day *Science* raised the possibility of a lab leak. He got an email from Kristian Andersen (soon to be the lead author of the proximal origin letter in *Nature*), which led to a conference call including Jeremy Farrar of the UK’s Wellcome Trust (a leading funder of genomic science). Andersen explained by email that he and E.C. Holmes had been looking at the genome of the virus and thought it was possible it had been manipulated in a lab. Andersen told Fauci that:

"the unusual features of the virus make up a really small part of the genome (<0.1%) so one has to look really closely at all the sequences to see that some of the features (potentially) look engineered." In a follow-up email the same day, Andersen added that he, Holmes, and some other leading scientists "all find the genome inconsistent with expectation from evolutionary theory...those opinions could still change." The next call, on Saturday, February 1, included Fauci, his boss Francis Collins, Andersen, Holmes, Farrar, and others. Someone took notes but what exactly was said was blacked out in the documents supplied to *BuzzFeed*. In a follow-up interview, Fauci told Alison Young of USA *Today* that people at the February 1 teleconference, which was supposed to be kept confidential, were on both side of the question—natural or engineered—but the conversation was more "heavily weighted" to natural origin.

Only three days later, according to Young, relying on emails obtained under a Freedom of Information application by US Right to Know, Andersen's opinion, or at least what he decided to say in public, had abruptly changed. Working with other scientists on a public letter to be sent by the National Academies of Sciences, Engineering and Medicine to the White House, he made it clear in an email that he wanted the letter to be strong enough to counter crackpot theories about lab manipulation. He also made it clear that the Academies' letter was in part aimed at structuring the narrative about the virus's origin, a way to quash the very possibility he'd raised himself to his main funder only three days earlier. Did the teleconference change his mind about the meaning of that minute portion of the virus's sequence? Or did Fauci and Collins point out the political danger of leading scientists suggesting in public that this could be a lab-altered virus?

"The main crackpot theories going around at the moment relate to this virus being somehow engineered with intent and that is demonstrably not the case," Andersen wrote. "Engineering can mean many things and could be done for either basic research or nefarious reasons, but the data conclusively show that neither was done."[411]

ANDERSEN SENT THAT email on February 4. He copied several people, including Stanley Perlman, Trevor Bedford and Ralph Baric. If you guessed the email was addressed to Peter Daszak, I've run out of brownie points to give you. Daszak was drafting the Academies' letter to the White House at the same time as he was drafting the *Lancet* statement.[412]

PRESIDENT BIDEN, WHO when running for office had been dismissive of the lab-leak idea, finally weighed in on May 26, 2021. He said he had asked the US intelligence community to re-examine the issue and come back with a report within 90 days. The intelligence community (there are 17 intelligence agencies in the US) had already produced a report for Biden in March which came to no conclusion. But a whole lot of information (gleaned by methods Edward Snowden informed us about), remained to be analyzed.[413]

President Biden said he'd make the report public.

I GOT MY second shot of the Pfizer-BioNTech vaccine on my birthday. It was a good present, and it was also an act of closure. I knew I had gone as far as I could investigating the origin of the pandemic. Following the competing narratives had consumed most of my waking moments for more than a year. I had dug my way through a mountain of secrets. I had read and re-read repellent prose, parsed muddled arguments, followed trails through thickets of undeclared interests, begged for information that I am entitled to by law but which bureaucrats withheld anyway, documented a national security scandal that had unfolded for years under the noses of three ministers of health and CSIS, fruitlessly banged on scientists' email inboxes, used every trick in the journalist's handbook to get to those who could explain and give context. I am not complaining. Frequent frustration is the lot of anyone working on a story whose lead characters don't wish to be brought to account, especially in a country such as Canada. Our governments display spectacular talent for keeping embarrassing secrets hidden despite laws to the contrary. In a way, I was grateful to them. Finding ways around all those obstacles, digging into the folkways of a scientific community that protects its interests like the old pols of Tammany Hall, helped keep despair at bay. It was dammed up by rage, which is a much more useful emotion. Rage gave me energy. It kept me going.

I was angry about the shameless cover-ups in the face of so much death and suffering. I was angry about crazy-dangerous experiments done in less than perfect labs operating in too many countries around the world, including my own. I was angry about the way a US charity acted as a cut-out to get American eyes and US government money into foreign labs where God-knows-what has been done that cannot be done under US rules. I was angry about the organized attempts to suppress all reasonable questions about a lab leak, about the way labels like *conspiracy theorist* or *racist* were used to make people afraid to raise them. I was angry at how the government of China took advantage of those in the West who wanted to help it advance, or at least become a lesser danger. I was angry about how the world's leading science journals lent themselves to the propaganda efforts of that government, which has no interest in truth, only in geopolitical advantage. I was angry about the gross corruption of the ideals of science. I was angry that important information about the origin of this pandemic only came to public attention due to the efforts of a group of volunteers, not thanks to the scientists we support with public funds to ask the right questions, no matter how discomfiting. I was more than angry over the failure of Western governments—Canada's governments—to protect the public from this pandemic.

Things would have been very different if Canadian bureaucrats had done what they'd been told to do after the SARS crisis: if they had properly maintained stockpiles of personal protective equipment; if they'd properly maintained the effective Global Public Health Intelligence Network (GPHIN) instead of stifling its ability to communicate. Things would have been very different even if they'd just watched television news like ordinary mortals and taken note of what was happening in Wuhan instead of relying on a dysfunctional algorithm; if they'd told Canadians the risk was high; if they'd recommended early on that Canadians use masks to reduce transmission; if they'd recommended that the prime minister shut Canada's borders and bring people home at the end of January.[414]

Instead of acting immediately, from January 31, 2020, WHO officials, health bureaucrats, and politicians who knew or should have known what was coming failed to act decisively for six more weeks.[415] As a result, by the end of May 2021, more than 25,000 Canadians had lost their lives. That's only half the number of Canadians who died in 1918–1919, but it's nothing to brag about. Modern medicine was in its infancy then: the only treatments available in Saskatchewan in the winter of 1919 were oranges, scotch, and aspirin.

In the US, President Trump was forced to take responsibility for his appalling failure to protect the American public, for the deaths of more than 600,000 Americans. He lost office, his top officials swept out with him. But in Canada almost no one has been held to account. No ministers of health or long-term care have stood up and said *I'm sorry, my department made serious mistakes, I resign, time for someone else to try*. Our politicians seem to have forgotten that in a parliamentary democracy, ministers are responsible for everything that goes on in their departments, the bad as well as the good. Somehow, we've allowed our political masters to arrange things so that ministers need not suffer political consequences for departmental failures. That makes departmental failures inevitable. If there are no political consequences, why would any minister probe what their bureaucrats have done or failed to do? It's safer to cover ears and eyes and practice deniability. So there is no accountability, just like there is no transparency. Without accountability and transparency there can never be good government.

PHAC, which reports to the minister of health, allowed Canada's only BSL-4 lab to be infiltrated and used by an unfriendly power for years. When found out, PHAC insisted, on the one hand, that such international collaborations are good for public health around the globe, while also claiming, on the other hand, that it could not explain, due to the constraints of national security, why Xiangguo Qiu and Keding Cheng were fired. They refused to hand over unredacted documents that would have shown what the minister and the top bureaucrats knew about the operations of Qiu and Cheng and when they knew it. When asked to stop the collaboration between PHAC and China, the prime minister told the Opposition to beware of racism, then maneuvered and stalled and finally sued the speaker of the House to keep the unredacted documents hidden until the House rose for the summer, putting off any possibility of accountability until after the next election.

PHAC's prime function is to protect public health. Though two of its top bureaucrats were shown the door early on in the pandemic as their failings became clear, the minister of health, their boss, retained her position, accepting no responsibility for PHAC's failure to prepare for a pandemic, or for allowing the government of China to use Canada's most important and secure microbiology facility as if it were its own. Similarly, Ontario's minister of long-term care, responsible for the system in which most of Ontario's deaths occurred, failed to resign in the face of abject failure. Not only did she not make sure long-term care homes in Ontario were prepared for the pandemic and their residents kept safe, her officials stopped inspecting the homes. While Quebec took care to hire more staff over the summer of 2020 to prepare for the second wave, Ontario's minister of long-term care didn't get around to it in time. Many more people died who shouldn't have.

In fact, the only Canadian politicians who lost their jobs during the pandemic of 2020–2021 were those stupid enough, entitled enough, to leave the country for winter vacations though the borders had been shut to all but essential travel and stay-at-home orders issued.

The premier of Ontario shuffled his cabinet just before the 2021 summer break. Instead of firing his minister of long-term care, he moved her to a new portfolio. He replaced her with his former minister of finance, Rod Phillips. Rod Phillips was one of those politicians forced to resign after being caught vacationing in the Caribbean over the Christmas period. He'd put up an image on his web page to make it look as if he'd remained in Ontario. When the premier was told he'd left town, the premier failed to tell him to come home. Until the story hit the front pages of the newspapers.

BY JUNE 2021, I knew what I would have said to my mother about the origin of this pandemic if she were still alive. I was pretty sure I'd marshalled enough evidence even to satisfy Stephen (though Stephen could always surprise me by pointing out some gaping hole in my thinking). In a civil action, truth about the events at issue is determined according to the preponderance of probabilities, not evidence beyond a reasonable doubt. There will always be a reasonable doubt about the origin of SARS-CoV-2 because the government of China has refused to permit any further investigation within its borders by the WHO or by anyone else. But the preponderance of the probabilities says that SARS-CoV-2 is the result of an accident following sloppy laboratory work either in the Wuhan CDC by Tian Jun-Hua's group, or at the Wuhan Institute of Virology by Shi's group (or by military researchers working secretly there).

I believe a precursor virus was either isolated from the

samples taken from those miners' lungs, or that its genome was assembled, the virus rescued and adapted further by passaging in human cells and humanized mice in Shi's lab. I think the result, SARS-CoV-2, may have been tested on macaques in the WIV's BSL-4 in 2019, perhaps after the WIV took its samples and sequences database offline in mid-September.[416] The leak could have happened in several ways: a problem with waste, a researcher inhaling something from a plate of cells, a bite by a rebellious macaque. The researcher may have remained asymptomatic, yet infectious. I believe some officials knew that an accident had occurred, though they may not have realized how serious it would become.[417] China's CDC, presided over by George Gao, had official reports by the last week in December that a pneumonia caused by a SARS-like, unknown coronavirus was circulating in Wuhan. Three commercial sequencing companies had reported those findings up the chain. They were ordered to destroy or return all samples to the authorities. No official told the WHO about this SARS-like problem until after Dr. Li's warning about SARS went viral on December 30, and ProMED posted about this new SARS-like pneumonia. That's when Shi Zhengli was told to come home and "deal with it." That's when the WHO started asking questions. That's when scientists all over the world started asking questions. The problem was no longer containable, so the government of China switched strategies: from total information suppression to cover-up and rewrite, including a heroic role for President Xi.

The leadership in China controlled and then orchestrated scientific publications on SARS-CoV-2, all the while scrubbing from the net anything that might undercut the new story. They must have worried that the name SARS-

CoV-2 brought the original SARS to mind, which brought cover-up to mind. In early February 2020 Shi Zhengli and three colleagues (including Shibo Jiang, co-author with Shi of that *EMI* paper submitted the same day as Shi's *Nature* paper introducing RaTG13) wrote to the international taxonomy subcommittee to ask that the virus's name be changed. They claimed that because SARS-CoV-2 employs an Arabic numeral, it's a hard name to remember, leading to the use of shorthand terms "such as 'Wuhan coronavirus'...[which would] stigmatize and insult the people in Wuhan, who are suffering this outbreak." She sent a copy of that letter to Ralph Baric.[418]

A successful cover-up entailed getting rid of any information that could point to a particular place of origin in Wuhan. The international members of the WHO-convened joint study team were told by their Chinese colleagues that no samples from flu sentinel hospitals were available for the last three weeks of December, that their Chinese colleagues needed complicated permissions to do retrospective tests for the virus in human blood donated during the last quarter of 2019. Even the date of the first known case was moved up eight days to avoid narrowing down when and where SARS-CoV-2 began to circulate in Wuhan.

Should we blame the Chinese scientists who participated in this chicanery? Zhihong Hu told Basil Arif that they have no choice. I'm sure that's true.

After Watergate, American journalists took to saying it's not the crime that gets you, it's the cover-up. But they were writing about the way things work in a democracy with multiple, competing centres of power, where the press is free to inquire and to publish and whistle-blowing has a long tradition. A cover-up and narrative rewrite perpetrated by the leadership of the Communist Party of China will have a different outcome. In China, competing centres of power have been stamped out. The press is required to produce propaganda when so instructed. Even CNKI, the database of all the learned journal articles and dissertations published by China's scholars, is partly supported by Communist Party of China's Propaganda Department. Social media are used as means of social control, not free expression. Whistle-blowers disappear or die. In a country where millions have been imprisoned and even tortured in re-education camps because terrorists were once among them, where people whose behaviour is deemed inappropriate can find themselves deprived of the right to board a fast train or a plane, where any business or institution or individual can be required to spy if the government so commands, where people are afraid to go out on their balconies to listen to a concert because the police said not to, who can blame scientists for failing to expose the smoking gun?

But there is no such excuse for the Western scientists, funding agencies, and leading science journals which participated in China's effort to write a false narrative instead of getting out the facts.

PEOPLE COMPLAINED TO the editors of *The Lancet* about the failure of the co-authors of the February 2020 statement to declare competing interests. The editors eventually wrote an addendum. It was attached to the statement on June 21, 2021.

"Some readers have questioned the validity of this disclosure, particularly as it relates to one of the authors, Peter Daszak," their addendum said. The editors explained that in accordance with the guidance of the International

Committee of Medical Journal Editors, medical journals ask authors to report "financial and non-financial relationships that may be relevant to interpreting the content of their manuscript." They had asked all 27 authors of the February statement to revisit their disclosures. They reported that Peter Daszak "has expanded on his disclosure statements for three pieces related to COVID-19 that he co-authored or contributed to." These were: the statement, a *Lancet* Commission statement, and a comment for the *Lancet* COVID-19 Commission.

Peter Daszak's new disclosure statement was longer than the editors' addendum. He maintained he had only ever been paid by EcoHealth Alliance (so he had no financial conflict of interest). He listed his organizations' activities in China. He claimed that all work done was in accordance with the rules and regulations of his funders, including the USAID and the NIH. He noted that the NIH had reviewed the "recombinant virus work" (which he described as modifying a small number of bat coronaviruses to study entry mechanisms) and that the NIH had "deemed it does not meet the criteria that would warrant further specific review by its Potential Pandemic Pathogen Care and Oversight (P3CO) committee." This was an attempt to rebut the July 8, 2020, letter the NIH sent to EcoHealth Alliance, made public in *Vanity Fair* in May 2021. That letter claimed that the WIV "has been conducting research at its facilities in China that pose serious bio-safety concerns and, as a result, create health and welfare threats to the public in China and other countries, including the United States."[419] Amazingly, Daszak's disclosure statement did not mention the Wuhan Institute of Virology or Shi Zhengli by name.

I wondered if the *Lancet* editors had been moved to act because of the public release of the email exchanges between Daszak and Ralph Baric as Daszak was drafting the *Lancet* statement.

On February 6, 2020, Daszak wrote to Baric.

> Re no need for you to sign the "statement" Ralph!!
>
> I spoke with Linfa last night about the statement we sent round. He thinks, and I agree with him, that you, me and him should not sign this statement, so it has some distance from us and therefore doesn't work in a counterproductive way. Jim Hughes, Linda Saif, Hume Field, and I believe Rita Colwell will sign it, then I'll send it round some other key people tonight. We'll then put it out in a way that doesn't link it back to our collaboration so we maximize an independent voice.

Baric wrote back the same day:

> I also think this is a good decision. Otherwise it looks self-serving and we lose impact.[420]

It doesn't get more damning than that.

I SCROLLED TO the bottom of the *Lancet* statement to see if any of the co-authors besides Daszak had declared competing interests. I saw none. I assumed I wasn't looking in the right place so I wrote to the *Lancet* press person for North America and asked where on their site I could find their statements. I mentioned the names of people who I thought would surely declare competing interests, such as Hume Field, who also works for EcoHealth Alliance, which funded Shi's lab, and Dennis Carroll who ran the PREDICT program, which funded Shi's lab through EcoHealth Alliance. The North American press person passed my email on to Emily Head, Media Relations Manager in the UK. She replied, "We do not have anything to add beyond the Addendum" and sent me the link to the original statement. In other words, none of the other co-authors had declared competing interests, including those who work for EcoHealth Alliance, like Field, or who served on his board, like Colwell, or who funded the WIV through EcoHealth Alliance, like Carroll. They remained as silent as the lambs.

On June 22, 2021, the *Lancet* Commission, which had created the task force into the origin of the virus and invited Daszak to be its chairman, put out a statement on its webpage. It said: "The Commission's technical work will be conducted by independent experts who were not themselves directly involved in US-China research activities that are under scrutiny. Dr. Peter Daszak has recused himself from the Commission's work on the origins of the virus."[421] On the task force's page, the word *recused* appeared in a bracket above Daszak's image instead of the word *chairman*. But recusal is not the same as resignation; a recused person keeps his position but does not take part in discussions or vote on issues in which he/she has interests. Daszak's recusal did not solve the task force's undeclared competing interest problem. It is Daszak's employer, EcoHealth Alliance, that funded experiments at the Wuhan Institute of Virology which might have resulted in the pandemic. Another task force member, Hume Field, one of the co-authors of the *Lancet* conspiracy theory statement, works for EcoHealth Alliance advising on China, and Daszak is his boss. Another active member is Danielle Anderson, who did her PhD under Linfa Wang, and then worked for him in Singapore for years, though she is now in Melbourne at the Peter Doherty Institute for Infection and Immunity. In fact, Anderson was doing research at the Wuhan Institute of Virology's BSL-4 on why Ebola and Nipah cause no illness in bats in November 2019, as the pandemic began. Her funding was courtesy of the WIV.

Though she had remained silent about her activities there for more than 18 months, she was suddenly featured in a Bloomberg story right after the announcement of Daszak's recusal. It described her as the only foreign scientist allowed to work at the WIV (which is not the case). She told the Bloomberg reporter that she didn't notice any staff getting sick or anything else untoward going on at the WIV while she was there. "If people were sick, I assume that I would have been sick—and I wasn't," she said. "I was tested for coronavirus in Singapore before I was vaccinated, and had never had it." The story appeared on June 27, 2021, one week after Daszak's recusal.[422]

Her timing was interesting. She obviously has a competing interest—she was paid by the WIV—and so her failure to recuse herself from the *Lancet* task force, as Daszak had done, is more interesting still. It means that the war to control the narrative of the pandemic's origin is not over. It won't be for many years.

The truth remains in a desperate state, splayed out in the narrative equivalent of an ICU.

What are we going to do about that?

Selected Bibliography

Baker, Nicholson. *Baseless: My Search for Secrets in the Ruins of the Freedom of Information Act*. New York: Penguin Press, 2020.

Endicott, Stephen, and Edward Hagerman. *The United States and Biological Warfare: Secrets from the Early Cold War and Korea*. Bloomington: Indiana University Press, 1998.

Gold, Hal. *Japan's Infamous Unit 731: Firsthand Accounts of Japan's Wartime Human Experimentation Program*. Tuttle Publishing, 2019.

Klein, Naomi. *The Shock Doctrine: The Rise of Disaster Capitalism*, Alfred A. Knopf Canada, 2007.

Kolata, Gina. *Flu: The Story of the Great Influenza Pandemic of 1918 and the Search for the Virus That Caused It*. New York: Atria Books, 2019.

Manthorpe, Jonathan. *Claws of the Panda: Beijing's Campaign of Influence and Intimidation in Canada*. Toronto: Cormorant Books, 2019.

Woodward, Bob. *Rage*. New York: Simon & Schuster, 2020.

Notes

1 Megan Ogilvie, Kenyon Wallace, and Jennifer Yang, "Shocking Rates Seen In Virus Hot Spots," *Toronto Star*, November 17, 2020.
2 Hope Reisenberg Richman, *Davidner Roots* (Browning Press, 1995).
3 Sol Sinclair. *Memories of Early Jewish Settlement at Lipton Saskatchewan* (unpublished).
4 "Prime Minister announces new actions under Canada's COVID-19 response," March 16, 2020; "Prime Minister announces temporary border agreement with the United States," March 20, 2020. See pm.gc.ca.
5 Donald E. Low, "SARS: Lessons from Toronto," in *Learning from SARS: Preparing for the Next Disease Outbreak: Workshop Summary*, eds. Stacy Knobler, Adel Mahmoud, Stanley Lemon, et al. (Washington: National Academies Press (US), 2004).
6 Kim Tingley, "Sewage Surveillance Might Have Helped Control the Pandemic Earlier—But Increasing Its Use Now Could Still Save Lives," *New York Times Magazine*, November 29, 2020.
7 Hua Guo et al., "Evolutionary Arms Race between Virus and Host Drives Genetic Diversity in Bat Severe Acute Respiratory Syndrome-Related Coronavirus Spike Genes," *Journal of Virology*, September 29, 2020.
8 Gail Dutton, "Compare Update: 2003 SARS Pandemic Versus 2020 COVID-19 Pandemic," *BioSpace*, September 7, 2020, biospace.com.
9 This is short for *angiotensin-converting enzyme 2*.
10 "Where Did This Coronavirus Originate? Virus Hunters Find Genetic Clues In Bats," NPR's *Short Wave*, April 15, 2020.
11 John M. Barry, *The Great Influenza: The Story of the Deadliest Pandemic in History* (New York: Penguin Books, 2005).
12 Jeronimo Cello, Aniko V. Paul, and Eckard Wimmer, "Chemical synthesis of poliovirus CDNA: generation of infectious virus in the absence of natural template," *Science* 297, no. 5583 (August 8, 2002): 1016–8.
13 Gina Kolata, *Flu: The Story of the Great Influenza Pandemic of 1918 and the Search for the Virus That Caused It* (New York: Atria Books, 2019), 254–315.
14 Douglas Jordan, "An Interview with Dr. Terrence Tumpey," CDC, May 24, 2018, www.cdc.gov.
15 Tran Thi Nhu Thao et al., "Rapid reconstruction of SARS-CoV-2 using a synthetic genomics platform," *bioRxiv*, February 21, 2020.
16 Wendong Li et al., "Bats are natural reservoirs of SARS-like coronaviruses," *Science* 310, no. 5748 (October 28, 2005): 676–9.
17 Siro Igino Trevisanato, "The 'Hittite plague', an epidemic of tularemia and the first record of biological warfare," *Medical Hypotheses* 69, no. 6 (2007): 1371–4.
18 Hal Gold, *Japan's Infamous Unit 731: Firsthand Accounts of Japan's Wartime Human Experimentation Program* (Tuttle Publishing, 2019).
19 A recent book on this subject is *Baseless*, by Nicholson Baker. In it, Baker describes the years he has spent trying to determine whether the US did use pathogens during the Korean War. He filed a vast number of Freedom of Information applications, which yielded very little hard evidence in the end. He also relied on a previous work by Canadian researchers Stephen Endicott and Edward Hagerman presented in a book called *The United States and Biological Warfare*, which relies on documents from China purporting to show evidence of such attacks. It is clear that Canada, the US, and Britain all have done significant offensive and defensive research on such weapons.

20 US Department of State, *Adherence to and Compliance with Arms Control, Non-proliferation and Disarmament Agreements and Commitments* (August 2019): 45. "Information indicates that the People's Republic of China (China) engaged during the reporting period in biological activities with potential dual-use applications which raises concerns regarding its compliance with BWC. In addition, the United States does not have sufficient information to determine whether China eliminated its assessed biological warfare (BW) program as required under Article II of the Convention.... Questions and concerns on its compliance with the Convention have been raised since the 1993 Report."

21 Karen Pauls, "Chinese researchers escorted from infectious disease lab amid RCMP investigation," *CBC News*, July 14, 2019.

22 World Health Organization, *SARS: How a global epidemic was stopped* (Western Pacific Region, 2006). See chapter 12 by Carolyn Abraham. Toronto was placed under a travel ban by the WHO. Ten thousand were quarantined, 250 became sick, 44 died. The economic cost was about $1 billion according to the Conference Board of Canada. See: "SARS Fallout to Cost Toronto Economy About $1 Billion: Conference Board," *CBC News*, December 4, 2003.

23 Gilles Demaneuf, "A review of SARS Lab Escapes from 2003–2004," *ResearchGate*, November 2020.

24 The Gang of Four were Party officials who came to power under Mao during the Cultural Revolution. They held on to power as Mao lay dying and continued to hold it for a for a short time after his death. They were led by Mao's last wife, Jiang Qing.

25 House Foreign Affairs Committee, Lead Republican Michael T. McCaul, *Final Report: The Origins of the COVID-19 Global Pandemic, Including the Roles of the Chinese Communist Party and the World Health Organization*, 116th Congress, September 21, 2020. See within the report a memo sent to the US State Department on April 19, 2018, from the American Consul, Wuhan: 72-75.

26 Ibid., 37.

27 Ibid., 38-39.

28 Eva Dou, "As China nears a coronavirus vaccine, bribery cloud hangs over drugmaker Sinovac," *Washington Post*, December 4, 2020. This story explains how the CEO of the second largest vaccine manufacturer in China, Sinovac, which trades on the NASDAQ, and sells its version of a SARS-CoV-2 vaccine called CoronaVac to the developing world, made a practice for years of bribing a Chinese drug regulator with cash gifts to fast-track applications for certification, as well as bribing various hospital officials to buy Sinovac vaccines. The officials were punished, but the CEO of Sinovac faced no prosecution.

29 Steven Chase, "Victims recount foreign state-sponsored harassment," *Globe and Mail*, November 27, 2020.

30 David Matas and Sarah Teich, "We must stop the genocide of Uighurs," *Toronto Star*, December 9, 2020.

31 The two pieces of legislation are the 2017 National Intelligence Law and the 2014 Counter-Espionage Law. See: Arjun Kharpal, "Huawei says it would never hand data to China's government. Experts say it wouldn't have a choice," *CNBC*, March 4, 2019.

32 Murray Brewster, "Canada warned of fallout on Five Eyes relationship if Huawei allowed on 5G," *CBC News*, November 23, 2019.

33 Kate Kelland and Se Young Lee, "Pneumonia outbreak in China may be linked to new virus from SARS, MERS Family: WHO," *Reuters*, January 8, 2020.

34 Jon Cohen and Dennis Normile, "World on alert for potential spread of new SARS-like virus found in China," *Science*, January 14, 2020.

35 Ibid.

36 Ibid.

37 In 1918 it was easy to suppress word of an epidemic, even a terrible pandemic.

During World War I there was censorship of the press due to concern about maintaining morale. President Wilson of the US never mentioned the pandemic in public. It became known as the Spanish flu because Spain was not a combatant and did not censor stories about it.

38 Gao Yu et al. "How early signs of the coronavirus were spotted, spread and throttled in China," *The Straits Times Asia*, February 28, 2020. Original from Caixin Global.

39 Raymond L. Zhong, Paul Mozur, and Aaron Krolik, with Jeff Kao, "Leaked Documents Show How China's Army of Paid Internet Trolls Helped Censor the Coronavirus," *New York Times* and *ProPublica*, December 19, 2020.

40 Yu et al, "How early signs of the coronavirus were spotted, spread and throttled in China."

41 Ibid.

42 Fan Wu et al. "A new coronavirus associated with human respiratory disease in China," *Nature* 579 (February 3, 2020): 265–9.

43 Ana Swanson and Alan Rappaport, "Trump Signs China Trade Deal, Putting Economic Conflict on Pause," *New York Times*, January 15, 2020. The pact kept most of $360 billion in US tariffs on Chinese goods in place to ensure compliance, with the threat of more if China failed to honour its obligations, which included purchase of $200 billion worth of US goods and services by 2021.

44 "China didn't warn public of likely pandemic for 6 key days," Associated Press, April 15, 2020.

45 Ibid.

46 Nathan VanderKlippe, "SARS-like virus has been transmitted from person to person, Chinese health authorities say," *Globe and Mail*, January 20, 2020.

47 James Bandler et al., "Inside the Fall of the CDC," *ProPublica*, October 15, 2020.

48 Chaolin Huang et al., "Clinical features of patients infected with 2019 novel coronavirus in Wuhan, China," *The Lancet* 395, no. 10223 (January 24, 2020): 497–506.

49 Xing-Yi Ge et al., "Isolation and characterization of a bat SARS-like coronavirus that uses the ACE2 receptor," *Nature* 503 (October 30, 2013): 535–8.

50 Ron Fouchier et al., "Transmission Studies Resume for Avian Flu," *Science* 339, no. 6119 (February 1, 2013).

51 Sander Herfst et al., "Airborne transmission of influenza A/H5N1 virus between ferrets," *Science* 336, no. 6088 (June 22, 2012): 1534–41. See also: Masaki Imai et al., "Experimental adaptation of an influenza H5 HA confers respiratory droplet transmission to a reassortant H5 HA/H1N1 virus in ferrets," *Nature* 486, no. 7403 (May 2012): 420–8.

52 Ed Yang, "Scientists create hybrid flu that can go airborne: H5N1 virus with genes from H1N1 can spread through the air between mammals," *Nature/News*, May 12, 2013.

53 Alison Young and Nick Penzenstadler, "Inside America's secretive biolabs," *USA Today*, May 28, 2015. See also: Alison Young, "Newly disclosed CDC biolab failures 'like a screenplay for a disaster movie'," *USA Today*, June 2, 2016.

54 Lynn C. Klotz and Gregory D. Koblentz, "New pathogen research rules: Gain of function, loss of clarity," *Bulletin of the Atomic Scientists*, February 26, 2018.

55 Young and Penzenstadler, "Inside America's secretive biolabs."

56 Jocelyn Kaiser, "Scientists call for limit on creating dangerous pathogens," *Science*, July 15, 2014. This piece announced the creation of the Cambridge Working Group, at Harvard. "'Laboratory creation of highly transmissible, novel strains of dangerous viruses, especially but not limited to influenza, poses substantially increased risk' that an accidental infection could lead to a global outbreak." Harvard epidemiologist Marc Lipsitch was among the signatories and founders of the group.

57 Vineet D. Menachery et al., "A SARS-like cluster of circulating bat coronaviruses

shows potential for human emergence," *Nature Medicine*, December 12, 2015.
58 Chen Wang et al., "A novel coronavirus outbreak of global health concern," *The Lancet* 395, no. 10233 (February 15, 2020): 470–3, published online on January 24, 2020.
59 Ibid.
60 Peng Zhou et al., "A pneumonia outbreak associated with a new coronavirus of probable bat origin," *Nature*, February 3, 2020.
61 David Quammen, "Did Pangolin Trafficking Cause the Coronavirus Pandemic?" *New Yorker*, August 31, 2020.
62 Botao Xiao and Lei Xiao, "The possible origins of the 2019-nCoV coronavirus," *ResearchGate*, February 6, 2020.
63 Peng Zhou et al. *Nature* February 3, 2020.
64 This journal is prepared for printing by a group in Shanghai. Its editor is an American who solicited and published an article in February 2020 declaring there is no credible evidence that the virus leaked from the WIV. Shibo Jiang, one of the co-authors of this paper, was also part of a group of scientists, led by Shi Zhengli, who tried to get the international virus taxonomy committee to change the name of SARS-CoV-2 to something without a link to the original SARS. See Biohazards Blog at USRTK.org.
65 Shibo Jiang, Lanying Du, and Zhengli Shi, "An emerging coronavirus causing pneumonia outbreak in Wuhan, China: calling for developing therapeutic and prophylactic strategies," *Emerging Microbes & Infections* 9, no. 1 (January 31, 2020): 275–7.
66 Joseph Wu, Kathy Leung, and Gabriel M. Leung, "Nowcasting and forecasting the potential domestic and international spread of the 2019-nCoV outbreak originating in Wuhan, China: a modelling study," *The Lancet* 395, no. 10225 (January 31, 2020): 689–97.
67 Zhou et al., "A pneumonia outbreak associated with a new coronavirus of probable bat origin."
68 Wu, Leung, and Leung, "Nowcasting and forecasting the potential domestic and international spread of the 2019-nCoV outbreak originating in Wuhan, China: a modelling study."
69 Seyi Samson Enitan et al., "The 2019 Novel Coronavirus Outbreak: Current Crises, Controversies and Global Strategies to Prevent a Pandemic," *International Journal of Pathogen Research* 4, no. 1 (March 7, 2020): 1–16.
70 Heath Kelly, "The classical definition of a pandemic is not elusive," *Bulletin of the World Health Organization* 89, no. 7 (2011): 540–1.
71 House Foreign Affairs Committee, *Final Report: The Origins of the COVID-19 Global Pandemic*, 46–7.
72 "Elections, ties with China shaped Iran's coronavirus response," *Reuters*, April 2, 2020.
73 Bob Woodward, *Rage* (New York: Simon & Schuster, 2020), xii–xv.
74 Lev Facher, "U.S. declares public health emergency over coronavirus, bans travel from China by foreign nationals," STAT, January 31, 2020.
75 Woodward, *Rage*, xii.
76 Michael Wines and Amy Harmon, "Biogen Conference May Have Spread Virus to 300,000," *New York Times*, December 11, 2020.
77 David Staples, part one of "The road to Canada's COVID-19 outbreak: timeline of federal government failure at border to slow the virus," *Edmonton Journal*, March 31, 2020, to April 28, 2020.
78 David Staples, part two of "The road to Canada's COVID-19 outbreak: timeline of federal government failure at border to slow the virus," *Edmonton Journal*, April 2, 2020, to April 12, 2020.
79 Marieke Walsh, Grant Robertson, and Kathy Tomlinson, "Federal emergency stockpile of PPE was ill-prepared for pandemic," *Globe and Mail*, April 30, 2020.

80 House Foreign Affairs Committee, *Final Report: The Origins of the COVID-19 Global Pandemic*, 47.
81 https://gphin.Canada.ca
82 Grant Robertson, "'Without early warning you can't have early response': How Canada's world-class pandemic alert system failed," *Globe and Mail*, July 25, 2020.
83 Grant Robertson, "Head of Canada's Public Health Agency resigns," *Globe and Mail*, September 18, 2020.
84 Grant Robertson, "Ottawa left door open to deadly pandemic: Bloodworth report," *Globe and Mail*, July 13, 2021.
85 Murray Brewster, "Public Health Agency was unprepared for the pandemic and 'underestimated' the danger, auditor general says," *CBC News*, March 25, 2021.
86 Cohen and Normile, "World on alert for potential spread of new SARS-like virus found in China."
87 Emily Baumgaertner, "As coronavirus outbreak worsens, China agrees to accept help from WHO," *Los Angeles Times*, January 28, 2020. See also: "How China blocked WHO and Chinese scientists early in coronavirus outbreak," Associated Press, June 2, 2020.
88 The State Council Information Office at the People's Republic of China, "Fighting COVID-19: China in Action," June 2020.
89 See the WHO coronavirus (COVID-19) dashboard at https://covid19.who.int.
90 COVID-19 confirmed and death case development in China 2020. https://www.statista.com.
91 https://www.statista.com.
92 Xingguang Li et al., "Evolutionary history, potential intermediate animal host, and cross-species analyses of SARS-CoV-2," *Journal of Medical Virology* 92, no. 6 (June 2020): 602–11.
93 Jon Cohen, "Mining coronavirus genomes for clues to the outbreak's origins," *Science*, January 31, 2020. See also: Kristian Andersen et al., "The proximal origin of SARS-CoV-2," *Nature Medicine* 26, no. 4 (April 2020): 450–5. Originally published online March 17, 2020.
94 See news stories in the *Globe and Mail* on December 21, 2020, regarding the new variant identified in the UK, also in Denmark and Australia, with 17 mutations that appear to make it more easily transmissible but not yet known to be more lethal or adapted to vaccine.
95 Lauren Pelley and Adam Miller, "'They were completely unaware': Why mass COVID-19 testing is key to stopping spread," *CBC News*, March 21, 2020.
96 Jian Shang et al., "Structural basis of receptor recognition by SARS-CoV-2," *Nature* 581 (May 12, 2020): 221–4.
97 Jian Shang et al., "Cell entry mechanisms of SARS-CoV-2," *PNAS* 117, no. 21 (May 26, 2020): 11727–34.
98 Himalaya Rose Garden Team. "[Exclusive] A University of Minnesota Professor, Fang Li, Is Named by Lude Media as a Key Figure Behind the CCP Virus," September 13, 2020.
99 I woke up early on the morning of January 1, 2021, to find my radio still on and blaring a podcast. Bill Gates was being interviewed so I listened for a while. He touted the virtues of digital IDs for everyone, especially in the context of the pandemic. These digital IDs could record who has had vaccines and who has not, making it easier to let the vaccinated cross borders at will while the possibly infectious remain locked down. I am sure any Gates conspiracy theorists who heard it banged their fists on their tables and shouted "I told you so," while privacy proponents went wild with worry.
100 Adam Taylor, "Experts debunk fringe theory linking China's coronavirus to weapons research," *Washington Post*, January 29, 2020.
101 Ben Westcott and Steven Jiang, "Chinese diplomat promotes conspiracy theory that US military brought coronavirus to Wuhan," *CNN*, March 13, 2020.

102 Nathan Allen and Inti Landauro, "Coronavirus traces found in March 2019 sewage sample, Spanish study shows," *Reuters*, June 26, 2020. See also: Gemma Chavarria-Miro, et al., "Sentinel surveillance of SARS-CoV-2 in wastewater anticipates the occurrence of COVID-19 tests" *medRxiv*. Last author Albert Bosch, a leading virologist. The group surveyed wastewater and found the telltale RNA in a sample taken January 15, 2020, 41 days in advance of the first confirmed case of COVID-19 in Spain, suggesting that wastewater is a good place to survey to warn of community transmission before other tests pick it up. They went back to search for the first occurrence, found nothing until they tested a March 2019 sample which held some of the RNA sequences of the virus. Unfortunately, this work could not be retested for accuracy as the sample was destroyed in the first test.

103 Taylor, "Experts debunk fringe theory linking China's coronavirus to weapons research."

104 David Cyranoski, "Inside the Chinese lab poised to study world's most dangerous pathogens," *Nature* 542 (February 22, 2017): 399–400.

105 CNAS supporters between October 1, 2019, and September 30, 2020. See https//www.cnas.org.

106 Elsa B. Kania and Wilson Vorndick. "Weaponizing Biotech: How China's Military Is Preparing for a 'New Domain of Warfare'," *Defense One*, August 14, 2019.

107 Monika Chansoria, "Is China Producing Biological Weapons? Look at Its Capabilities and International Compliance," JAPAN *Forward*, March 5, 2020.

108 Nicholson Baker, *Baseless: My Search for Secrets in the Ruins of the Freedom of Information Act* (New York: Penguin Press, 2020), 190–1.

109 Corey Pfluke, "Biohazard: A Look at China's Biological Capabilities and the Recent Coronavirus Outbreak," *Wild Blue Yonder*, March 2, 2020.

110 Amy Harmon, "Lab Severs Ties with James Watson, Citing 'Unsubstantiated and Reckless' Remarks," *New York Times*, January 11, 2019.

111 Elaine Dewar, *Bones: Discovering the First Americans* (Toronto: Random House of Canada, 2001).

112 Kay Prüfer et al., "The complete genome sequence of a Neandertal from the Altai Mountains," *Nature* 505, no. 7481 (January 2, 2014): 43–9. First published digitally on December 18, 2013.

113 Hugo Zeberg and Svante Pääbo, "The major genetic risk factor for severe COVID-19 is inherited from Neandertals," *bioRxiv*, July 3, 2020.

114 Ibid. Published in *Nature* on September 30, 2020.

115 Xiao and Xiao, "The possible origins of the 2019-nCoV coronavirus."

116 Ibid.

117 Matthew Tye, *US-China Today*. https://uschinatoday.org

118 "Coronavirus and China," May 21, 2020, https://www.kiwiblog.co.nz.

119 Jim Geraghty, "The Trail Leading Back to the Wuhan Labs," *National Review*, April 3, 2020.

120 Cohen, "Mining coronavirus genomes for clues to the outbreak's origins."

121 Ibid.

122 Li et al., "Bats are natural reservoirs of SARS-like coronaviruses," 676–9.

123 Please see Vincent Racaniello's podcast interview with Peter Daszak: "TWiV 615: Peter Daszak of EcoHealth Alliance," *This Week in Virology (TWiV)*, May 19, 2020, https://www.microbe.tv/twiv/twiv-615.

124 See emails between Daszak and Colwell obtained from a Freedom of Information request by a group called US Right to Know, available at https://usrtk.org/biohazards-blog-index.

125 Andersen et al., "The proximal origin of SARS-CoV-2."

126 Charles Calisher et al., "Statement in support of the scientists, public health professionals, and medical professionals of China combatting COVID-19," *The Lancet* 395, no. 10226 (March 7, 2020). First published digitally on February 19, 2020.

127 Andersen et al., "The proximal origin of SARS-CoV-2."
128 Menachery et al., "A SARS-like cluster of circulating bat coronaviruses shows potential for human emergence." Corrected April 6, 2016, and again May 22, 2020. Interestingly, the first correction is about the failure of the article to mention that the USAID/PREDICT program had funded Shi Zhengli's work on the project through EcoHealth Alliance. The second correction was in regard to the sequence of the mouse-adapted SHC015-MA15 virus (a chimera). The original article claimed it had already been deposited to GenBank. The actual deposit wasn't made until 2020. Claiming a sequence has been deposited when it has not been seems to be a pattern with the Shi group.
129 Andersen et al., "The proximal origin of SARS-CoV-2." See endnote number 23.
130 Joint Prevention and Control Mechanism of the State Council in Response to the Novel Coronavirus Pneumonia Scientific Research Group, March 3, 2020.
131 Memo to the Offices of the China Center for Disease Control and Prevention. Memo 202, no. 16, of the Science and Technology Department. "On the Supplementary Regulations on Strengthening the Management of Science and Technology During the Emergency Response to the Novel Coronavirus Pneumonia."
132 Dake Kang, Maria Cheng, and Sam McNeil, "China clamps down in hidden hunt for coronavirus origins," Associated Press, December 29, 2020. This story includes screenshots of the issued notices above.
133 Andersen et al., "The proximal origin of SARS-CoV-2."
134 Donald G. McNeil Jr., "Scientists Were Hunting for the Next Ebola. Now the US Has Cut Off Their Funding," *New York Times*, October 25, 2019. Funding was partially restored for a short period after this story came out and the pandemic began. McNeil would later declare himself to be a friend of Peter Daszak's.
135 Sisi Wei, Annie Waldman, and David Armstrong, "Dollars for Profs: Dig Into University Researchers' Outside Income and Conflicts of Interest," *ProPublica*, December 6, 2019.
136 Edward C. Holmes, Andrew Rambaut, and Kristian G. Andersen, "Pandemics: spend on surveillance, not prediction," *Nature* 558, no. 7709 (June 14, 2018): 180–2.
137 Calisher et al., "Statement in support of the scientists, public health professionals, and medical professionals of China combatting COVID-19."
138 Ibid.
139 Biohazards Blog, https://usrtk.org/biohazards-blog-index.
140 Karl Sirotkin and Dan Sirotkin, "Might SARS-CoV-2 Have Arisen via Serial Passage through an Animal Host or Cell Culture?" *BioEssays* 42, no. 10 (2020).
141 Fred Guterl, Naveed Jamali, and Tom O'Connor, "The Controversial Experiments and Wuhan Lab Suspected of Starting the Coronavirus Pandemic," *Newsweek*, April 27, 2020.
142 Minnie Chan and William Zheng, "Meet the major general on China's coronavirus scientific front line," *South China Morning Post*, March 3, 2020. And "Profile: Chen Wei, military medical scientist marching toward vaccine," *Xinhua*, October 8, 2020.
143 Sarah Owermohle, "Trump cuts US research on bat-human virus transmission over China ties," *Politico*, April 27, 2020.
144 House of Commons, Canada, *Report on Canadian Security Intelligence Service Director Richard Fadden's Remarks Regarding Alleged Foreign Influence of Canadian Politicians*, Eighth Report of the Standing Committee on Public Safety and National Security, 40th Parliament, 3rd session, March 2011.
145 Robert Fife and Steven Chase, "Trudeau attended cash-for-access fundraiser with Chinese billionaires," *Globe and Mail*, November 22, 2015.
146 Arthur Cockfield, "The high price of Chinese money laundering in Canada," *Globe and Mail*, February 8, 2019.

147 Alex Joske, "The party speaks for you," *Australian Strategy Policy Institute*, June 9, 2020.
148 See Elections Canada website: Annual Financial Report, Papineau, Quebec, fiscal year ends for 2014, 2015, 2016, 2018, and 2019.
149 Robert Fife and Steven Chase, "Chinese envoy says Canada's acceptance of Hong Kong refugees jeopardizes Canadians in former British colony," *Globe and Mail*, October 15, 2020.
150 Perry Diaz, "Is COVID-19 a sinister plot to destroy America?" *Mindanao Gold Star Daily*, March 27, 2020.
151 See Falun Gong's Wikipedia listing.
152 Brandy Zadrozny and Ben Collins, "Trump, QAnon and an impending judgment day: Behind the Facebook-fueled rise of the Epoch Times," *NBC News: Tech News*, August 18, 2019.
153 Jim Waterson, "Trump pardons fraudster Conrad Black after glowing biography," *Guardian*, May 15, 2019.
154 For a list of columns go to https://www.theepochtimes.com.
155 "National High-tech R&D Program (863 Program)," Consulate General of the People's Republic of China in New York, March 5, 2016, http://newyork.china-consulate.org.
156 Captain Scott Tosi, "Steal the Firewood from Under the Pot," *Military Review: Army University Press*, September-October 2020.
157 Diaz, "Is COVID-19 a sinister plot to destroy America?"
158 Dan Hu et al., "Genomic characterization and infectivity of a novel SARS-like coronavirus in Chinese bats," *Emerging Microbes and Infections* 7, no. 154 (2018).
159 Mark Moore, "54 NIH scientists reportedly fired or resigned during espionage probe," *New York Post*, June 17, 2020.
160 Nidhi Subbaraman, "US investigations of Chinese scientists expand focus to military ties," *Nature* 585, no. 7824 (September 3, 2020): 170–1.
161 Ibid.
162 Edward Wong and Julian E. Barnes, "US to Expel Chinese Graduate Students With Ties to China's Military Schools," *New York Times*, May 28, 2020. Also: Humeyra Pamuk, "US revokes more than 1,000 visas of Chinese nationals, citing military links," *Reuters*, September 9, 2020.
163 Evan Dyer, "Experts call on Canadian universities to close off China's access to sensitive research," *CBC News*, September 15, 2020.
164 Rowan Jacobsen, "The Non-Paranoid Person's Guide to Viruses Escaping from Labs," *Mother Jones*, May 14, 2020.
165 Heather Mongilio, "CDC inspection findings reveal more about USAMRIID research suspension," *The Frederick News-Post*, November 23, 2019.
166 Young and Penzenstadler, "Inside America's secretive biolabs."
167 Jacobsen, "The Non-Paranoid Person's Guide to Viruses Escaping from Labs." By September 2020, the number of people suffering from brucellosis had risen to 3,245 with 1,401 more suspected, and the infections had spread to the province next door. See: Jessie Yeung and Eric Cheung, "Bacterial outbreak infects thousands after factory leak in China," *CNN*, September 17, 2020.
168 Interview with Stefan Wagener, former administrator directing lab safety at the NML.
169 Katie Pedersen, "Workers may have been exposed to Ebola, HIV and TB at Winnipeg lab, reports reveal," *CBC News: CBC Investigates*, February 2, 2017.
170 Dylan Robertson, "TB sample lost on floor for year: Winnipeg microbiology lab documents," *Winnipeg Free Press*, June 6, 2019.
171 "Frank Plummer opens up about his alcoholism—and the experimental surgery he says saved him," CBC Radio's *As It Happens*, December 16, 2019.
172 Scott Shane and Eric Lichtblau, "Scientist's Suicide Linked to Anthrax Inquiry," *New York Times*, August 2, 2008.

173 Dylan Robertson, "Misery infects virology lab: deadly pathogens, toxic workplace," *Winnipeg Free Press,* September 9, 2019. Story includes extensive documentation. See also: Jef Akst, "Union Says National Lab in Canada Is a Toxic Workplace," *The Scientist*, September 30, 2019.

174 See Human Pathogens and Toxins Act, https://laws.justice.gc.ca/eng/AnnualStatutes/2009_24/page-1.html

175 See Human Pathogens and Toxins Act, section 38, (2) Information and (4) Excluded information, https://laws.justice.gc.ca/eng/AnnualStatutes/2009_24/page-4.html#h-16

176 D. Choucrallah et al., "Surveillance of laboratory exposures to human pathogens and toxins: Canada 2018," *CCDR* 49, no. 9 (September 5, 2019): 244–51.

177 Amanda Lien et al., "Surveillance of laboratory exposures to human pathogens and toxins, Canada 2019," *Public Health Agency of Canada* 46–9, Force Health Protection (September 3, 2020).

178 See Access to Information Act, RSC, 1985, c. A-1, 4 (1), https://laws-lois.justice.gc.ca/eng/acts/a-1/

179 Bryn Williams-Jones, Elise Smith, and Catherine Olivier, "Governing 'Dual-Use' Research in Canada: A Policy Review," *Science and Public Policy* 41, no. 1 (2014): 76–93.

180 Ron Fouchier et al., "Pause on Avian Flu Transmission Research," *Science* 335, no. 6067 (January 27, 2012).

181 Ryan S. Noyce, Seth Lederman, and David H. Evans, "Construction of an infectious horsepox virus vaccine from chemically synthesized DNA fragments," *Plos One,* January 19, 2018.

182 Ibid.

183 See Access to Information Act, RSC, 1985, c. A-1, https://laws-lois.justice.gc.ca/eng/acts/a-1/

184 "Expert who quit Canada's Vaccine Task Force says more transparency needed to ensure public trust," CBC Radio's *The Current,* September 28, 2020.

185 "Dr Xiangguo Qiu & Dr Gary Kobinger: Against all odds they discovered a treatment for Ebola," *Governor General's Innovation Awards* website, May 18, 2018.

186 Dylan Robertson, "Microbiology lab researcher suspended by U of M," *Winnipeg Free Press,* July 15, 2019.

187 Karen Pauls, "Ouster of researchers from National Microbiology Lab still a mystery," *CBC News,* July 23, 2019.

188 Dylan Robertson, "Protocol ignored in pathogens transfer from Winnipeg's microbiology lab: sources," *Winnipeg Free Press,* July 31, 2019.

189 Karen Pauls, "Canadian lab's shipment of Ebola, Henipah viruses to China raises questions," *CBC News,* August 2, 2019.

190 Karen Pauls, "Canadian government scientist under investigation trained staff at Level 4 lab in China," *CBC News,* October 3, 2019.

191 Karen Pauls, "Head of National Microbiology Lab resigns, takes bio-research job in UK," *CBC News,* May 15, 2020.

192 "Former head of National Microbiology Lab resigned for family reasons, he says," *CBC News,* May 20, 2020.

193 Karen Pauls, "Canadian scientist sent deadly viruses to Wuhan lab months before RCMP asked to investigate: Documents show concerns about Ebola shipment from National Microbiology Lab, no relation to COVID-19," *CBC News,* June 14, 2020.

194 Bryan D. Griffin et al., "North American deer mice are susceptible to SARS-CoV-2," *bioRxiv,* July 26, 2020.

195 Pauls, "Canadian scientist sent deadly viruses to Wuhan lab months before RCMP asked to investigate."

196 Megan Benedictson, "Frank Plummer, former head of National Microbiology Lab, dead at 67," *CTV News*, February 4, 2020.

197 "Charged with smuggling in US, ex-Winnipeg researcher cops a plea," *CBC News*, May 22, 2009.

198 "Canadian charged with smuggling Ebola," *Nature* 459, no. 311 (May 20, 2009).

199 Han Xia et al., "Biosafety Level 4 Laboratory User Training Program, China," *EID Journal* 25, no. 5 (May 2019).

200 Both of these cables may be viewed in the House Foreign Affairs Committee's *Final Report: The Origins of the COVID-19 Global Pandemic*, 69, including the roles of the Chinese Communist Party and the World Health Organization.

201 http://english.whiov.cas.cn/

202 The Wuhan Institute of Virology website also lists members of an International Executive Committee of Performance Evaluation of WIV/CAS. Members include Linfa Wang, said to be affiliated with the Australian Animal Health Laboratory in Geelong, Australia, though he is now in Singapore; Chris Le of the University of Alberta; and Stefan Wagener, who is mentioned elsewhere as an international advisor. Wagener was the chief administrative officer of the NML in Winnipeg from 2000 to 2013 until he was told his services were no longer required. He moved on to the Canadian Grain Commission and is now an international consultant in bio-risk based in Winnipeg.

203 The WIV was then the only known BSL-4 operating in China. However, a veterinary BSL-4 was about to open in Harbin, the place where the Japanese conducted horrific biological warfare experiments on Chinese prisoners during World War II. As the US House of Representatives final minority staff report on the origin of SARS-CoV-2 makes clear, the French were asked to provide dozens of pressure suits, many more than could be used at the WIV, suggesting there could be another BSL-4 lab either up and running or about to be in China. The minority report says that that until 2003, all BSL-3 labs in China were controlled by the PLA, and that the person in charge of the WIV's new BSL-4 until January 2020 was Yuan Zhiming, President of the CCP Committee within the Wuhan branch of the Chinese Academy of Sciences. Yuan was replaced by Major General Chen Wei "the PRC's top biowarfare expert," according to the Report.

204 Access to Information Act, RSC, 1985, c. A-1, https://laws-lois.justice.gc.ca/eng/acts/a-1/

205 See https://www.ourcommons.ca/content/Committee/432/CACN/WebDoc/WD11405787/11405787/PublicHealthAgencyOfCanada-Revised-e.pdf

206 Ibid., email to Xiangguo Qiu, May 30, 2018, 10:58 am, copy to Mike Drebot.

207 Ibid.

208 Those who got the fragments later became seriously ill and were killed in the name of science ethics with an intracardiac shot of pentobarbital sodium. The nonhuman primates used in this study were likely macaques, who are very intelligent, well able to feel fear and anticipate suffering.

209 Hualei Wang et al., "Equine-Origin Immunoglobulin Fragments Protect Nonhuman Primates from Ebola Virus Disease," *Journal of Virology* 93, no. 5 (March 2019).

210 Yongkun Zhao et al., "Equine immunoglobulin F(ab')2 fragments protect mice from Rift Valley fever virus infection," *International Immunopharmacology* 64 (November 2018): 217–22.

211 Shao-Li Hong et al., "Cellular-Beacon-Mediated Counting for the Ultrasensitive Detection of Ebola Virus on an Integrated Micromagnetic Platform," *Analytic Chemistry* 90, no. 12 (June 2018): 7310–7.

212 Xiangguo Qiu et al., "Reversion of advanced Ebola virus disease in nonhuman

primates with ZMapp," *Nature* 514, no. 7520 (October 2, 2014): 47–53.

213 To read of these experiments involving lethal doses of these viruses given to imprisoned macaques, who invent and use tools in the wild and have interesting societies, is awful. They are infected with the virus, then given the antibodies according to an experimental regime in order to determine the optimal combination of antibodies. Those who show signs of significant illness are killed (the word used in the papers is *euthanized*). In other words, in labs like the NML, "nonhuman primates" live to die for us.

214 I am indebted to Evan Osnos for his biography of Xi Jinping: "Born Red," which appeared in the *New Yorker*, March 30, 2015.

215 Yuwen Wu, "China's Class of 1977: I took an exam that changed China," BBC *News*, December 14, 2017.

216 Keding Cheng et al., "Leukemia cell line inhibition of cell proliferation, autophosphorylation, kinase activity, and induction of apoptosis by tyrphostin AG1112 and AG957," *Blood* 10 suppl., part 1-2 (1996).

217 Keding Cheng et al., "Reduced FAK and paxillin phosphorylation in BCR-ABL transfected 3T3 cells," *Blood* 88 suppl., part 1-2 (1996): 2297.

218 Surender Kharbanda et al., "Translocation of SAPK/JNK to mitochondria and interaction with Bcl-x(L) in response to DNA Damage," *Journal of Biological Chemistry* 275, no. 1 (January 7, 2000).

219 Moshe Talpaz et al., "Autoantibodies to Abl and Bcr proteins," *Leukemia* 14 (2000): 1661–6.

220 Keding Cheng et al., "Reduced focal adhesion kinase and paxillin phosphorylation in BCR-ABL-transfected cells," *Cancer* 95, no. 2 (July 15, 2002).

221 Oleg Krokhin et al., "Mass spectrometric characterization of proteins from the SARS virus: a preliminary report," *Molecular & Cellular Proteomics* 2, no. 5 (May 2003): 346–56.

222 Oleg Krokhin et al., "Mass spectrometric based mapping of the disulfide bonding patterns of integrin alpha chains," *Biochemistry* 42, no. 44 (October 17, 2003).

223 Ningjun Li et al., "The role of angiotensin converting enzyme 2 in the generation of angiotensin 1-7 by rat proximal tubules," *American Journal of Physiology: Renal Physiology* 288, no. 2 (October 5, 2004).

224 Adam Burgener et al., "Identification of differentially expressed proteins in the cervical mucosa of HIV-1-resistant sex workers," *Journal of Proteome Research* 7, no. 10 (2008): 4446–54.

225 Keding Cheng, "MS-H: a novel proteomic approach to isolate and type the E. coli H antigen using membrane filtration and liquid chromatography-tandem mass spectrometry (LC-MS/MS)," *Public Library of Science One* 8, no. 2 (2013): e57339.

226 Chris Druar et al., "Analysis of the expressed heavy chain variable-region genes of Macaca fascicularis and isolation of monoclonal antibodies specific for the Ebola virus' soluble glycoprotein," *Immunogenetics* 57 (2005): 730–8.

227 Natalie Salat, "Waging War on Infectious Disease," *Legion Magazine*, November 1, 2005.

228 "High-tech Canadian lab spills wastewater," CBC *News*, July 13, 1999. See also: Roberta Rampton, "Lab criticized for leak," *The Western Producer*, July 22, 1999.

229 Robert Fife and Steven Chase, "Canadian spy chief calls China strategic threat," *Globe and Mail*, A7, February 10, 2021.

230 Qiu et al., "Reversion of advanced Ebola virus disease in nonhuman primates with ZMapp."

231 Jonathan Audet et al., "Molecular Characterization of the Monoclonal Antibodies Composing ZMAb: A Protective Cocktail Against Ebola Virus," *Scientific Reports* 4, no. 6881 (November 6, 2014).

232 Xiangguo Qiu et al., "Establishment and characterization of a lethal mouse

model for the Angola strain of Marburg virus," *Journal of Virology* 88, no. 21 (online October 6, 2014): 12703–14.

233 Gary Wong et al., "Development and Characterization of a Guinea Pig-Adapted Sudan Virus," *Journal of Virology* 90, no. 1 (December 17, 2015): 392–9.

234 Gary Wong et al., "Pathogenicity Comparison Between the Kikwit and Makona Ebola Virus Variants in Rhesus Macaques," *Journal of Infectious Diseases* 214, suppl. 3 (October 15, 2016): S281–9.

235 Demin Duan, "Nanozyme-strip for rapid local diagnosis of Ebola," *Biosensors and Bioelectronics* 74 (December 15, 2015): 134–41.

236 Shao-Li Hong et al., "Cellular-beacon-mediated counting for the ultrasensitive detection of ebola virus on an integrated micromagnetic platform," *Analytical Chemistry* 90, no. 12 (May 24, 2018): 7310–7.

237 Zhen Wu et al., "Ultrasensitive ebola virus detection based on electroluminescent nanospheres and immunomagnetic separation," *Analytic Chemistry* 89, no. 3 (December 26, 2016): 2039–48.

238 Shipo Wu et al., "An Adenovirus Vaccine Expressing Ebola Virus variant Makona Glycoprotein Is Efficacious in Guinea Pigs and Nonhuman Primates," *Journal of Infectious Diseases* 214, suppl. 3 (October 4, 2016): S326–32.

239 Ibid.

240 Alex Cooke, "Canadian COVID-19 clinical trial scrapped after China wouldn't ship potential vaccine," *CBC News*, August 26, 2020.

241 Aw Cheng Wei, "ST Asians of the Year: 'People's Hero' worked on virus treatment at Ground Zero," *The Straits Times*, December 5, 2020.

242 Ruili Huang et al., "Massive-scale biological activity-based modeling identifies novel antiviral leads against SARS-CoV-2," *bioRxiv*, July 27, 2020, later published as "Biological activity-based modeling identifies antiviral leads against SARS-CoV-2," *Nature Biotechnology* 39, no. 6 (February 23, 2021): 747–53.

243 Logan Banadyga et al., "Atypical Ebola Virus Disease in a Nonhuman Primate following Monoclonal Antibody Treatment Is Associated with Glycoprotein Mutations within the Fusion Loop," *mBio* 12, no. 1, (January 17, 2021).

244 Robert Fife and Steven Chase, "Infectious disease scientists at Canada's high-security lab collaborated with China," *Globe and Mail*, May 20, 2021. This story identified Feihu Yan as being with the People Liberation Army's Academy of Military Medical Sciences.

245 Pengei Fan et al., "Potent neutralizing monoclonal antibodies against Ebola virus isolated from vaccinated donors," *mAbs* 12, no. 1 (January–December 2020): 1742457. Published digitally on March 26, 2020.

246 Ourcommons.ca.

247 Tu Thanh Ha, "At one Quebec nursing home, staff were ordered to remove masks before COVID-19 outbreak," *Globe and Mail*, June 8, 2021.

248 Erika Kinetz, "Takeaways: AP investigation of China COVID-19 disinformation," *AP News*, February 15, 2021.

249 Shan-Lu Liu et al., "No credible evidence supporting claims of the laboratory engineering of SARS-CoV-2," *Emerging Microbes and Infections* 9, no. 1 (February 26, 2021): 505–7. Received February 13, accepted February 13, 2020, and published online February 26, 2021.

250 Sainath Suryanarayanan, "Chinese-linked journal editor sought help to rebut COVID-19 lab origin hypothesis," *USRTK.org*, April 7, 2021.

251 Menachery et al., "A SARS-like cluster of circulating bat coronaviruses shows potential for human emergence."

252 Yi Fan et al., "Bat Coronaviruses in China," *Viruses* 11, no. 3 (March 2019).

253 Hongying Li et al., "Human-animal interactions and bat coronavirus spillover potential among rural residents in Southern China," *Biosafety Health* 1, no. 2, (September 2019): 84–90. This paper was published online November 9, 2019,

just before the first infections were believed to have happened not in Yunnan or Guangxi or Guangdong, but in Wuhan. This paper was supported by PREDICT and the National Institute of Allergy and Infectious Diseases of the National Institutes of Health.

254 Ben Hu et al., "Discovery of a rich gene pool of bat SARS-related coronaviruses provides new insight into the origin of SARS coronavirus," *PLOS Pathogens*, November 30, 2017. Authors on this paper included Peter Daszak and Linfa Wang, and some funding came from NIH and USAID/PREDICT. This study involved taking samples over a five-year period from bats in a single Yunnan cave, successfully isolating viruses, creating chimeras and testing to see whether they could infect human cells. And they could.

255 Carroll also told me that "gain-of-function dual use is the biggest biological threat we have. COVID-19 and SARS is mother nature. Dual use is where we expedite our own demise." He also went on to decry the sloppiness of certain labs that do this work in less than adequate containment, especially a lab he visited in Indonesia, and the ease with which synthetic biology may now be done, predicting people will be able to order sequences "off Amazon," which one day will lead to high school kids building viruses in the garage.

256 Wang explained that after the paper was submitted to *Science*, with no response, he was called by a journalist at *Science Magazine* to comment on another group's paper on a similar theme that had just been published elsewhere. He told the journalist that he'd sent in his own paper on the same subject to *Science* and had heard nothing. The journalist found out that the sub editor Wang had sent the paper to had taken three weeks off and so the paper was just sitting on a desk unread. The journalist intervened with the editor, and boom, history.

257 Wei Quan, Bikun Chen, and Fei Shu, "Publish or impoverish: An investigation of the monetary reward system of science in China (1999–2016)," *Aslib Journal of Information Management* 69, no. 5 (September 18, 2017): 1–17.

258 James T. Areddy, "China Bat Expert Says Her Wuhan Lab Wasn't Source of New Coronavirus," *Wall Street Journal*, April 21, 2020.

259 Sarah Zheng, "Chinese virologist at centre of 'coronavirus came from a laboratory' claim denies defecting," *South China Morning Post*, May 2, 2020.

260 Qiang Zhang et al., "SARS-CoV-2 neutralizing serum antibodies in cats: a serological investigation," *bioRxiv*, April 3, 2020. The article appeared with a new title, "A serological survey of SARS-CoV-2 in cat in Wuhan," in the peer-reviewed journal *Emerging Microbes and Infections*, volume 9, issue 1, the same journal that published the article with Shi's name on it that said the closest known viral sequences to SARS-CoV-2 were two sequences discovered by military researchers. This paper was submitted July 24, accepted August 28, and published online September 17 2020. The co-authors added to the original article by following two cats with antibodies for 120 days to see how long they stayed in the cats' systems. But the rest of the article was the same as the first, repeating the line that perhaps the stray cats picked it up from a polluted environment.

261 Jane Qiu, "How China's 'Bat Woman' Hunted Down Viruses from SARS to the New Coronavirus," *Scientific American*, June 1, 2020. First published digitally on March 11, 2020.

262 See chapter 11. The addendum is attached to the original *Nature* article of February 3, 2020.

263 Qiu, "How China's 'Bat Woman' Hunted Down Viruses from SARS to the New Coronavirus."

264 Dailymail.com Reporter, "Did coronavirus leak from a research lab in Wuhan? Startling new theory is 'no longer being discounted' amid claims staff 'got infected after being sprayed with blood'," *Daily Mail*, April 5, 2020.

265 Ibid.
266 Glenn Owen, "REVEALED: US government gave $3.7 million grant to Wuhan lab at center of coronavirus leak scrutiny that was performing experiments on bats from the caves where the disease is believed to have originated," *Daily Mail*, April 12, 2020.
267 Qiu, "How China's 'Bat Woman' Hunted Down Viruses from SARS to the New Coronavirus."
268 Josh Rogin, "Opinion: State Department cables warned of safety issues at Wuhan lab studying bat coronaviruses," *Washington Post*, April 14, 2020.
269 Jon Cohen, "Wuhan coronavirus hunter Shi Zhengli speaks out," *Science* 369, no. 6503 (July 31, 2020): 487–8.
270 Science News Staff, "Nobel laureates and science groups demand NIH review decision to kill coronavirus grant," *Science*, May 21, 2020.
271 Cohen, "Wuhan coronavirus hunter Shi Zhengli speaks out."
272 See chapter 15.
273 Hu et al., "Discovery of a rich gene pool of bat SARS-related coronaviruses provides new insight into the origin of SARS coronavirus." Shi was last author, but co-authors included Linfa Wang and Peter Daszak. The way the article is presented, it is as if none of the above had demonstrated that SARS likely arose in a horseshoe bat in Yunnan, rather than in other parts of China, such as Hubei. The last part of the paper describes in detail the making of chimeras by making copies of the spike sequences from sequences pulled from bat caves, inserting the sequences into a virus already isolated by Shi's lab, and in the end, making two of the chimeras infect Vero 6 cells and HeLa (human) cells via the ACE2 receptor.
274 "Reply to *Science Magazine*," scim.ag/ShiZhengli.
275 Josh Rogin, "In 2018, Diplomats Warned of Risky Coronavirus Experiments in a Wuhan Lab. No One Listened," *Politico*, March 8, 2021. Excerpt from Josh Rogin's *Chaos Under Heaven: Trump, Xi, and the Battle for the Twenty-First Century* (HMH Books & Media, 2021).
276 House Foreign Affairs Committee's *Final Report: The Origins of the COVID-19 Global Pandemic*, 39. The report points out that early in 2020, Yuan was replaced as head of the BSL-4 lab by Chen Wei, aka the Major General of the PLA and the leading Ebola and biowarfare expert in China. The WIV website does not list her as the director of the BSL-4.
277 House Foreign Affairs Committee's *Final Report: The Origins of the COVID-19 Global Pandemic*.
278 By March 11, 2021, one year after the WHO at last declared the existence of a pandemic, more than 118 million infections would be confirmed around the world and 2,622,375 million deaths. Canada would endure 902,234 cases and 22,330 deaths while the land of the free would suffer 29,155,279 infections and over 529,269 deaths, more than died in World Wars I and II and Vietnam combined, and almost as many as died from flu in 1918. *Johns Hopkins Coronavirus Resource Center*, http://coronavirus.jhu.edu.
279 Nathan VanderKlippe, "Thousands of Uyghur workers in China are being relocated in an effort to assimilate Muslims, documents show," *Globe and Mail*, March 2, 2021.
280 Dean Bengston, "All journal articles evaluating the origin or epidemiology of SARS-CoV-2 that utilize RaTG13 bat strain genomics are potentially flawed and should be retracted," *ResearchGate*, April 2020.
281 Occam's razor is named for William of Ockham, a theologian of the late 13th century from Surrey, educated at Oxford. The razor refers to getting rid of the extraneous bits of an argument. Ockham was a very influential logician and natural philosopher.

282 Aksel Fridstrøm, "The fight for a controversial article," *Minerva*, July 13, 2020.
283 Birger Sørensen, Andres Susrud, and Angus George Dalgleish, "Biovacc-19: A Candidate Vaccine for Covid-19 (SARS-CoV-2) Developed from Analysis of its General Method of Action for Infectivity," QRB *Discovery* 1, no. e6 (2020): 1–11.
284 Hong Zho et al., "A Novel Bat Coronavirus Closely Related to SARS-CoV-2 Contains Natural Insertions at the S1/S2 Cleavage Site of the Spike Protein," *Current Biology* 30, no. 11 (June 3, 2020): 2196–203.
285 Peng Zhou et al., "Fatal swine acute diarrhoea syndrome caused by an HKU2-related coronavirus of bat origin," *Nature* 556 (April 12, 2018). Other authors on this paper include Shi Zhengli, Linfa Wang, and Peter Daszak. Peng Zhou, the lead author, is with the Wuhan Institute of Virology's Laboratory of Special Pathogens and Biosafety, and is a long-time collaborator of Shi Zhengli. Funding came from various Chinese agencies, including the Strategic Priority Research Program of the Chinese Academy of Sciences, as well as USAID's PREDICT and the NIH/NIAID.
286 Birger Sørensen, Angus Dalgleish, and Andres Susrud, "The Evidence which Suggests that This Is No Naturally Evolved Virus: A Reconstructed Historical Aetiology of the SARS-CoV-2 Spike." This article may be found via a link in Sørensen, Susrud, and Dalgleish, "Biovacc-19: A Candidate Vaccine for Covid-19 (SARS-CoV-2) Developed from Analysis of its General Method of Action for Infectivity," referenced above.
287 Sophie Tanno, "Ex-head of MI6 Sir Richard Dearlove says coronavirus 'is man-made' and was 'released by accident' - after seeing 'important' scientific report," *Daily Mail*, June 4, 2020. The podcast is *The Telegraph's Planet Normal*.
288 Shing Hei Zhan, Benjamin E. Deverman, and Yujia Alina Chan, "SARS-CoV-2 is well adapted for humans. What does this mean for re-emergence?" *bioRxiv*, May 2, 2020.
289 Later, in a preprint titled "Single source of pangolin CoVs with a near identical Spike RBD to SARS-CoV-2," Yujia Alina Chan and Shing Hei Zhan demolished a series of papers published by different groups (with one shared author) in *Nature*, *Current Biology*, and PLOS *Pathogens* purporting to show that sick, smuggled pangolins found in different areas of China carried the same virus and were the probable intermediate animal from which it jumped to humans. Chan and Shing Hei Zhan showed that all of the papers referred to a single small group of pangolins smuggled from Malaysia and that they probably caught the virus from human handlers. "Our observations highlight the importance of requiring authors to publish their complete genome assembly pipeline and all contributing raw sequence data...in order to empower peer-review and independent analysis of the sequence data."
290 Jason Lemon, "Scientists Shouldn't Rule Out Lab As Source of Coronavirus, New Study Says," *Newsweek*, May 17, 2020.
291 Rowan Jacobsen, "Could COVID-19 Have Escaped from a Lab? The world's preeminent scientists say a theory from the Broad Institute's Alina Chan is too wild to be believed. But when the theory is about the possibility of COVID being man-made, is this science or censorship?" *Boston Magazine*, September 9, 2020. The preprint did not in fact argue that COVID was man-made: it merely said the question should not be ignored.
292 Jonathan Latham and Allison Wilson, "The Case Is Building That COVID-19 Had a Lab Origin," *Independent Science News*, June 1, 2020.
293 Jonathan Latham and Allison Wilson, "A Proposed Origin for SARS-CoV-2 and the COVID-19 Pandemic," *Independent Science News*, July 15, 2020.
294 Li Xu, "The analysis of 6 patients with severe pneumonia caused by unknown viruses" (MSc thesis (original in Mandarin), Emergency Department and EICU, The 1st Affiliated Hospital of Kunming Medical University, Kunming, 2013). https://oversea.cnki.net/KCMS/detail/detail.aspx?dbcode=CMFD&db-

name=CMFD201401&filename=1013327523.nh&v=kF6UVNY%25mmd2FoD%25mmd2F1El1IDANNWHFuucyWAXVDjm5luXgLcCgwxB5FlzUqb-jD5dixWZHYA

295 Rossana Segreto and Yuri Deigin, "Is considering a genetic-manipulation origin for SARS-CoV-2 a conspiracy theory that must be censored?" April 2020, https://www.researchgate.net/publication/340924249. Their article was eventually published in a peer-reviewed journal under the title "The genetic structure of SARS-CoV-2 does not rule out a laboratory origin: SARS-CoV-2 chimeric structure and furin cleavage site might be the result of genetic manipulation," *BioEssays* 43, no. 3 (March 2021): e2000240

296 Monali C. Rahalkar and Rahul A. Bahulikar, "Understanding the Origin of 'BatCovRATG13,' a Virus Closest to SARS-CoV-2," *Preprints*, May 24, 2020. This was later published in a peer-reviewed journal under the title "Lethal Pneumonia Cases in Mojiang Miners (2012) and the Mineshaft Could Provide Important Clues to the Origin of SARS-CoV-2," *Frontiers in Public Health* 8 (October 20, 2020): 581569. The authors say in the *Frontiers* version that they were alerted to the existence of the master's thesis by someone with the Twitter handle @seeker268. As I later learned, that person had a very important role in driving interest in the lab escape theory. His work found an audience thanks to DRASTIC.

297 When RATG13 raw reads were finally uploaded in May 2020, they revealed that the sequencing had been done in 2018, not "recently" as stated in *Nature*. The authors pointed out that synthesis of SARS-like genomes with pandemic potential was part of EcoHealth Alliance's grant application: they argued that the Shi lab already had expertise in that regard, so a lab outbreak was not unlikely.

298 DRASTIC stands for *Decentralized Radical Autonomous Search Team Investigating COVID-19*. See chapter 15.

299 Anonymous OSINT Contributor, Billy Bostickson, and Gilles Demaneuf, "An investigation into the WIV databases that were taken offline," *ResearchGate*, February 2021. See also: Qiu, "How China's 'Bat Woman' Hunted Down Viruses from SARS to the New Coronavirus."

300 CNKI is the name of the site. The English reader version can be found at https://global.cnki.net. On the home page it is described as "a key national research and information publishing institution in China, led by Tsinghua University and supported by PRC Ministry of Education, PRC Ministry of Science, Propaganda Department of the Chinese Communist Party and PRC General Administration of Press and Publication."

301 Flinders University, "Researchers Find COVID-19 Virus Was 'Highly Human Adapted' – Exact Origins Still a Mystery," *SciTechDaily*, June 25, 2021. The paper was originally published as a preprint, and finally made it into *Nature's* Scientific Reports. Authors included Nikolai Petrovsky of Flinders University and Professor David Winkler of La Trobe University.

302 Xing-Yi Ge et al., "Coexistence of multiple coronaviruses in several bat colonies in an abandoned mineshaft," *Virologica Sinica* 31, no. 3, (2016): 31–40. This abstract says, "we conducted a surveillance of coronaviruses in bats in an abandoned mineshaft in Mojiang County, Yunnan Province, China, from 2012–2013." They found six bat species, including horseshoe bats, and sequenced the RdRp sequences of diverse coronaviruses found in different bat species and found two "unclassified beta coronaviruses, one new strain of SARS-like coronavirus, and one potentially new betacoronavirus species. Furthermore, coronavirus co-infection was detected in all six bat species." The abstract said that bats are natural reservoirs and potential sources of zoonotic viral pathogens. Yet the paper made no mention of the men who got sick and died in the mine where these viruses were found. Shi's lab trapped hundreds of bats, beginning in August 2012, while the miners were still in hospital, and came back at

least three more times, in September 2012, in April, and in July 2013. Most of the grant money came from China but a grant from NIH/NIAID is acknowledged. Neither Peter Daszak nor Linfa Wang are co-authors of this paper.

303 From the EcoHealth Alliance Award abstract re: NIAID award number R01AI110964. The grant title is Understanding the Risk of Bat Coronavirus Emergence. This grant was first awarded in 2014 and was continued without competition right through 2019 until the Trump administration refused to pay out what had been promised for 2019, retracting $336,819. The total amount paid out under that grant over five years was $3,748,715.

304 The doctoral thesis described how several groups went to the mine to investigate the cause of the miners' illness, including the Center for Disease Control and Prevention of Chengdu Military Region, which sampled blood and tested for a whole range of infectious viruses. Results were negative. Another group from Beijing found a Henipah-like virus in rats in the mine. In their report they mentioned the mine and the fact that three miners died there, but nothing about the WIV or other groups looking for coronaviruses in the same place. The resulting paper was written about in *Science* in 2014, a story about finding a possible new SARS. Zhong Nanshan's lab tested blood from the miners and did not find anything SARS-like. The group from the Center for Disease Control and Prevention, under the direction of George F. Gao and including Canping Huang, trapped and dissected 87 bats and gathered feces. They discovered a novel virus that appeared to be the product of recombination between two different viral species. Canping Huang's thesis work produced two peer-reviewed papers, both of which listed George Gao as the last author, one including Edward C. Holmes as a co-author. Neither mentioned the mine as the source of these viruses. The doctoral thesis recorded that in addition to bat dissection products and samples from patients' lungs, blood was sent to the WIV for analysis: the WIV found that four patients' blood showed antibodies to something SARS-like. Canping Huang's dissertation was submitted to the Center for Disease Control and Prevention on October 20, 2016, signed May 24, 2017. The two papers appeared in *Biochemistry* and *PLOS* in 2016 before the dissertation was accepted. The paper from Shi's lab describing finding various viruses in individual bats, published in *Virologica Sinica*, did not mention the miners' illnesses or deaths.

305 They published later a much more detailed argument against a zoonotic spillover, quoting many of the more recent papers describing the unique structure and biochemistry of the SARS-CoV-2 spike: Segreto and Deigin, "The genetic structure of SARS-CoV-2 does not rule out a laboratory origin."

306 Among many of the questions raised, they asked if any of the groups that went to the mine between 2012 and 2014 could have become infected themselves. They wanted to know why the miners' disease was not reported to the WHO. They wanted to know why information about these miners had not appeared in any of Shi's lab's papers and why a WIV database that carried the datasets from the WIVs expedition to the mine had been taken offline. The doctoral dissertation reported that Shi's lab found antibodies to SARS or something SARS-like in four of the patients' blood samples, but they wanted to know what else Shi might have learned and whether the samples still existed at the WIV.

307 "Transcript: Matt Pottinger on 'Face the Nation,' February 21, 2021," *CBS News*, February 21, 2021.

308 Elaine Okanyene Nsoesie et al., "Analysis of hospital traffic and search engine data in Wuhan China indicates early disease activity in the Fall of 2019," June 8, 2020, http://nrs.harvard.edu/urn-3:HUL.InstRepos:42669767

309 Ken Dilanian et al., "Report says cellphone data suggests October shutdown at Wuhan lab, but experts are skeptical," *NBC News*, May 8, 2020. Click through NBC site for mention of original report called "MACE E-PAI COVID-19 ANALYSIS" available at www.documentcloud.org.

310 "Second China-US Workshop on the Challenges of Emerging Infections, Laboratory Safety and Global Health Security," *Study Resource*, https://studyres.com/doc/21384953/second-china-u.s.-workshop-on-the-challenges-of-emerging
311 The *National Post* carried a story by Tom Blackwell in the summer of 2019 about the viruses sent from the NML in Winnipeg to China. The story detailed that Nipah is considered a likely bioweapon because it could be engineered for mass dissemination and it has a death rate from 50 percent to 100 percent. Johns Hopkins ran an exercise that involved a version of Nipah engineered to pass between people more easily. See Tom Blackwell, "Bio-warfare experts question why Canada was sending lethal viruses to China," *National Post*, August 8, 2019.
312 Wang Yiwei, "How China's Bat Caves Hold the Secret to Preventing Epidemics," *Sixth Tone*, May 24, 2018.
313 Keoni Everington, "WHO inspector's denials of bats in Wuhan lab contradicted by facts," *Taiwan News*, February 18, 2021.
314 Ibid.
315 The interview with Daszak is "TWiV 615: Peter Daszak of EcoHealth Alliance" and may be found at https://www.microbe.tv/twiv/twiv-615.
316 Lekshmy Sreekumar, "Singapore Conference Explored Global Threat of Nipah Virus," *Duke Today*, December 11, 2019.
317 Return of Organization Exempt from Income Tax. EcoHealth Alliance, for the fiscal year ended June 2019, submitted September 1, 2020.
318 The Consortium for Conservation Medicine worked in concert with the Wildlife Preservation Trust International Inc., set up in the US in 1976 (a similar organization was set up in Canada in 1986).
319 Glenn McDonald, "The Global Virome Project is Hunting Hundreds of Thousands of Deadly Viruses," *Seeker*, March 1, 2018.
320 http://www.globalviromeproject.org/leadership-team.
321 I'd never heard of the Cosmos Club so I looked it up. It's a private club in Washington, DC, launched in 1878 as a leading place for the very best minds in science, arts, and letters, including among its current and former members many Nobel laureates and some US presidents and vice presidents. However, it had to be forced, under threat of losing its licences to operate, to permit women to join in 1988.
322 http://www.globalviromeproject.org. Peter Daszak PhD. Accessed April 5, 2021.
323 The award is R01-TW05869. "Ecology of Infectious Diseases" from the NIH/NSF John E. Fogarty International Center and the V. Kann Rasmussen foundation. It described sequencing a full SARS-related virus from bat fecal sample. The sequence is called SL-CoVRP3. The sequence proved to be between 92 and 94 percent identical to SARS-Tor, a sample taken from a person with SARS in Toronto, establishing bats as a reservoir for SARS. Interestingly, the grant does not say which of the regions sampled produced that positive genome sequence, Guangxi, or Hubei.
324 This was a reference to her 2013 *Nature* letter reporting genome sequences pulled from horseshoe bat feces, very similar to SARS, that also use the ACE2 receptor for entry into human, bat, and civet cells. It reported the successful isolation of one virus later used to make chimeras in the Baric lab's gain-of-function experiments two years later.
325 To examine EcoHealth Alliance's 501(c)(3) annual returns, please go to EcohealthAlliance http://projects.propublica.org. Their non-profit provides PDFs of the annual returns by year.
326 Sam Husseini, "Peter Daszak's EcoHealth Alliance has hidden Almost 40 million in Pentagon and Militaried Pandemic Science," *Independent Science News*, December 16, 2020.
327 See: letter to Drs. Chimura and Daszak of July 8, 2020 from Michael S. Lauer, NIH Deputy Director for Extramural Research, Department of Health &

Human Services, National Institutes of Health, National Institute of Allergy and Infectious Diseases. In: Eban, Katherine. "The Lab-Leak Theory: Inside the fight to Uncover COVID-19's Origins." *Vanity Fair*, May 2021.

328 Ibid, Eban.

329 Yu Shang, et al. "Construction and Rescue of a Functional Synthetic Baculovirus," ACS *Synthetic Biology*, 2017, 6, 1393-1402.

330 As Arif explained later, there are two types of baculoviruses. One is used as a vector to carry genetic instructions to make certain proteins into insect cells, the other is used to control insect pests. The one that can be used as a vector "abundantly expresses a variety of foreign proteins. It does not make sense to let the virus synthesize so many proteins that do not contribute to its capacity as an expression vector. Therefore, if the genes expressing these 'unnecessary' proteins are deleted, it is likely we can improve its efficiency as an expression vector. A baculovirus was used to synthesize the vaccine against HPV that causes cervical cancer and is now being used to make a vaccine against the SARS-CoV-2." The vaccine for SARS-CoV-2 that uses a baculovirus as a vector is made by Novavax. The virus "expresses the gene encoding the spike protein of the coronavirus." It carries it into insect cells, and in the process the protein is carried to the surface of the insect cells and harvested "in the form of 'nanoparticles', purified, incorporated with an adjuvant and used for immunization." Dr. Arif believes that this vaccine is 90 percent effective and he had hoped to be immunized with it himself but Pfizer's vaccine came along first.

331 "GLFC launches 75th anniversary celebrations: a look at some of the pioneering scientists." e-Bulletin. The Great Lakes Forestry Centre Issue, 40, June 2020.

332 For those too young to remember, the George W. Bush administration insisted Saddam Hussein had acquired weapons of mass destruction capable of reaching the United States and so regime change was required. This led to Gulf War II. The plan was to depose and kill Hussein, and replace him with a democracy that would change the balance of power in the Middle East. There were no weapons of mass destruction found and the invasion killed hundreds of thousands of Iraqis as well as Saddam Hussein, tilting the balance of power in the Middle East toward Iran, an avowed enemy of the US.

333 David Helwig, "Baghdad-born Saultite pleads for peace," *SooToday News*, March 9, 2003.

334 Hedi Zhou et al., "Accidental discovery and isolation of Zika virus in Uganda and the relentless epidemiologist behind the investigations," *Virologica Sinica* 31, no. 4, August 2016: 357–362.

335 Monali C. Rahalkar and Rahul A. Bahulikar, "Lethal Pneumonia Cases in Mojiang Miners (2012) and the Mineshaft Could Provide Important Clues to the Origin of SARS-CoV-2," *Frontiers in Public Health*, October 20, 2020.

336 Peng Zhou et al., "Addendum: A pneumonia outbreak associated with a new coronavirus of probably bat origin," *Nature* 588, E6, November 17, 2020.

337 Mikkel Winther Pedersen et al. "Environmental genomics of late Pleistocene black bears and giant short-faced bears. *Current Biology*, 31, pp. 1-9. Published online April 19, 2021.

338 Benjamin Vernot et al., "Unearthing Neandertal population history using nuclear and mitochondrial DNA from cave sediments," *Science*, April 15, 2021.

339 Xingyi Ge et al, "Metagenomic analysis of viruses from bat fecal samples reveals many novel viruses in insectivorous bats in China," *Journal of Virology*, February 15, 2012. This article shows that the Solexa sequencing method was used to produce a large number of reads and some sequences as long as 1000 base pairs. It was done at the Beijing Institute of Genomics of the Chinese Academy of Sciences.

340 The article found no benefit and some harms from various forms of chloroquine but it was based on datasets not available to the peer reviewers or to more

than one of the authors. The authors did not declare conflicts though at least one author had enjoyed many personal fee payments from a host of drug companies. See: Mandeep R. Mehra, Frank Ruschitzka, and Amit N. Pater, "Retraction—Hydroxychloroquine or chloroquine with or without a macrolide for treatment of COVID-19: a multinational registry analysis," *The Lancet*, June 5, 2020.

341 Naomi Klein, *The Shock Doctrine: The Rise of Disaster Capitalism* (Alfred A. Knopf Canada, 2007).

342 Elaine Dewar, *Cloak of Green* (James Lorimer & Company, 1995).

343 See: Jeffrey D. Sachs, https://csd.columbia.edu.

344 Elizabeth Schulze, "Geopolitical cold war with China would be a dangerous mistake, economist Jeffrey Sachs says," CNBC, August 10, 2020.

345 Katrina Northrup, "Jeffrey Sachs on Not Pointing Fingers," *The Wire China*, January 24, 2021, is-org.cdn.armproject.org.

346 Jeffrey D. Sachs, William Schabas. "The Xinjiang Genocide Allegations are Unjustified," *Project Syndicate*, April 20, 2021.

347 *Al Jazeera* Staff, "Amnesty says China has created 'dystopian hellscape' in Xinjiang: Rights group alleges 'crimes against humanity' being perpetrated against Uighurs, other Muslim minorities," *Al Jazeera*, June 10, 2021.

348 And more brownie points if you can guess who Daszak put on this task force. They include a fellow employee at EcoHealth Alliance, Hume Field, Stanley Perlman one of those Daszak rounded up to sign the statement of support published in *The Lancet*. Task force members also include Danielle Anderson, who until recently directed a BSL-3 lab at Linfa Wang's Duke-NUS in Singapore and Linda Saif, distinguished professor at Ohio State, who was also involved in creating *The Lancet* statement. The task force support staff either work for EcoHealth Alliance or for Sach's Center for Sustainable Development at Columbia. See: *The Lancet* COVID-19 Commission Task Force Statement, December 2020, COVID19Commission@unsdsn.org. The early February 2020 letters between Saif, Perlman, and Daszak about the statement may be seen at USRTK.org, the result of a US Freedom of Information application concerning Saif.

349 Daszak's work organizing the *Lancet* statement included correspondence with Ralph Baric and a few others. Baric in the end was not a signatory. Others involved in drafting the statement include Linda Saif and Stanley Perlman who were appointed by Daszak to the *Lancet* task force. The correspondence and the many drafts of that statement were discovered thanks to a Freedom of Information application regarding Ralph Baric's correspondence. This group also found letters written by Shi Zhengli to the official taxonomy commission that named SARS-CoV-2. These emails and drafts may be read at USRTK.org under the title "FOI documents on origins of SARS-CoV-2, hazards of gain-of-function research and biosafety labs."

350 The Scripps Research Institute, which employs Andersen, had serious financial difficulties some years ago which led to major investments in the Institute by Chinese based corporations. See: Bradley J. Fikes, "Scripps Research CEO outlines path to financial sustainability," *San Diego Union-Triibune*, February 23, 2018. On November 27, 2019, Scripps entered a research collaboration with Shenzhen Bay Laboratory and Peking University in China. See: Scripps Research at htpps://www.scripps.edu.

351 Karl Sirotkin and Dan Sirotkin, "Might SARS-CoV-2 Have Arisen via Serial Passage through an Animal Host or Cell Culture?" *BioEssays* 42, no. 10 (August 12, 2020)

352 "Scientists outraged by Peter Daszak leading enquiry into possible COVID lab leak," *GMwatch*, September 23, 2020.

353 "WHO-convened Global Study of the Origins of SARS-CoV-2: Terms of Reference for the China Part," *World Health Organization*, November 5, 2020.

354 "WHO-convened Global Study of the Origins of SARS-CoV-2: China Part," *Joint*

WHO-*China Study*, January 14–February 10, 2021. Members of the international team: Professor Thea Fisher, Denmark; Professor John Watson, UK; Professor Marion Koopmans, Netherlands; Professor Dominic Dwyer, Australia; Vladimir Dedkov, Russia; Hung Nguyen-Viet, Vietnam; Fabian Leendertz, Germany; Peter Daszak, US; Farag El Moubasher, Qatar; Professor Ken Maeda, Japan.

355 William Karesh, the executive vice president of EcoHealth Alliance, is president of the World Organization for Animal Health Working Group on Wildlife Diseases.

356 Eban, "The Lab-Leak Theory: Inside the fight to Uncover COVID-19's Origins."

357 "WHO-convened Global Study of Origins of SARS-CoV-2: China Part": 12.

358 Alice Fabbri et al., "The Influence of Industry Sponsorship on the Research Agenda: A Scoping Review," *The American Journal of Public Health* 108, no. 11, November 2018: 9–16. See also: Simon Young, "Bias in the research literature and conflict of interest: an issue for publishers, editors, reviewers and authors, and it is not just about the money," *Journal of Psychiatry and Neuroscience* 34, no. 6 (November 1, 2009: 412–417)

359 "WHO-convened Global Study of Origins of SARS-CoV-2: China Part": 12.

360 The *Lancet* COVID-19 Commission, "Task Force on the Origins and Early Spread & One Health Solutions to Future Pandemic Threats," *The Lancet*, December 1, 2020, http://covid19commission.org.

361 "WHO team investigating coronavirus origins denied entry into China," *the journal.ie*, January 6, 2021.

362 Helen Regan and Sandi Sidhu, "WHO team blocked from entering China to study origins of coronavirus," CNN, January 6, 2021.

363 As of the last quarter of 2019, China contributed only .2 percent of the WHO budget. Canada contributes 1.7 percent. The Democratic Republic of Congo contributes more than China. Pakistan contributes more than China. The Wellcome Trust contributes more. The US contributes 15.5 percent of the WHO budget and the second largest funder is the Bill & Melinda Gates Foundation. So: why is the WHO so careful to please China? Is it the prizes? Or is it the risk? Source: *World Health Organization*, Contributors by name. https://open.who.int.

364 Dake Kang, Maria Cheng, and Sam McNeil, "China clamps down in hidden hunt for coronavirus origins," *AP News*, December 30, 2020.

365 "COVID: WHO team probing origin of virus arrives in China," *BBC News*, January 14, 2021.

366 Office of the Spokesperson, "Fact Sheet: Activity at the Wuhan Institute of Virology," *US Department of State*, January 15, 2021.

367 "China Mission VPC 09 Feb 2021," *World Health Organization*, February 9, 2021.

368 "All hypotheses remain open on COVID origins: WHO chief," *Al Jazeera*, February 12, 2021.

369 Javier C. Hernández and James Gorman, "On WHO Trip, China Refuses to Hand Over Important Data," *New York Times*, February 12, 2021.

370 "WHO experts slam NYT report on China trip, say they were 'misquoted,'" *CGTN*, February 14, 2021.

371 "Statement by National Security Advisor Jake Sullivan, Statements and Releases," *Briefing Room*, February 13, 2021.

372 Open Letter. "Call for a Full and Unrestricted International Forensic Investigation into the Origins of COVID-19," March 4, 2021. http://www.wsj.com.

373 Sanjay Gupta, "Autopsy of a pandemic: 6 doctors at the center of the US COVID-19 response," CNN, March 28, 2021.

374 Michaeleen Doucleff and Suzette Lohmeyer, "WHO Report: Wildlife Farms, Not Market, Likely Source of Coronavirus Pandemic," NPR, March 29, 2021.

375 "WHO-Convened Global Study of Origins of SARS-CoV-2: China Part." This is available on the WHO website at: https://www.who.int. All quotes and precis in this chapter come from that document unless otherwise endnoted.
376 Li-Meng Yan et al., "Unusual Features of the SARS-CoV-2 Genome Suggesting Sophisticated Laboratory Modification Rather Than Natural Evolution and Delineation of Its Probable Synthetic Route," *Rule of Law Society and Rule of Law Foundation*, 2020. See also: Li-Meng Yan et al., "SARS-CoV-2 Is an Unrestricted Bioweapon," *Rule of Law Society & Rule of Law Foundation*, 2020.
377 Gilles Demaneuf, "The Good, the Bad and the Ugly: a review of SARS Lab Escapes," *Medium*, November 16, 2020. May be obtained from Gilles Demaneuf at contact@demaneuf.com.
378 "WHO-convened Global Study of Origins of SARS-CoV-2: China Part." Annexes.
379 "Joint Statement on the WHO-Convened COVID-19 Origins Study," *Global Affairs Canada*, March 30, 2021.
380 Elaine Dewar, "The True North Strong and Free," *The Canadian*, September 16, 1978.
381 See: Luke Harding, *The Snowden Files: The Inside story of the World's Most Wanted Man*. Guardian Books, 2014., and also: Glenn Greenwald, *No Place to Hide: Edward Snowden, the NSA and the US Surveillance State*, Metropolitan Books, 2014, and watch *Citizenfour* (2014), Laura Poitras' film on surveillance which won Best Documentary at the Oscars in 2015.
382 Contributor, Bostickson, and Demaneuf, "An investigation into the WIV databases that were taken offline." https://www.researchgate.net/publication/349073738_An_investigation_into_the_WIV_databases_that_were_taken_offline.
383 Hua Guo et al., "Identification of a novel lineage bat SARS-related coronavirus that use bat ACE2 receptor," *bioRxiv*, May 21, 2021.
384 Dan Sirotkin, "Logistical and Technical Exploration into the Origins of the Wuhan Strain of Coronavirus," *Harvard to the Big House*, January 31, 2020. This site is styled as a place where criminology, behavioral economics and evolutionary anthropology make babies. See harvard2thebighouse.com.
385 Later a peer-reviewed paper authored by Karl and Dan Sirotkin was published called "Might SARS-CoV-2 Have Arisen via Serial Passage through an Animal Host or Cell Culture?"
386 Ian Birrell, "The COVID dissidents taking on China: Beijings science stooges are being unmasked by an international team of online sleuths," *UnHerd.com*, April 13, 2021.
387 Yuri Deigin, "Lab-Made? SARS-CoV-2 Genealogy Through the Lens of Gain-of-Function Research," *Medium*, April 22, 2020.
388 Robert Fife and Steven Chase, "CSIS warned Ottawa about two scientists at Winnipeg disease Lab," *Globe and Mail*, May 12, 2021.
389 Fife and Chase, "Infectious disease scientists at Canada's high-security lab collaborated with China," *Globe and Mail*, May 20, 2021.
390 Fife and Chase, "Infectious disease scientists at Canada's high-security lab collaborated with China." The proper name is "The Convention on the Prohibition of the Development, Production and Stockpiling of Bacteriological (Biological) and Toxin Weapons and on their Destruction," http://www.un.org. Listed organisms include pathogens like smallpox and tularemia. Coronaviruses are not listed.
391 Christopher Reynolds, "Cries of racism in Commons over questions about Chinese researchers turfed from Winnipeg disease lab," *Canada's National Observer*, June 3, 2021.
392 John Ibbitson, "Questioning government policy on China is not fomenting racism, Prime Minister," *Globe and Mail*, May 31, 2021.

393 Robert Fife and Steven Chase, "Commons votes to demand Liberals reveal reasons for firing of two federal scientists from infectious-disease lab," *Globe and Mail*, June 2, 2021. See also: "Business of Supply," *Journals* no. 109, June 2, 2021, www.ourcommons.ca

394 Robert Fife and Steven Chase, "Liberals breached parliamentary privilege over documents on fired scientists, House Speaker rules," *Globe and Mail*, June 16, 2021.

395 Robert Fife and Steven Chase, "Liberals censured over files in case of fired scientists," *Globe and Mail*, June 18, 2021.

396 Karen Pauls, "'Wake-up call for Canada': Security experts say case of 2 fired scientists could point to espionage," *CBC News*, June 10, 2021.

397 Robert Fife and Steven Chase, "Ottawa takes Speaker to court over records," *Globe and Mail*, June 24, 2021.

398 Tom Blackwell, "Fired Winnipeg lab scientist listed as co-inventor on two Chinese government patents," *Calgary Herald*, June 23, 2021.

399 See: Glenn Kessler, "Timeline: How the Wuhan lab-leak theory suddenly became credible," *Washington Post*, May 25, 2021. See also: David Leonhardt, "The Lab-Leak Theory: We have an explainer," *New York Times*, May 27, 2021.

400 Nicholson Baker, "The Lab-Leak Hypothesis," *New York Magazine*, January 4, 2021.

401 Nicholas Wade, "The origin of COVID: Did people or nature open Pandora's box at Wuhan?" *Bulletin of Atomic Scientists*, May 5, 2021, https:// thebulletin.org. See also: Jesse D. Bloom et al., "Investigate the origins of COVID-19," *Science* 372, no. 6543, May 14, 2021. And: Michael R. Gordon, Warren P. Strobel, and Drew Hinshaw, "Intelligence on Sick Staff at Wuhan Lab Fuels Debate on COVID-19 Origin," *Wall Street Journal*, May 23, 2021.

402 Wade, "The origin of COVID: Did people or nature open Pandora's box at Wuhan?"

403 See: "Transcript: The Origins of COVID-19: Policy Implications for the Future," *Hudson Insitute*, March 12, 2021. Participants included David Asher, Dr. Miles Yu, a China expert who served under Pompeo as a China policy advisor and Jamie Metzl of the Atlantic Council, who is an expert advisor to the WHO's Committee on Human Genome Editing. Metzl was concerned that the independent joint study, convened by the WHO, dealt so poorly with the lab leak thesis that it would tar the WHO. He organized a public letter, signed by several members of the DRASTIC group, to say so. It's called the Paris Letter.

404 Eban, "The Lab-Leak Theory: Inside the fight to Uncover COVID-19's Origins."

405 Bloom et al. "Investigate the origins of COVID-19."

406 Jeremy Page, Betsy McKay, and Drew Hinshaw, "The Wuhan Lab Leak Question: A Disused Chinese Mine Takes Center Stage," *Wall Street Journal*, May 24, 2021.

407 Aisla Chang, "Many Scientists Still Think the Coronavirus Came From Nature." NPR's *Goats and Soda*, May 28, 2021.

408 John Wayne Ferguson, "In Galveston, a cautious perspective on lab leak theories," *Daily News*, June 25, 2021.

409 Staff. Mint, May 24, 2021.

410 Peter Wehner, "NIH Director: We Need an Investigation Into the Wuhan Lab-Leak Theory," *The Atlantic*, June 2, 2021.

411 Alison Young, "'I remember it very well': Dr. Fauci describes a secret 2020 meeting to talk about COVID origins," *USA Today*, June 17, 2021.

412 "US Right to Know" email from Kristian Andersen to Peter Daszak, copies to Andrew Pope, Ralph Baric, Trevor Bedford, Stanley Perlman, etc. Sent February 4, 2020 at 12:05:54 p.m. re: drafting of letter to the White House from the National Academies of Sciences, Engineering and Medicine.

413 Kate Sullivan, Donald Judd, and Phil Mattingly, "Biden tasks intelligence

community to report on COVID origins in 90 days," CNN, May 27, 2021. See also: Julian E. Barnes, "Biden says he will release the results of an inquiry into the virus' origin," *New York Times*, May 27, 2021.

414 Doug Saunders, "Canada's pandemic plan ought to make the next pandemic non-existent," *Globe and Mail*, June 2, 2021. This piece notes that data tracked in the UK by the Oxford COVID-19 Government Response Tracker shows that the governments that succeeded in controlling the pandemic without having to resort to lockdowns instituted strict quarantine in quarantine centres at their borders by the end of January 2020 for everyone, including citizens returning home.

415 Grant Robertson, "Early days in COVID-19 fight marked by lack of focus, wrong data: panel," *Globe and Mail*, May 28, 2021.

416 On March 10, 2021, a UK think tank called Chatham House held an event called "COVID-19 Briefing: The WHO-China Mission." It was held well before the actual release of the joint study report. Peter Daszak and Marion Koopmans were featured. Koopmans answered a question sent in by someone in the online audience who wanted to know why the WIV had taken down the database listing all its samples and sequences in September 2019. Koopmans said she had been told by people at the WIV that there had been over 3,000 hacking attempts. She did not explain why there were so many hacking attempts before the pandemic began. www.chathamhouse.org.

417 Linfa Wang told me that it wasn't obvious that anything serious was afoot when he visited in Wuhan in the middle of January 2020. Everything seemed normal, he said, nobody on the street was wearing a mask. Which is an interesting story, but his close colleagues at the WIV would have been talking about the urgent need for a vaccine by then.

418 Sainath Suryanarayanan, "Chinese scientists sought to change name of deadly coronavirus to distance it from China," *USRTK*, February 17, 2021, https://usrtk.org/biohazards. See: Ralph Baric letters. Shi Zhengli sent a copy of the letter sent to the subcommittee to Ralph Baric on February 13, 2020.

419 See letter disclosed to Katherine Eban, "The Lab-Leak theory: Inside the fight to Uncover COVID-19's Origins," from Michael S. Lauer, NIH Deputy Director for Extramural Research, July 8, 2020, to Peter Daszak and Aleksei Chmura, EcoHealth Alliance.

420 usrtk.org/wp-content/uploads/2021/02/Baric-Emails-2.17.21.pdf

421 Statement by the *Lancet* COVID-19 Commission, June 22, 2021, https://covid-19commission.org.

422 Michelle Fay Cortez, "The Last—and Only—Foreign Scientist in the Wuhan Lab Speaks Out," *Bloomberg*, June 27, 2021.

Acknowledgements

THIS IS THE first non-fiction book I've done that did not start with a written proposal and a lot of preliminary research. Instead, it began with a phone call from publisher Dan Wells as the first wave of the pandemic receded. He said he was starting a new series of short books called Field Notes and asked if I would like to write about pandemic profiteering. I said no: I don't want to do another business book, and besides, I'm fascinated by the origin of SARS-CoV-2. He asked why. I said "because it's such a smart virus." I mentioned the possibility of a lab leak and gain-of-function experiments. I explained there might be a connection to Chinese researchers suspended from the National Microbiology Laboratory in Winnipeg. He said, "Okay, let's do it," and my fate was sealed. I asked him to edit it because he does not have a science background and a book about science for general readers needs an editor who asks: *What does this mean? What's that about?* His editorial comments were fair, considered, his questions important. His demands for endnotes made me crazy, but he was right to ask for more and when I looked for them, I turned up more tidbits that I would have missed otherwise. He made this a much better book than it might have been without his care and attention.

Many other people helped along the way. My agent, Sam Hiyate was, as always, enthusiastic and supportive: he may be the last enthusiastic literary agent in Canada. The Ontario Arts Council gave me a small grant, which was a great help. The media, major and minor, provided reliable, steady reporting on what was going on in China before and after the start of the pandemic, as well as explaining the politics that drove the behaviour of the WHO. Unlike the flu pandemic of 1918–1919, when censorship prevailed in all countries involved in World War I, the information dug out with such effort by so many reporters around the world was vitally important to the public and to policy makers. Without their efforts, this book could not have been done. Most of all, I am indebted to my friends and family. Marc Côté, publisher of Cormorant Books, sent me the crucial Latham/Wilson publication, which I would have missed otherwise. My niece, Samantha Landa, the family genealogist, passed on to me the memoirs of Hope Richman and Sol Sinclair. My daughters, Anna Dewar Gully and Danielle Dewar, listened patiently to endless expositions about what I was finding out. So did my sons-in-law, Timothy Gully and Brandon Smollet. My lifelong friend Larry Gelmon helped me with important contacts. My friend, the brilliant journalist Marci McDonald, sent cards and emails wishing me courage. My friend of many years, Charles Greene, sent links to important stories, helped me think through the meaning of what I was learning, and then read and commented on an early draft. Dawn Deme, also a friend of many years and my former editor, did the same, hearing me out, countering thinking errors, making suggestions that were so helpful. Both warned me that the original title didn't capture what this book is about, and so I changed it. Their insights are gratefully appreciated.

Author and media lawyer Mark Bourrie read the manuscript and made useful comments. He too sent links to stories he thought I should know about. Editor John Metcalf read it too and he and his wife Myrna informed me of an important development just as I was finishing the final version.

Meghan Desjardins, Emily Donaldson, Theo Hummer, and John Sweet, under the watchful eye of managing editor Vanessa Stauffer, scrubbed out my grammatical and spelling errors and made this work much more readable.

All their comments helped in so many ways. But any errors are my own.

ELAINE DEWAR—AUTHOR, JOURNALIST, television story editor—has been honoured by nine National Magazine awards, including the prestigious President's Medal, and the White Award. Her first book, *Cloak of Green*, delved into the dark side of environmental politics and became an underground classic. *Bones: Discovering the First Americans*, an investigation of the science and politics regarding the peopling of the Americas, was a national bestseller and earned a special commendation from the Canadian Archaeological Association. *The Second Tree: of Clones, Chimeras, and Quests for Immortality* won Canada's premier literary non-fiction prize from the Writers' Trust, and *The Handover: How Bigwigs and Bureaucrats Transferred Canada's Best Publisher and the Best Part of Our Literary Heritage to a Foreign Multinational* was nominated for the Governor General's Award for Non-fiction. Called "Canada's Rachel Carson," Dewar aspires to be a happy warrior for the public good.